Logic & Memory Experiments Using TTL Integrated Circuits—

Book 1

by

David G. Larsen and Peter R. Rony

Also Published as
Logic & Memory Experiments Using TTL Integrated Circuits
by E & L Instruments, Inc.

Howard W. Sams & Co., Inc.
4300 WEST 62ND ST. INDIANAPOLIS, INDIANA 46268 USA

FIRST EDITION
SECOND PRINTING—1980

International Standard Book Number: 0-672-21542-X
Library of Congress Catalog Card Number: 78-57209

Printed in the United States of America.

Preface

Welcome to the "wonderful world of digital electronics." The exciting new advances in digital electronics will profoundly influence the way that we shall all live. We hope that this book will in some way give you insight into what the excitement is all about. This book and Book 2 represents a somewhat novel approach to the study of electronics. Rather than start you, as is customarily done in an introductory electronics course, with experiments using electronic *components* such as *resistors, capacitors, diodes,* and *transistors,* we instead introduce you immediately to *integrated-circuit chips.* We also introduce you to the concepts of *logic switches, lamp monitors, pulsers,* and *displays;* show you how to use such auxiliary functions; and provide you with numerous experiments that are based on connections between integrated-circuit chips and such devices. We believe that *digital electronics* is a simpler subject to grasp by the novice in electronics than is *analog electronics.* Success with digital electronics should give you a need to know that will make your study of analog electronics much more interesting and meaningful.

We would like to acknowledge Mr. Murray Gallant and E & L Instruments, Inc. for making it possible to publish the Blacksburg Continuing Education Series™ and for their enthusiasm and vigorous support for the entire LR OUTBOARDS® approach to digital electronics. Dr. Jonathan A. Titus, president of Tychon, Inc. in Blacksburg, participated significantly in the design and implementation of the actual LR OUTBOARDS® hardware. If you like the individual Outboards®, you have Jon to thank. Dr. Roberts A. Braden provided many superb suggestions concerning how to make this book more useful to you. The lists of objectives and the emphasis on you, the reader, were but two of his significant recommendations.

Mr. Bu Up Kim, a sophomore chemical engineering student, performed every experiment in this book early in the summer of 1974 without the benefit of any prior experience in electronics. We would like to acknowledge his efforts to ensure the accuracy of our experiments. Mr. Harry Goldman, another student at VPI&SU, worked very hard on the teaching equipment for our course in digital electronics; we gratefully acknowledge his efforts. Finally, our sincere thanks go to Mr. Bob Veltri, who did an excellent job of photographing the various versions of the Outboards®; to Mr. James Ruggiero, for helpful comments on the software. Our greatest thanks, however, go to our wives, Nancy and Myriam, who have been very patient and understanding.

The pin configurations, block diagrams, and digital waveforms are courtesy of the Texas Instruments Incorporated "TTL Data Book for Design Engineers." They are a vital part of this book and we are very grateful to TI for allowing us to use them. We also would like to thank Signetics Corporation and Fairchild Semiconductor for permission to use one or two pages from their publications.

As this revision goes to press, we have had the opportunity to use the books to personally teach hundreds of individuals, ranging from a freshman engineer to professional engineers and scientists with Ph.D. or M.D. degrees. Even our eleven-year-old sons have been able to wire moderately complicated circuits. Such experience justifies our conviction that digital electronics is easier to grasp than analog electronics.

David G. Larsen
Peter R. Rony

Contents

CHAPTER 4

CHAPTER 5

APPENDIX

1

Breadboarding Digital Circuits

THE ELECTRONICS REVOLUTION

In this book, you will set up and analyze experiments that demonstrate principles, concepts, and applications of *integrated-circuit "chips."* In doing so, you will develop skills that will enable you to participate in one of the most exciting of all of the recent technological revolutions: "the electronics revolution." As the editors of the magazine *Electronics* has stated in their article titled, "The Great Takeover:"

> "How else can it be described? Never before have such young, breathtaking aggressive, technologically sophisticated industries become so pervasive so quickly: scarcely a sector of our lives is untouched. New, gigantic industries have risen, replacing older ones based on other technologies. And more and more, human existence and electronics are becoming inter-twined."
>
> . . . *Electronics,* October 25, 1973

You are already able to purchase a $9.95 pocket calculator that can add, subtract, multiply, and divide six-digit floating-point numbers; a $29.95 pocket slide rule that makes the popular mechanical version obsolete; a $295 pocket-sized programmable computer-calculator that accepts small preprogrammed magnetic memories, a tiny computer-on-a-chip, no larger than the size of a postage stamp, that can, with the addition of several similar-sized chips, perform most of the arithmetic operations that the enormous state-of-the-art vacuum-tube computers could perform back in the mid '50s. And this is just a beginning. In the future, you can look forward to tiny computers called *microcomputers* in the home, in automobiles, at work, and at play. Large flat television screens will be the center of home computer/communication terminals that will interact with the outside world via cable television. Such terminals, along with small electronic devices, will

- Control home heating and lighting systems.
- Obtain emergency aid.
- Check all doors and windows to ensure that they are locked at night.
- Perform sophisticated calculations required in homework problems or in the pursuit of your profession.
- Reduce the necessity for you to be physically present somewhere else in order to accomplish something, such as purchase groceries, take a test, check out a book from the library, obtain information, listen to a lecture, etc.

Perhaps the most exciting aspect of the "electronics revolution," certainly for the novice, is that it is getting easier to understand and use the new integrated circuits that are daily coming to the marketplace. A complicated circuit that could be wired in a week by an electrical engineer during the mid '60s can today be wired in an afternoon by a high school student who has only a modest background in digital electronics. In fact, *digital electronics is becoming easier, not harder, as a result of the electronics revolution.*

In this book, we hope that we can convey some of the excitement and vitality of the "new electronics," as the electronics revolution has been called by many individuals. We shall make no attempt to explain why the remarkable devices called *integrated circuits* work they way they do. Such information is available in a number of texts and reference books, which can be consulted by the interested reader. Instead, we shall provide you with an extensive series of experiments that, when completed, will allow you to (a) conceive your own circuits that incorporate integrated-circuit chips; (b) know how to purchase the necessary components that you would require for such circuits; (c) wire the circuits using the purchased chips and components; and finally, (d) demonstrate the successful operation of the circuits. In short, we seek to make you independent and creative; independent of this book and creative in the sense that you are able to conceive and execute your own integrated circuit designs.

Good luck!

INTRODUCTION

In Chapter 2, you will embark upon an extensive series of experiments that will help you to develop competence in a wide area of *digital electronics.* All experiments will be performed with inexpensive electronic building blocks—a *breadboard* or two, *auxiliary functions* called OUTBOARD® modules, wire, a few *resistors* and *capacitors,* a *battery* or *power supply,* and integrated-circuit "chips" —all of which are both lightweight and portable, two properties that will enable you to perform these experiments almost anywhere: in the home, at school, at work, or in a motel room while on the road. In this chapter, we shall provide you with hands-on experience in the use of the OUTBOARD® modules. The description should be sufficient to enable you to successfully wire all of the circuits given in subsequent chapters. Most of these experiments can be performed on any type of digital logic trainer.

OBJECTIVES

At the completion of this chapter, you will be able to do the following:

- Set up any experiment given in subsequent chapters by following the steps and schematic diagrams provided.
- Demonstrate the operation of the principal OUTBOARD® modules and associated equipment, including
  - battery or power supply
  - breadboard
  - logic switches
  - lamp monitors
  - pulsers
  - clock
  - seven-segment LED display
  - a typical integrated-circuit chip
  - capacitors
- Identify the schematic symbols *employed in this book* for the following auxiliary functions,
  - battery or power supply
  - logic switches
  - lamp monitors
  - pulsers
  - clock
  - seven-segment LED display
  - zero/one LED display

  as well as for the following components,
  - resistors
  - capacitors
  - diodes
  - integrated circuits

- Make wire connections between the above auxiliary functions and components using a breadboard.
- Handle the above auxiliary functions, components, and breadboard with sufficient care that none will be destroyed during the entire series of experiments given in this book.
- Recognize that the majority of simple digital electronic circuits are based upon two states, a "logic 0" state and a "logic 1" state, which correspond to voltages of approximately 0 volt and +5 volts with TTL integrated circuits.

DEFINITIONS

One of the most serious barriers that you will encounter in your study of digital electronics is a jargon barrier. In each chapter, we shall provide you with definitions for important terms and concepts. The definitions for this chapter are as follows. Superscript numbers are for references listed in the Appendix.

alligator clip—A spring-loaded metal clip with long, narrow meshing jaws, used for making temporary electrical connections.[1]

auxiliary function—Any separate electronic device that is required to make operational a digital electronic circuit consisting of integrated circuits, resistors, capacitors, etc., that is wired on a breadboard.

battery—A dc voltage source, consisting of two or more cells, which converts chemical energy into electrical energy.[1]

breadboard—Any aid used to temporarily wire together a circuit to prove the feasibility of a circuit, device, system, etc.

breadboarding—The action of using a breadboard to temporarily wire together an electrical circuit.

capacitance—In a capacitor or a system of conductors and dielectrics, that property which permits the storage of electrically separated charges when potential differences exist between the conductors. The capacitance of a capacitor is defined as the ratio between the electric charge that has been transferred from one electrode to the other *and* the resultant difference in potential between the electrodes. The value of this ratio is dependent on the magnitude of the transferred charge. The unit of capacitance is a Farad.[1]

$$C[\text{farads}] = Q[\text{coulombs}]/V\,[\text{volts}]$$

capacitor—A device consisting essentially of two conducting surfaces separated by an insulating material or dielectric such as air, paper, mica, glass, plastic film, or oil. A capacitor stores electrical energy, blocks the flow of direct current, and permits the flow of alternating current to a degree dependent essentially on the capacitance and the frequency.[1]

chip—See integrated circuit.

circuit—The interconnection of a number of devices in one or more closed paths to perform a desired electrical or electronic function.[1]

clock—(1) Any device that generates at least one clock pulse. (2) A timing device in a system; usually it provides a continuous series of timing pulses.[1]

clock pulse—A complete logic cycle, i.e., a transition from logic 0 to logic 1 and back to logic 0, *or* a transition from logic 1 to logic 0 and back to logic 1.

complement—The binary opposite of a variable or function. The complement of 1 is 0 and the complement of 0 is 1; thus, for example, the complement of 011010 is 100101.[1]

component—An essential functional part of a subsystem or of equipment. It may be any self-contained element with a specific function, or it may consist of a combination of parts, assemblies, accessories, and attachments.[1]

connection—(1) The attachment of two or more component parts so that conduction can take place between them. (2) The point of such attachment.[1]

current—The movement of electrons through a conductor. It is measured in amperes and its symbol is I.[1]

DCBA—(Or ABCD) Binary or binary-coded decimal (BCD) representation. $A = 2^0$ position; $B = 2^1$ position; $C = 2^2$ position; and $D = 2^3$ position. This nomenclature is frequently found in truth tables and on specifications for auxiliary functions and integrated-circuit chips.

decade counter—A logic device that has 10 stable states and may be cycled through these states by the application of 10 clock or pulse inputs. A decade counter usually counts in a binary sequence from state 0 through state 9 and then cycles back to 0. Sometimes referred to as a divide-by-10 counter.[1]

device—(1) A single discrete conventional electronic part such as a resistor or transistor, or a microelectronic circuit. (2) A mechanical, electrical, and/or electronic contrivance intended to serve a specific purpose.[1]

digit—One of the symbols, 0, 1, 2, 3, 4, 5, 6, 7, 8, and 9, used in numbering in the scale of 10.[1]

digital—(1) Using numbers expressed in digits and in a certain scale of notation to represent all the variables that occur in a problem. (2) Of or pertaining to the class of devices or circuits in which the output varies in discrete steps (i.e., pulses or "on-off" operation). (3) Of or pertaining to an element or circuit the output of which is utilized as a discontinuous function of its input.[1]

digital data—Data represented in discrete, discontinuous form, as contrasted with analog data represented in continuous form. Digital data are usually represented by means of coded characters (e.g., numbers, signs, symbols, etc.).[1]

digital electronics—Pertaining to that branch of electronics that is concerned with digital components, devices, and circuits.

digital output—An output signal which represents the size of a stimulus or input signal in the form of a series of discrete quantities which are coded to represent digits in a system of numerical notation. This type of output is to be distinguished from one which provides a continuous rather than a discrete output signal.[1]

digital signals—Discrete or discontinuous signals whose various states are discrete intervals apart.[1]

diode—A two-electrode semiconductor device that makes use of the rectifying properties of a pn junction (junction diode) or a sharp metallic point in contact with a semiconductor material (point-contact diode). Also called crystal diode, rectifier diode, and semiconductor diode.[1]

display—A device that provides a visual presentation of an electronic signal.[1]

dual-in-line package (DIP)—A type of housing for integrated circuits. The standard form is a molded plastic container about 3/4 inch long and 1/3 inch wide, with two rows of pins spaced 0.1 inch between centers. Abbreviated DIP.[1]

electronic—(1) Pertaining to that branch of science which deals with the motion, emission, and behavior of currents of free electrons, especially in vacuum, gas, or phototubes and special conductors or semiconductors. This is contrasted with electric, which pertains to the flow of large currents in metal conductors. (2) Of or pertaining to devices, circuits, or systems using the principle of electron flow through a conductor—for example, electronic control, equipment, instrument, circuit.[1]

frequency—Symbolized by f (or by ν). The number of recurrences of a periodic phenomenon in a unit of time. Electrical frequency is specified as so many hertz.[1]

ground—(1) A metallic connection with the earth to establish ground potential. (2) The voltage reference point in a circuit. There may or may not be an actual connection to earth, but it is understood that a point in the circuit said to be at ground potential could be connected to earth without disturbing the operation of the circuit in any way. (3) In digital electronic circuits, generally the logic 0 state.[1]

hookup wire—The wire used in coupling circuits together. It may be solid or stranded, and is usually tinned and insulated No. 18 to No. 26 soft-drawn copper.[1]

integrated circuit—Abbreviated IC. (1) A combination of interconnected circuit elements inseparably associated on or within a continuous substrate. (2) Any electronic device in which both active and passive elements are contained in a single package. In digital electronics, the term chiefly applies to circuits containing semiconductor elements.[1]

integrated-circuit chip—Slang term generally used to mean a complete, operating semiconductor integrated circuit.

interconnection—The physical wiring between components (outside a module), between modules, between units, or between larger portions of a system or systems.[1]

logic—(1) The science dealing with the basic principles and applications of truth tables, switching, gating, etc. (2) Also called symbolic logic. A mathematical approach to the solution of complex situations by the use of symbols to define basic concepts. The three basic logic symbols are AND, OR, and NOT. When used in Boolean algebra, these symbols are somewhat analogous to addition and multiplication. (3) In computers and information-processing networks, the systematic method that governs the operations performed on information, usually with each step influencing the one that follows. (4) The systematic plan that defines the interactions of signals in the design of a system for automatic data processing.[1]

logic 0—One of the two values of a binary digit signal. If the signal is a voltage, logic 0 usually is the lower of the two voltage levels.[1] Frequently GROUND.

logic 1—The opposite of logic 0. Frequently +5 volts.

logic circuit—A circuit (usually electronic) that provides an input-output relationship corresponding to a Boolean-algebra logic function.[1]

logic devices—Digital components that perform logic functions. They can gate or inhibit signal transmission in accordance with the application, removal, or combination of input signals.

logic diagram—In logical design, a diagram representing the logic elements and their interconnections, but not necessarily their construction and engineering details.[1]

logic element—A device that performs a logic function; a gate or flip-flop, or in some cases, a combination of these devices treated as a single entity.[1]

logic function—One of the Boolean-algebra functions, AND, OR, and NOT, or a combination of these.[1] Important combinations in digital electronics include NAND, NOR, EXCLUSIVE OR, AND-OR-INVERT, etc.

logic level—One of two possible states, 0 or 1.[1]

logic switch—(1) A mechanical device that generates a logic 0 or logic 1 state. (2) A diode matrix or other switching arrangement that can direct an input signal to a selected one of several outputs.[1]

logic symbol—A symbol used to represent a logical element on a graph.[1]

logic system—A group of interconnected logic elements that act in combination to perform a relatively complex logic function.[1]

logical decision—The ability of a computer to make a choice between two alternatives; basically, the ability to answer yes or no to certain fundamental questions concerning equality and relative magnitude.[1]

logical design—The synthesizing of a network of logical elements to perform a specified function.[1] In digital electronics, these logical elements are digital electronic devices, such as gates, flip-flops, decoders, counters, etc.

logical element—In a computer or data-processing system, the smallest building blocks which operators can represent in an appropriate system of symbolic logic. Typical logical elements are the AND gate and the "flip-flop."

light-emitting diode (LED)—Abbreviated LED. A pn junction that emits light when biased in the forward direction.[1]

lamp—A device for producing light.[1] A device that emits visible light when power is applied.

lamp monitor—A digital electronic device that generally is lit in the logic 1 state and unlit in the logic 0 state.

medium-scale integration (MSI)—Integrated circuits that function as simple, self-contained logic systems, such as decade counters or five-bit shift registers. Such chips may contain up to 100 gates. Abbreviated MSI.[1]

microelectronics—The entire body of electronic art which is connected with or applied to the realization of electronic systems from extremely small electronic parts.[1]

pin connections—Connections made to pins in an integrated-circuit chip.

power supply—A unit that supplies electrical power to another unit. It changes ac to dc and maintains a constant voltage output within limits.[1]

pulse—(1) The variation of a quantity having a normally constant value. This variation is characterized by a rise and a decay of a finite duration (2) An abrupt change of a voltage, either positive or negative, which conveys information to a circuit. (3) A clock pulse.

pulser—A logic switch that generates a clock pulse.

resistance—A property of conductors which—depending on their dimensions, material, and temperature—determines the current produced by a given difference of potential; that property of a substance which impedes current and results in the dissipation of power in the form of heat. The practical unit of resistance is the ohm. It is defined as the resistance through which a difference of potential of one volt will produce a current of one ampere.[1]

resistor—A device connected into an electrical circuit to introduce a specified resistance.[1]

schematic diagram—A written representation of electronic components, auxiliary functions, devices, etc., that are wired in some useful manner.

seven-segment display—An electronic display that contains seven lines or segments spatially arranged in such a manner that the digits 0 through 9 can be represented through the selective lighting of certain segments.

small-scale integration (SSI)—Integrated circuits that only function as simple gates or flip-flops. Such chips generally contain less than ten to twenty gates. Abbreviated SSI.

state—The conditions at the output of a circuit that represents logic 0 or logic 1.[1]

symbol—A simplified design representing a part in a schematic circuit diagram.[1]

truth table—A tabulation that shows the relation of all output logic levels of a digital circuit to all possible combinations of input logic levels so as to characterize the circuit functions completely.[1]

volt—The unit of measurement of electromotive force. It is equivalent to the force required to produce a current of 1 ampere through a resistance of 1 ohm. Abbreviated V.[1]

+5 volts—In digital electronics, generally the logic 1 state.

voltage—(1) Electrical pressure—i.e., the force which causes current to flow through an electrical conductor. (2) Symbolized by E. The greatest effective difference of potential between any two conductors of a circuit.[1]

wire—A solid or stranded group of solid cylindrical conductors having a low resistance to current flow, together with any associated insulation.[1] Useful wire sizes for breadboarding of digital electronic circuits are the No. 22 and No. 24 wire sizes.

SYMBOLS FOR AUXILIARY FUNCTIONS AND COMPONENTS

What Is an Auxiliary Function?

We shall consider an *auxiliary function* to be any separate electronic device that is required to make operational a digital electronic circuit consisting of integrated circuits, resistors, and capacitors that are wired on a breadboard. Devices that fall within the definition of auxiliary function include the following:

batteries
power supplies
logic switches
lamp monitors
pulsers
clocks
displays
counters
function generators
and others.

Such devices will be individually described as they are introduced in this book.

Why Symbols and Schematic Diagrams?

A *symbol* is defined by the Oxford English Dictionary as "a written character or mark used to represent something; a letter, figure, or sign conventionally standing for some object, process, etc." In electronics, symbols are used to represent electronic *components* such as resistors, capacitors, batteries, integrated-circuit chips, diodes, transistors, light-emitting diodes (LEDs), etc. In this book, we shall use the accepted symbols for electronic components, but in addition we shall also use symbols to represent the auxiliary functions described immediately above. A *schematic diagram* is simply a printed representation of electronic components and auxiliary functions that are wired up in some useful manner; *symbols* are used in this representation.

As discussed in *How To Read Schematic Diagrams* (Donald E. Herrington, Howard W. Sams & Co., Inc.), symbols are used in electronic diagrams because experience has shown us that they are the quickest and easiest way to convey the needed information. Simple symbols are a form of electronic shorthand. With them, a digital electronic circuit can be wired from a schematic diagram in a very short time. In general, the use of symbols enables all of the necessary circuit information to be given in a relatively small written diagram, as will be seen later in this chapter. Such a diagram is much easier to follow than a photograph of the completely assembled circuit. Further, because electronic symbols and circuit diagrams are generally standardized, they can be easily interpreted by others. On the following pages a summary of symbols for electronic components and auxiliary functions is provided.

Voltage Source

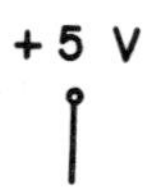

A voltage source will be indicated by this symbol and should be assumed to be referenced to ground unless otherwise indicated.

Power Supply

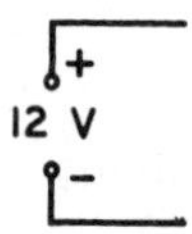

A power supply or voltage source may be represented in different ways. In this book, a power supply will be indicated by this symbol. The polarity and the voltage will usually be given.

Battery

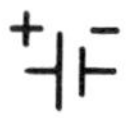

This is the customary symbol for a single-cell battery. The short bar represents the negative terminal, while the longer bar represents the positive terminal. The (−) and (+) designations may not always be shown. Occasionally, the voltage will be given.

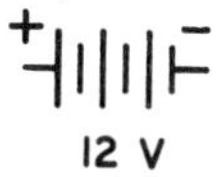

A multiple cell battery will be represented with additional pairs of bars. Usually the voltage will be given.

Ground

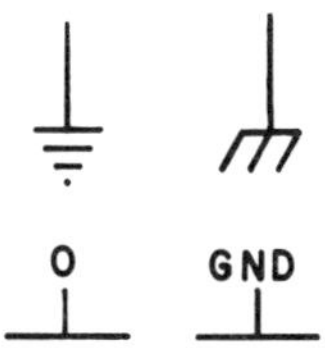

These symbols indicate GROUND, which is also abbreviated as GRD or GND. All points shown in a circuit with this symbol must be assumed to be at the same electrical potential. Generally, this potential can be assumed to be 0 volts.
These word symbols also indicate GROUND. They will be frequently used in this book.

Resistor

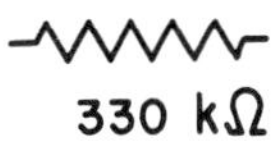

This symbol indicates a fixed value resistor. Usually the value of the resistance is given in units of ohms (Ω), kilohms (kΩ), or megohms (MΩ), where kilo = 1000 and meg = 1,000,000.

Capacitor

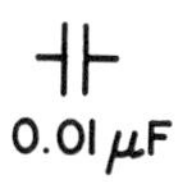

This symbol indicates a capacitor of fixed value. Usually the value of the capacitance is given in units of microfarads (μF), nanofarads (nF), or picofarads (pF), where micro = 10^{-6}, nano = 10^{-9}, and pico = 10^{-12}.

Diode

This symbol indicates a diode. The direction of the arrow and the bar indicate its polarity in the circuit. Significant current will only flow through the diode when it is *forward biased,* with the *anode* connected to a positive potential with respect to the *cathode,* as shown in the symbol.

Logic Switches

This symbol represents a group of four logic switches, which serve as vital auxiliary functions in almost any digital circuit. The letters ABCD represent the four *bits* in a 4-bit *binary word,* with A being the *least-significant bit* and D being the *most-significant bit.*

Lamp Monitors

This symbol represents a group of four lamp monitors, which also serve as vital auxiliary functions in any digital circuit. As with the logic switches, the letters ABCD represent the four bits of a 4-bit binary word.

Clock

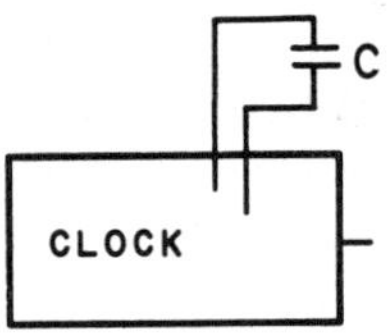

This symbol represents a clock, an auxiliary function that produces a square wave whose voltage ranges between 0 volt and approximately +5 volts. Usually, the value of the clock frequency is given, in units of hertz (Hz) or kilohertz (kHz). One hertz is a unit of frequency corresponding to one cycle/second. A series of *clock pulses* produced by a clock is frequently called a *train of clock pulses.*

Pulser

This symbol represents a pulser, an auxiliary function that allows you to apply a single *pulse* to a digital electronic circuit. The two outputs are the complements of each other; one output is at a logic 1 state while the other is at a logic 0 state, and vice versa. Unless otherwise indicated, the pulser is *debounced,* a term that shall be explained shortly.

Seven-Segment LED Display

This symbol represents a seven-segment light-emitting diode (LED) display, an auxiliary function that has the ability to display decimal numbers from 0 to 9 as well as a group of six rather unusual symbols that represent the decimal numbers 10, 11, 12, 13, 14, and 15. As was the case with the logic switches and lamp monitors, the letters ABCD represent the four bits in a 4-bit binary word.

Zero/One Display

This symbol represents a zero/one LED display, an auxiliary function made from a seven-segment LED display that has the ability to display a decimal 0 or 1 depending upon whether input A is at 0 volt or at +5 volts, respectively.

Integrated-Circuit Chip

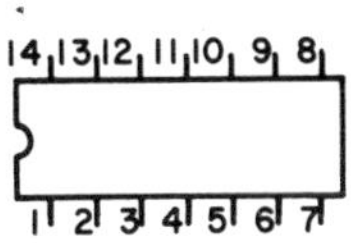

This is one of several related but different types of symbols that represent an integrated-circuit chip. Two types of information must be conveyed, if possible, by the symbol: the pin numbering system and the integrated-circuit function corresponding to each of the different pins that exist. Owing to the variety of symbols that exist, we shall discuss the subject of integrated-circuit chip symbols as a separate topic in Chapter 2.

Programmable Timer

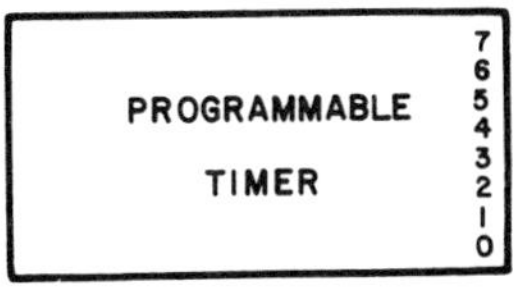

This symbol represents the output from a programmable timer, in which clock frequencies can be generated in multiples of 1, 2, 4, 8, 16, 32, 64 and 128 of a fundamental clock frequency. Output "0" is the lowest frequency and output "7" the highest frequency.

THE OUTBOARD® MODULES

Purpose

The purpose of the OUTBOARD® modules is to provide you with practical experience in the wiring of digital electronic circuits using integrated circuits, auxiliary functions, and electronic components. A remarkable aspect of the teaching of digital electronics is the fact that an individual can gain considerable understanding concerning the construction of digital circuits with the aid of only very simple equipment, including some logic switches, lamp monitors, pulsers, and integrated circuits as well as a breadboard, clock, and seven-segment LED display.

With the breadboarding system, you will be able to wire most of the experiments described in this book with the aid of readily available integrated circuits. The experiments range from simple circuits that require no integrated-circuit chips to complex circuits that require up to five or six chips and approximately 30 to 40 wire interconnections between them. We hope that the OUTBOARD® modules approach to digital electronics will convey the vitality and excitement of digital technology to you.

Breadboard

The breadboard is designed to accommodate the many experiments that you will perform in subsequent units in this book. Integrated circuits, resistors, capacitors, and auxiliary functions as well as the power supply all connect or tie in directly to the breadboard.

In Fig. 1-1, we show a top view of the breadboard, which contains 64 × 2 sets of five electrically connected solderless terminals that straddle both sides of a narrow center groove, and 8 sets of twenty-

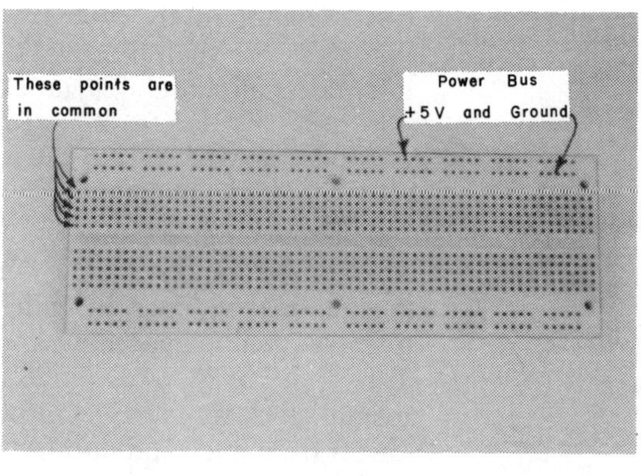

Fig. 1-1. The E & L Instruments, Inc. SK-10 breadboarding socket.

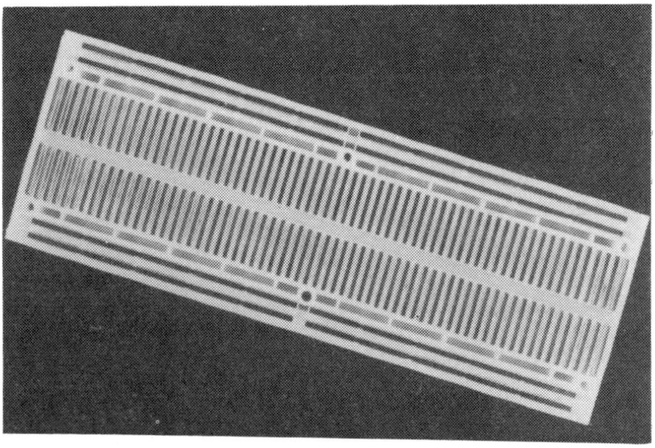

Fig. 1-2. Bottom view of solderless breadboard SK-10 socket showing 8 sets of connected terminals along the edges.

five electrically connected terminals along the edges as shown in Fig. 1-2. The center groups of five electrically connected terminals accommodate integrated circuits and permit up to four additional connections to be made at each pin of the smaller 14-pin and 16-pin integrated circuits. The groups of twenty-five electrically connected terminals at the edges are used to provide power to the integrated-circuit chips as well as to the auxiliary function modules, which are called OUTBOARD® modules. A sample of a very densely packed breadboard, without wires, is shown in Fig. 1-3. Note that there are six 14-pin integrated circuits and five auxiliary function modules.

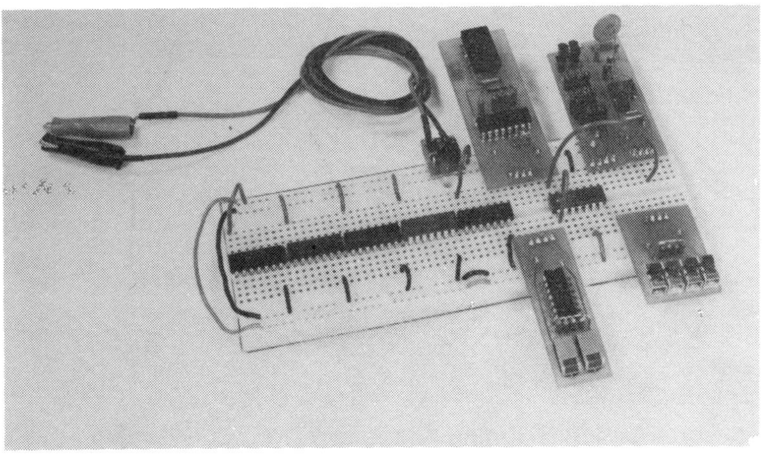

Fig. 1-3. An example of a densely packed SK-10 socket; wires are omitted for clarity.

The breadboard comes in a precision molded high impact plastic housing that has a total of 840 noncorrosive high-tensile–strength contacts, which have an expected contact life of over 10,000 wire insertions. The solderless terminals accept wires between the gauges of No. 22 and No. 26. Study the arrangement of the breadboard so that you will be familiar with it when wiring up circuits for the experiments described later.

The LR-1 Power Outboard Module

Any source of dc power can be used with the breadboarding system as long as the source provides no less than +4.75 volts and no more than perhaps +6.00 volts to the outer four groups of twenty-five electrically connected terminals along the two edges of the breadboard. We recommend that the outer groups of terminals be wired together in the manner shown in Fig. 1-21.

A portable power source is a heavy-duty +6-volt lantern battery, that can be connected to the breadboard with the aid of the *LR-1 Power Outboard* module (Fig. 1-4) which is all that is needed to apply power to 50 solderless terminals. The remaining 50 terminals are tied to the first 50 via jumper wires, *which should always remain in place.* The LR-1 provides a remarkable amount of control over and information about the dc power applied to the breadboard. The light-

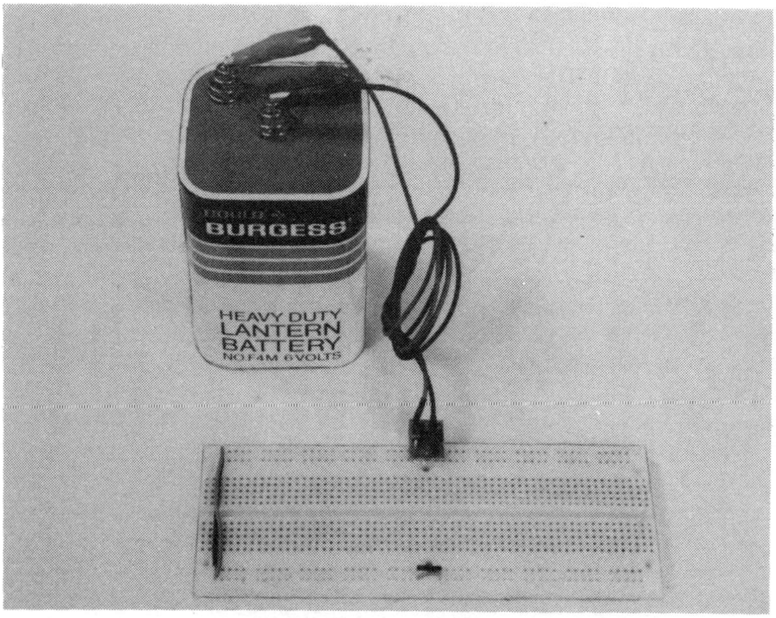

Fig. 1-4. A lantern battery for applying power to the breadboard.

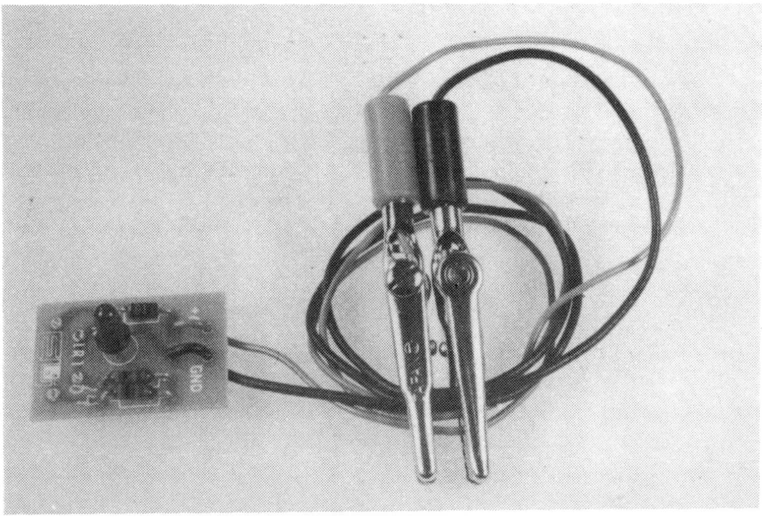

Fig. 1-5. A close-up view of the LR-1 power module.

emitting diode (LED) on the LR-1 module in Fig. 1-5 indicates when there is dc power that is applied to the breadboarding system, when there is no dc power, when the alligator clips have been connected to the wrong lantern-battery or power-supply terminals (the LED will be unlit), and finally, when the lantern-battery voltage is low (the LED will be dimly lit). The rectifier diode on the power module prevents adverse consequences resulting from the improper connection of the alligator clips to the battery terminals. If such clips are connected improperly, no voltage will be impressed on the breadboarding system and the LED will not light.

The LR-2 Logic Switch Outboard Module

A *logic switch* is a mechanical device that applies either a logic 0 or a logic 1 state at its output terminal. The *LR-2 Logic Switch Outboard* module, shown in Fig. 1-6, provides four logic switches that switch between GROUND potential (logic 0) and +5 volts (logic 1). The module can be located on either side of the breadboard provided that the outer two rows of terminals have the proper jumper wires.

The LR-4 Seven-Segment LED Display Outboard Module

A *display* is a device that provides a visual presentation of an electronic signal.[1] A very popular and effective kind of display is the *seven-segment LED display,* which contains seven segments spatially arranged in such a manner that the digits 0 through 9 can be

Fig. 1-6. A close-up view of the LR-2 logic switch module.

represented through the selective lighting of certain segments. The *LR-4 Seven-Segment LED Display Outboard module* (Fig. 1-7) contains, in addition to an Opcoa SLA-1 seven-segment LED display, seven current-limiting resistors, and a 7447 BCD-to-seven-segment

Fig. 1-7. A close-up view of the LR-4 seven-segment LED display module.

decoder/driver integrated circuit chip. The four inputs, ABCD, to the 7447 chip are used to generate 0 through 9, five symbols, and a blank condition. To the right of the 7447 chip in Fig. 1-7 are three inputs which are, from top to bottom, the BLANKING INPUT (pin 5), BLANKING OUTPUT (pin 4), and LAMP TEST (pin 3).

The LR-5 Clock Outboard Module

A *clock* is any device that generates at least one *clock pulse,* which is a complete logic cycle, *i.e.*, a transition from logic 0 to logic 1 and back to logic 0, or a transition from logic 1 to logic 0 and back to logic 1. The *LR-5 Clock Outboard* module generates a sequence of clock pulses, called a *train of clock pulses,* whose frequency depends on the value of the timing capacitor that is inserted into the two socket pins that bracket the letter C on the lower right-hand side of the Outboard module, as shown in Fig. 1-8. The negative terminal of

Fig. 1-8. A close-up view of the LR-5 clock module.

an electrolytic capacitor is inserted in the socket pin below the letters LR5. The clock frequency is generated with the aid of a 555 Timer, three resistors, and the timing capacitor. An on-board light-emitting diode (LED) gives a visual indication that the clock is operating properly. External components may be added to utilize the full 300-kHz to 1-MHz frequency range of the 555 Timer. A 0.33-microfarad timing capacitor provides a clock frequency of 0.7 Hz. Your clock frequency for an identical timing capacitor may vary within ±20% of this value. The proper output pin from this Outboard module is labeled "CLK."

The LR-6 LED Lamp Monitor Outboard Module

A *LED lamp monitor* is a light-emitting diode (LED) that is lit in the "logic 1" state and unlit in the "logic 0" state. It is used as a display for binary digital signals. The *LR-6 LED Lamp Monitor Outboard* module contains four LED lamp monitors arranged in a row at one end of the Outboard module. The use of transistor lamp drivers reduces the current to each lamp monitor to a level of only 1.5 mA. The module is shown in Fig. 1-9.

Fig. 1-9. A close-up view of the LR-6 LED lamp monitor module.

The LR-7 Debounced Dual Pulser Outboard Module

A *pulser* is a logic switch that generates a single clock pulse. In a mechanical logic switch, contact bounce—the uncontrolled making and breaking of contact when the switch contacts are opened or closed—is a common occurrence. In some digital applications, such bouncing—which produces a series of clock pulses—is not a problem. However, in most digital circuits, it is imperative that the output from a pulser be bounce free, thus permitting its use in flip-flops, counters and counter circuits, memories, shift registers, and related digital circuits that require clock pulses. The *LR-7 Dual Pulser Outboard* module uses four 2-input NAND gates to produce a pair of *debounced* pulsers, i.e., pulsers whose outputs do not exhibit contact bounce. Complementary "0" and "1" outputs are provided for each pulser on the LR-7 Outboard module, which is shown in Fig. 1-10.

The LR-26, LR-27, and LR-28 Latch/Display Outboard Modules

The *LR-26, LR-27,* and *LR-28 Latch/Display Outboard* modules are based on the remarkable 5082-7300 series of numeric and hexadecimal indicators manufactured by the Hewlett-Packard Company. The indicators are eight-pin displays, either decimal or hexadecimal, that contain a built-in *decoder/driver* and *latch.* In contrast to the LR-4 Outboard module, the indicator display is a 4 × 7 dot matrix that is very easy to read. This dot matrix permits the 5082-7340 indicator chip to exhibit *hexadecimal characters,* i.e., A, B, C, D, E, and F. The numeric displays exhibit blanks for such characters. Binary codes, including both octal and hexadecimal, are discussed in Chapter 5.

The *LR-26 Single Latch/Display Outboard* module contains a single numeric or hexadecimal indicator. Five input pins to the Outboard permit you to connect the A, B, C, or D inputs to the indicator. A strobe input pin is also provided. The *LR-28 Three Digit Latch/Display Outboard* module contains three numeric or hexadecimal indicators and three sets of A, B, C, and D input pins to the Outboard. The strobe inputs are tied together and provided as a single Outboard input. The *LR-27 Octal Latch/Display Outboard* module contains three numeric indicators that have their D inputs grounded. Two sets of indicators have A, B, and C inputs, whereas the third set has the C input also grounded. The strobe inputs are tied together as in the LR-28 Outboard module. This Outboard module is specifically for use as a bus monitor for an 8-bit data bus,

Fig. 1-10. A close-up view of the LR-7 dual pulser module.

Fig. 1-11. A close-up view of the LR-26 single latch/display module.

such as would occur with an 8-bit microcomputer. With it, you are able to read eight bits of parallel data as three octal digits. These three modules are shown in Figs. 1-11 through 1-13.

The LR-25 Breadboarding Station Outboard Module

The *LR-25 Breadboarding Station Outboard* module consolidates the LR-1, LR-2, LR-5, LR-6, LR-7, and LR-26 individual Outboard

Fig. 1-12. A close-up view of the LR-27 octal latch/display module.

Fig. 1-13. A close-up view of the LR-28 three-digit latch/display module.

modules into a single unit that serves as a complete breadboarding station for the experiments in this book. You should be able to perform all experiments, except those that require additional numeric displays or logic switches, with this module. The module contains four logic switches, two debounced pulsers, eight lamp monitors, an 8-pin socket for a Hewlett-Packard numeric or hexadecimal indicator, a diode-protected power input, and a power-on indicating LED, and is shown in Figs. 1-14, 1-15, and 1-16.

Labeling of the LR Outboard Modules

In the preceding pages, we have completed our description of the OUTBOARD® modules. The input/output pin connections on the LR-4, LR-5, and LR-7 Outboard modules are such that the only correct way to label the pins is as shown in Fig. 1-17.

The labeling for the LR-2 and LR-6 Outboards is optional and is omitted from the face of the two modules. You will discover that it is advantageous to change labeling schemes on these two Outboards when they are on opposite sides of the breadboarding socket. Some possible labeling schemes include the following (Fig. 1-18).

RULES FOR SETTING UP EXPERIMENTS

In this and all following chapters, you will be using the OUTBOARD® modules to perform a variety of experiments in digital

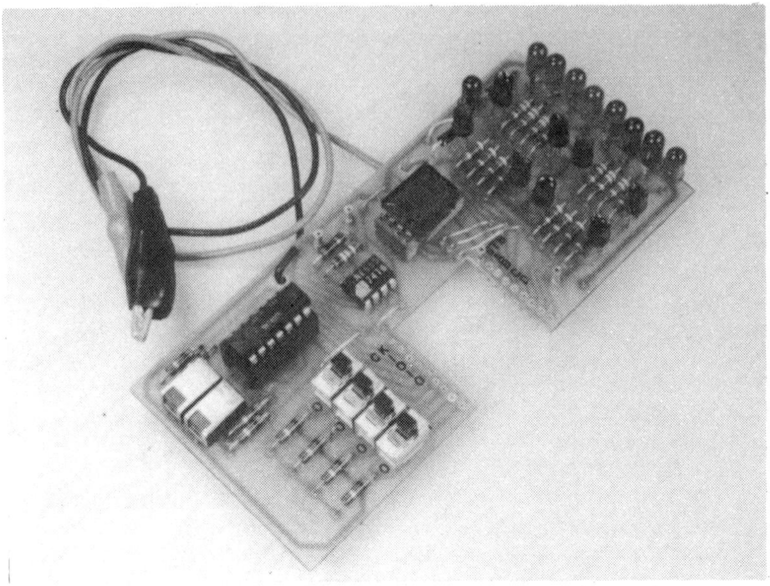

Fig. 1-14. A close-up view of the LR-25 breadboarding station module.

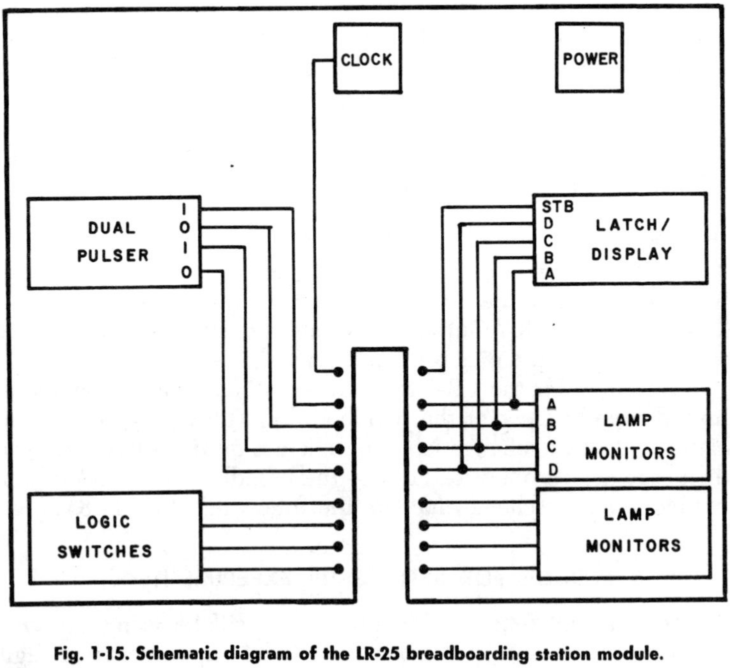

Fig. 1-15. Schematic diagram of the LR-25 breadboarding station module.

Fig. 1-16. The LR-25 breadboarding station attached to an SK-50 half socket; system can accomodate three integrated-circuit chips.

electronics. Before you set up any experiment, we recommend that you do the following:

1. Plan your experiment beforehand. Know what kinds of results you are expected to observe.
2. Disconnect or turn off all power to the breadboard and auxiliary functions. The source of power will usually be a +6-volt battery or +5-volt power supply.

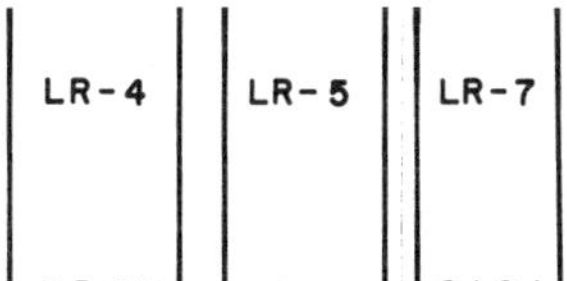

Fig. 1-17. The correct way to label pins in the LR-4, LR-5, LR-7, and LR-9 modules.

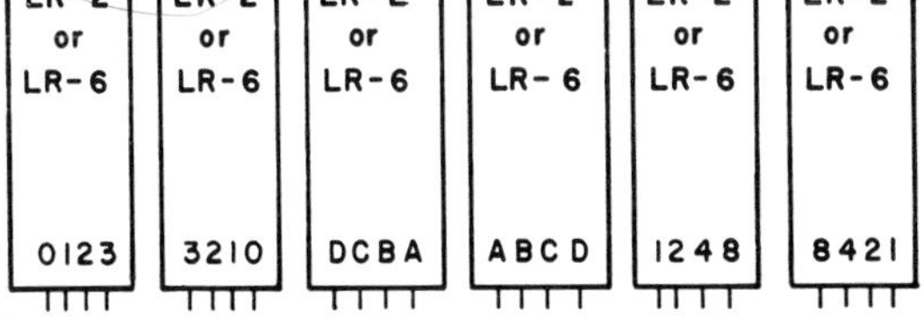

Fig. 1-18. Alternative ways to label a set of four logic switches or lamp monitors, such as the LR-2 and LR-6 modules.

3. Clear the breadboard of all wires and components from previous experiments.
4. Carefully wire up the digital electronic circuit, using whatever integrated circuits, auxiliary functions, and components are required. Read over the "helpful hints" section given later in this chapter for suggestions on the use of pliers, integrated-circuit sockets, "zero-insertion" integrated-circuit sockets, and small wire lengths.
5. Pay particular attention to the specific location of the various components and auxiliary functions on the breadboard. The judicious location of logic switches, lamp monitors, and seven-segment LED displays can frequently minimize the "jungle" of wires that you create on the breadboard.
6. Check the wired circuit to make certain that it is correct. Are any integrated-circuit chips plugged in backwards on the breadboard? Does each integrated circuit have both a GROUND and a +5 Volts connection?
7. Connect or turn on the power to the breadboard LAST. If available, use a current-limiting setting on the power supply to minimize the potential of burning out an improperly wired integrated-circuit chip. If the operation of the circuit appears satisfactory under a "current-limiting" condition, switch over to the normal current setting on the power supply.
8. Make certain that you disconnect or turn off all power to the breadboard when you are finished for the day.
9. If you encounter any unfamiliar terms, go to the beginning of the chapter and try to find them in the list of definitions before proceeding any further.

EXPERIMENT INSTRUCTIONS FORMAT

The instructions for each experiment are presented in the format described below.

Purpose

The material presented under this heading states the purpose of your performing the experiment. It is well for you to have this intended purpose in mind as you conduct the experiment.

Pin Configurations of Integrated-Circuit Chips

Pin configurations, shown with permission of Texas Instruments Incorporated, are given under this heading for all of the integrated-circuit chips used in the experiment. The pin configuration of the 555 Timer is shown in this book with the permission of Signetics Corporation.

Schematic Diagram of Circuit

Under this heading is a schematic diagram of the completed circuit that you will wire up in the experiment. You should analyze this diagram in an effort to obtain an understanding of the circuit *before* you proceed further with the experiment. We shall almost exclusively rely on the use of schematic diagrams in lieu of photographs, figures showing the wiring on the SK-10 socket, and detailed explanations on how to wire up the circuit. *We encourage you to become oriented toward the use of schematic diagrams as your chief tool in wiring digital circuits.* Pay particular attention to the pin numbers on the integrated-circuit chips. Don't forget pins 7 and 14 (or 8 and 16, or 12 and 24, or 5 and 10, etc.), the power connections to the integrated circuits.

Steps

Under the heading of each sequential step, i.e., Step 1, Step 2, Step 3, Step 4, etc., are detailed instructions for setting up portions of the experiment. Questions are also asked at different points in this material. You should answer the questions in the space provided *at the time that you are performing that step of the experiment.* After you have written down your answer, determine whether the correct answer is provided in the text following the space. If it is and if the two answers do not agree, be certain that you fully understand the discrepancy (and correct your answer, if possible) before you continue further in the experiment. Not all answers to questions are provided directly in the text of this book.

Questions

We finish each experiment with one or more questions that probe (a) your understanding of the experiment that you have just completed, (b) your ability to anticipate future experiments or problems, (c) your ability to correlate textual material with experimentally determined material, or (d) your ability to make wild guesses or suggest reasonable explanations to more general questions covering material to which you haven't been previously exposed. The number of questions given will depend on the nature of the experiment and how far you are in the book. Fewer questions are asked in later chapters.

Definitions

Definitions for new terms and concepts are given at the beginning of each chapter.

HOW MANY EXPERIMENTS DO YOU PERFORM?

In this book, there are at least two and perhaps three times as many experiments as you need to perform in order to gain a sound under-

standing of elementary digital electronics! In other words, do not feel it necessary to perform every experiment given in each chapter. Be selective. Choose about one-third to one-half of the experiments given and perform them carefully. When an experiment is particularly difficult, we state that such is the case. "Difficult" experiments may take two or three hours to wire correctly. Sufficient experiments are given for a full semester course in digital electronics based on a single three-hour laboratory session per week.

HELPFUL HINTS AND SUGGESTIONS

A wide variety of hints and suggestions concerning the purchase and handling of integrated-circuit chips, auxiliary functions, breadboards, and other electronic components are given below.

Tools

- Only three tools are really necessary for all of the experiments given in this book.

  a pair of pliers (We currently prefer the *Xcelite* 5-inch chain nose plier with spring, No. 72CG.)

  a screwdriver (Any inexpensive screwdriver that is approximately 4 to 5 inches long and has a tip width of 1/8 inch or a little less is quite satisfactory.)

  a wire stripper/cutter (We prefer the *Xcelite* wire stripper/cutter, No. 101-S.)

  We don't recommend that any other tools be purchased.
- The pliers are used to straighten out the bent ends of hookup wire that is used to wire up circuits on the breadboard.
- The pliers are also used to straighten out or bend into position the pins on integrated-circuit chips.
- The pliers are also used to bend resistor and capacitor leads to the proper position so that they can be conveniently inserted on the breadboard.
- The wire stripper/cutter is used to cut hookup wire to size and to strip the ends to a distance of approximately 3/8 inch.

- The screwdriver is an excellent integrated-circuit chip removal tool. Simply insert the screwdriver in the center groove of the SK-10 breadboard and gently ease out the integrated-circuit chip from the center terminals.

- The screwdriver can also be used to help remove the auxiliary function OUTBOARD modules from the edges of the breadboard. We recommend a small length of wood or plastic as being less likely to damage the soldering job on the underside of the printed-circuit board.

- The pliers can also be used to insert and remove hookup wire from the breadboard terminals. We have found this to be particularly useful when we have a "jungle" of wires with a complex digital circuit.

Wire

- Three wire sizes can be employed with the SK-10 solderless breadboarding socket,

No. 22
No. 24
No. 26

- We prefer No. 24 wire. As a second choice, we recommend No. 22 wire.
- An excellent source of pre-stripped No. 24 wire is Storm Products Company, 112 South Glasgow Avenue, Inglewood, California 90301. We purchased one thousand pieces each of the following sizes and colors, with each wire stripped 1/4 inch from each end:

red wire, 3-in. length
black wire, 3-in. length
yellow wire, 6-in. length
green wire, 10-in. length

We have found other colors to be useful, but have not changed the basic wire sizes originally chosen. These colors include

white wire, 3-in. length
blue wire, 6-in. length
brown wire, 6-in. length
purple wire, 6-in. length

- The hook-up wire used for making connections on the breadboard must be solid, not stranded.
- The ends of the wire should be stripped to a length of 1/4 inch or 5/16 inch from the ends. These are the optimum lengths of bare wire for the SK-10 breadboard.
- At least four to six different hook-up wire colors should be available for breadboarding applications. We recommend that the very short wires be red and black; these are the standard colors for +5 Volts and GROUND connections in our system.

- We have found it useful to cut in half the Storm Products Company 3-inch lengths of black and red wire. They permit us to make low profile connections between integrated-circuit chips and the power buses.
- Hookup wire that we have also used has been Alpha No. 7161/24 irradiated PVC hookup wire, No. 24 solid tinned copper, 0.010-inch wall, 80 °C, 1000 ft. Alpha No. 7161, which is No. 22 solid tinned copper wire, is a suitable second choice.

Breadboarding

- We recommend the judicious use of colors and wire lengths when wiring more complex digital circuits. There are numerous ways of doing this, and we will leave it up to your imagination to determine which way is the "best."
- Never insert too large a wire, pin, or device into a breadboard.
- Never insert a bent wire into a breadboard terminal. Straighten out the end with a pair of pliers before inserting it. The insertion of a bent wire is the easiest way to damage a solderless terminal.
- Try to maintain a low wire profile so that you can better see the pins on the integrated circuits.
- Try to avoid producing a "jungle" of wires of uncoordinated size and color.
- Arrange the integrated circuits and auxiliary function Outboard modules on the breadboard so that only short wire interconnections are necessary.
- Never insert an integrated-circuit chip with one or more bent pins into the breadboard. Straighten out the pins first with a pair of pliers.
- Use *integrated-circuit sockets* for 40-pin or expensive integrated-circuit chips.
- The most useful wire connections are those that lie flat on the breadboard without blocking unused terminals. The 1-inch and 1/8-inch wire lengths are especially handy for connecting +5 Volts and GRD pins on integrated-circuit chips to bus terminals on the SK-10 socket.
- The outer group of 100 terminals on the edge of the breadboard are the +5 Volts bus terminals. Make certain that all 100 of the +5 Volts terminals are connected together.
- The inner group of 100 terminals on the edge of the breadboard are the GROUND bus terminals. Make certain that all 100 GROUND terminals are connected together.
- The breadboards are not too expensive. You may wish to have more than one breadboard available so that you do not have to destroy one digital circuit in order to wire another.

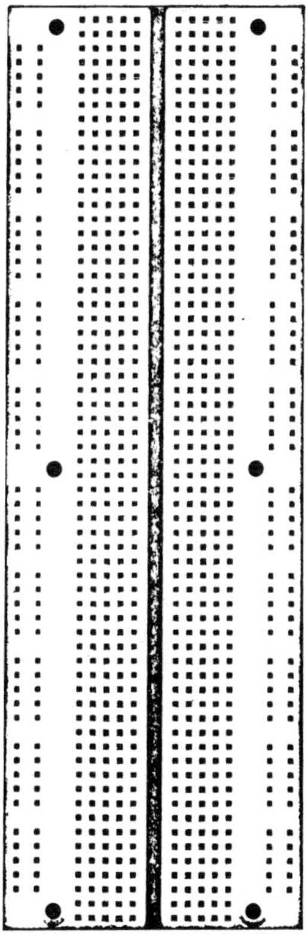

Fig. 1-19. Reproduced layout sheet made by placing SK-10 socket on flat surface of recent model IBM duplicating machine and laying four sheets of white paper around the socket edges for a clear impression.

- Plan the construction of your digital circuit on a layout sheet before you breadboard it. Layout sheets can be made by placing an SK-10 socket on the flat surface of an IBM or Xerox copying machine (Fig. 1-19), as we have done for this book.
- The socket holes on the SK-10 socket may be initially quite stiff. Use caution when you insert auxiliary functions into a new socket.
- Before you disconnect a circuit, check the following experiment to see if the same basic circuit is used.

Auxiliary Functions

- Handle them carefully! They cost an average of $15 each.
- To remove an auxiliary function Outboard from the SK-10 socket, use a flat wooden or plastic instrument, if possible. An acceptable instrument would be a small screwdriver that has its tip taped to minimize the danger of damaging the printed circuit on the underside of the Outboard module.

- Your circuit need not be limited simply to one lamp monitor, logic switch, pulser, or seven-segment LED display Outboard module. Use as many such Outboards as you require for the circuit.
- The labeling of the LR-2 Logic Switch and LR-6 Lamp Monitor Outboard modules is arbitrary. Label the four input or output pins on these two Outboard modules ABCD, or DCBA, or 0123, or 1248, or whatever you choose.
- Locate the Outboard modules opposite each other on the SK-10 socket to conserve valuable socket space.
- We have found it advantageous to locate all of the displays on one side of the SK-10 socket and the pulsers and switches on the opposite side of the socket. The clock Outboard modules can be located on either side.
- The connecting pins to the SK-10 socket from the Outboard modules are somewhat fragile. You may wish to increase the amount of solder on the underside of the modules to increase the strength of their attachment.
- Do not locate an integrated-circuit chip immediately opposite the input or output pins on an Outboard module. Chaos will result. You will burn out either the chip or the Outboard module or both! Locate integrated-circuit chips on one side of or between adjacent Outboard modules that are spaced at least seven or eight solderless terminals apart.
- By the end of this book, you will understand how all of the Outboard modules operate electronically.
- The LR-7 Dual Debounced Pulser Outboard module is a very useful teaching aid. It permits you to apply one or more clock pulses at irregular time intervals, thus allowing you to observe your displays and write down your observations between clock pulses.
- Be careful not to bend or break the two power pins on each auxiliary function Outboard module. They are more vulnerable than the four input/output pins on each module. We replace broken power pins with wire-wrap pins that are cut in half.

Components and Other Electronic Devices

A variety of different resistors will be required for circuits in this book. They include the following:

1000-ohm pull-up resistors for open collector outputs on integrated-circuit chips
150- to 300-ohm current-limiting resistors for LEDs and seven-segment LED displays
resistors of different values for use in setting RC time constants

A variety of different capacitors will be required for circuits in this book to set RC time constants in timers and monostable multivibrators. Useful values include:

100 pF
200 pF
500 pF
0.001 μF
0.002 μF
0.005 μF
0.01 μF
0.02 μF
0.05 μF
0.10 μF, Mylar
0.22 μF, Mylar
0.50 μF, Mylar or electrolytic
1.0 μF, electrolytic
5.0 μF, electrolytic
10. μF, electrolytic
20. μF, electrolytic
50. μF, electrolytic

Over the past year, we have obtained most of our fixed resistors and capacitors from James Electronics, P. O. Box 822, Belmont, California 94002. Red, green, and yellow light-emitting diodes (LEDs) are also available from them at modest cost.

We like the Opcoa SLA-1 seven-segment LED display, which is available from several different sources. The cost is reasonable and the light intensity per segment is good. We prefer it to the Litronix Data Lit 704. The Hewlett-Packard five-digit seven-segment LED readout is an excellent display.

Integrated Circuits

Most 14-pin and 16-pin integrated circuits can generally be inserted directly into the SK-10 breadboard. Make sure that the pins are not bent and that they are vertical. Use a pair of pliers or a hard flat surface to gently bend a group of 7 or 8 pins into a vertical position. Expensive integrated circuits (those exceeding $5 in price) or those that have more than 24 pins should be placed in an *integrated-circuit socket* prior to insertion on the breadboard. This action will prevent pin breakage and integrated-circuit chip destruction.

Zero-force insertion sockets can be used whenever it is necessary to repeatedly insert and remove integrated circuits from the breadboard, as would be the case when integrated-circuit chips are being tested for malfunctions. Such sockets work extremely well.

Integrated circuits in the 7400-series TTL logic family are relatively inexpensive today. They can be obtained from many different suppliers. Shop wisely and you can save money. Our favorite supplier is James Electronics, which markets the regular LS, S, H, and 5400-series of TTL chips. Order only dual-in-line package (DIP) integrated circuits. The "flat-pack" integrated-circuit chips, which are not commonly available, cannot be conveniently used on the SK-10 solderless breadboard.

All integrated circuits come with the type of chip and date code printed on the top; some chips also have numbers on the underside. Do not rub out, dissolve, scratch, or otherwise obliterate these numbers. Develop some fail-safe system so that you can readily identify chips whose numbers have been partially obliterated.

Chip burnout is usually caused by improper power connections (+5 Volts to GROUND input pin and GROUND to +5 Volts input pin). Be careful!

The pins on some integrated circuits are pointed at the ends, whereas the pins on others have flat ends. Naturally, ICs with pointed pins are considerably easier to insert into a solderless breadboard. Correct pin function identification on integrated circuits is an important part of wiring any digital electronic circuit. You will find the following manufacturer's specification handbooks to be exceedingly useful in your experiments:

The TTL Data Book for Design Engineers
Texas Instruments Incorporated
P. O. Box 5012
Dallas, Texas 75222

Signetics Digital, Linear, and MOS Handbook
Signetics Corporation, Marketing Communications Dept.
811 East Arques Avenue
Sunnyvale, California 94086

Digital Integrated Circuits
National Semiconductor Corporation
2900 Semiconductor Drive
Santa Clara, California 95051

We consider such handbooks to be our "Bibles" as we wire up new digital electronic circuits. In general, it is not necessary to have all three. The first two handbooks appear to be the most comprehensive for the "54/74 series" of integrated circuits.

The storage of integrated circuits can occasionally be a problem. They should generally not be kept loose in a pile, since pin damage might result. We recommend the use of properly labeled foamed polystyrene, small glass or plastic jars, special plastic cabinets, etc., as containers.

Prices for integrated circuits vary, but the following should give you a feel for their cost as of mid 1978:

Small-scale integration (SSI)	*Price per IC*
simple gates	$0.10 to $0.25
flip-flops	$0.30 to $0.80
Medium-scale integration (MSI)	
counters	$0.65 to $1.50
decoders	$0.75 to $1.25
data selectors/multiplexers	$0.95 to $1.25
shift registers	$0.90 to $2.25
decoder/drivers	$0.75 to $1.15
arithmetic units	$0.75 to $3.55
random access memories (RAMs)	$2.00 up

The dimensions of typical plastic dual-in-line package (DIP) integrated circuits are given in Fig. 1-20. Note that the dimensions are in metric units.

The numbering of the 7400-series integrated circuits can be confusing on occasion. On the top of each chip, you will usually find three kinds of information:

The manufacturer's name, symbol, or initials
The manufacturer's date code
The 7400-series number of the chip

Only the 7400-series number is important to you. The leading manufacturers of the 7400-series chips include

Fairchild Semiconductor
ITT Semiconductors (ITT)
Motorola Semiconductors
National Semiconductor (NS)
Raytheon Semiconductor
Signetics (S)
Sprague Electric Company (Sprague)
Stewart Warner Corp. (SW)
Texas Instruments (Small map of Texas)

The date code on the integrated circuit gives the year and the week of that year during which the chip was manufactured. Thus, a date code of 7106 indicates that the IC was manufactured in the sixth week of 1971. Be alert for confusion with chips that were manufactured in 1974. The number 7410 could mean the 7410 triple 3-input NAND gate or the fact that the chip in question was manufac-

TTL INTEGRATED CIRCUITS MECHANICAL DATA

N plastic dual-in-line packages

These dual-in-line packages consist of a circuit mounted on a 14-, 16-, or 24-lead frame and encapsulated within an electrically nonconductive plastic compound. The compound will withstand soldering temperature with no deformation and circuit performance characteristics remain stable when operated in high-humidity conditions. These packages are intended for insertion in mounting-hole rows on 7.62-mm (or 15.24-mm) centers. Once the leads are compressed and inserted, sufficient tension is provided to secure the package in the board during soldering. Silver-plated leads (–00) require no additional cleaning or processing when used in soldered assembly.

14-PIN N PLASTIC DUAL-IN-LINE PACKAGE OUTLINE

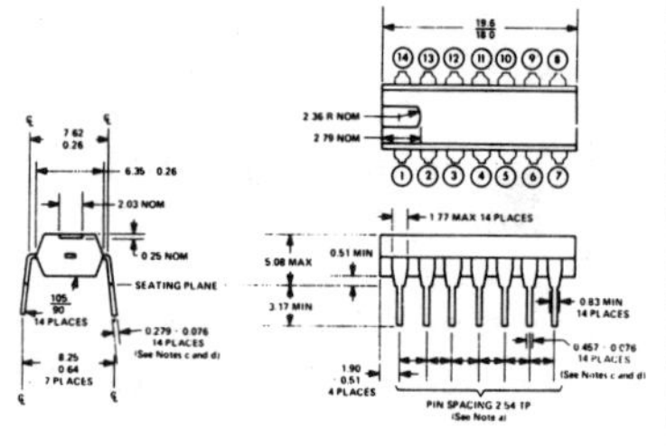

Falls Within JEDEC TO-116 and MO-001AA Dimensions

16-PIN N PLASTIC DUAL-IN-LINE PACKAGE OUTLINE

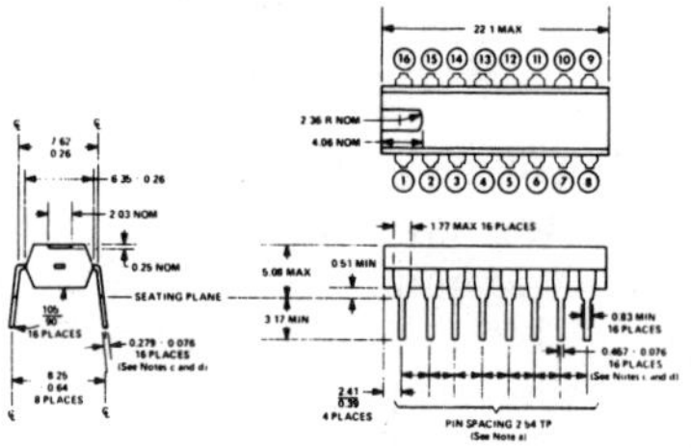

24-PIN N PLASTIC DUAL-IN-LINE PACKAGE OUTLINE

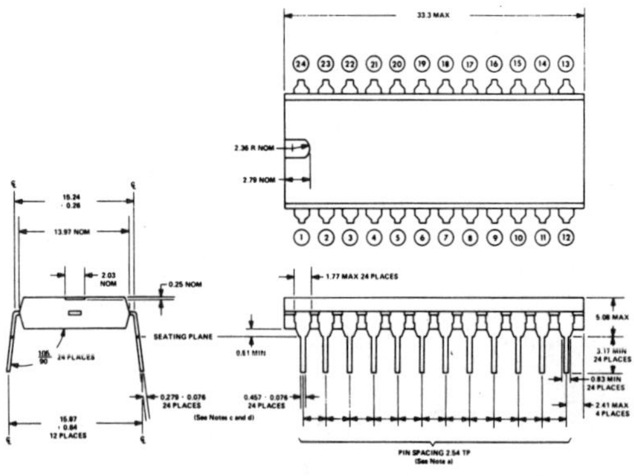

NOTES:

a. Each pin centerline is located within 0.26 mm of its true longitudinal position.

b. All dimensions are in millimeters unless otherwise noted.

c. This dimension does not apply for solder-dipped leads.

d. When solder-dipped leads are specified, dipped area of the lead extends from the lead tip to at least 0.5 mm above the seating plane.

Courtesy Texas Instruments Incorporated

Fig. 1-20. Data sheet.

tured during the tenth week in 1974. To illustrate the identification of integrated circuits, consider the following examples:

1. Sprague 7106 US7400A: This is a 7400 quad 2-input NAND gate manufactured during the sixth week of 1971.
2. SW 7400-N 20231 7216: This is also a 7400 quad 2-input NAND gate manufactured by Stewart Warner Corp. in the sixteenth week of 1972.
3. S7319 N7400 AH: This is a third 7400 quad 2-input NAND gate, this time manufactured by Signetics Corporation during the nineteenth week of 1973.
4. ITT 7420J 7307: A 7402 quad 2-input NOR gate manufactured by ITT during the seventh week of 1973.
5. SW 7490-N 7304: A 7490 decade counter manufactured by Stewart Warner Corp. in the fourth week of 1973.
6. NS 351 DM7489N: This is a 7489 64-bit RAM manufactured by National Semiconductor during the 51st week of 1973.
7. 7324 SN7440N Map of Texas: This is a 7440 dual 4-input positive NAND buffer manufactured by Texas Instruments during the 24th week of 1973.

In the following pages, we have summarized the "pin configurations" for most of the integrated circuits that we shall use in this book.

INTRODUCTION TO THE EXPERIMENTS

The following experiments are designed to demonstrate the operation of the principal OUTBOARD® modules. To conduct these experiments, you will need the following:

- LR-1 power module
- LR-2 logic switch module
- LR-4 seven-segment LED display module
- LR-5 clock module
- LR-6 LED lamp monitor module
- LR-7 dual pulser module

Alternatively, you can perform most of the experiments in this chapter and in the rest of the book using the

- LR-25 Breadboarding Station module

which contains the LR-1, LR-2, LR-4 (in the form of an LR-26 latch/display), LD-5, LR-6, and LR-7 digital functions on a single printed-circuit board. In addition to the above functions, you will also need No. 24 hookup wire in different colors and lengths, capacitors in the vicinity of 0.03 to 1.0 μF, and the 7490 integrated-circuit chip.

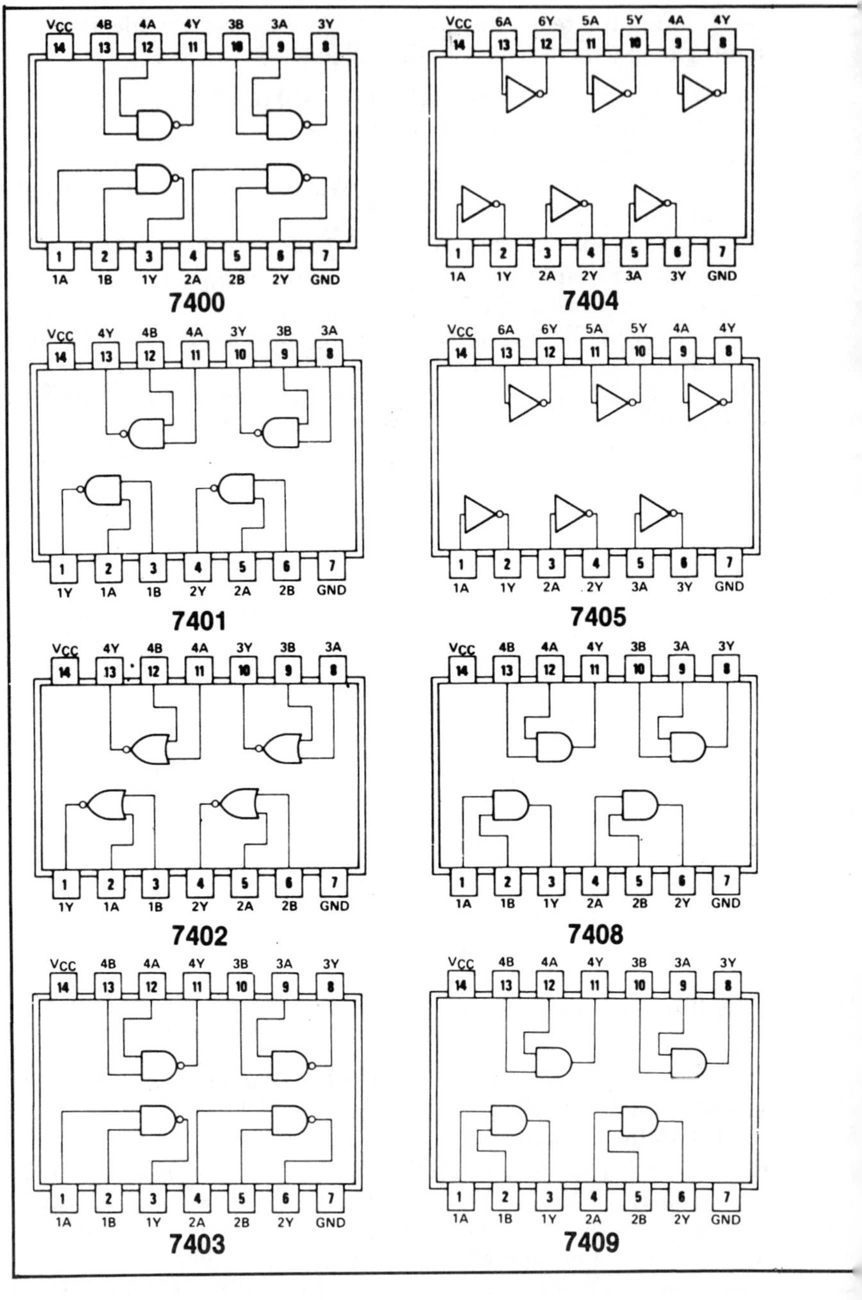
VCC 4B 4A 4Y 3B 3A 3Y
14 13 12 11 10 9 8
1 2 3 4 5 6 7
1A 1B 1Y 2A 2B 2Y GND
7400
VCC 6A 6Y 5A 5Y 4A 4Y
1A 1Y 2A 2Y 3A 3Y GND
7404
VCC 4Y 4B 4A 3Y 3B 3A
1Y 1A 1B 2Y 2A 2B GND
7401
VCC 6A 6Y 5A 5Y 4A 4Y
1A 1Y 2A 2Y 3A 3Y GND
7405
VCC 4Y 4B 4A 3Y 3B 3A
1Y 1A 1B 2Y 2A 2B GND
7402
VCC 4B 4A 4Y 3B 3A 3Y
1A 1B 1Y 2A 2B 2Y GND
7408
VCC 4B 4A 4Y 3B 3A 3Y
1A 1B 1Y 2A 2B 2Y GND
7403
VCC 4B 4A 4Y 3B 3A 3Y
1A 1B 1Y 2A 2B 2Y GND
7409

Fig. 1-21. Pin

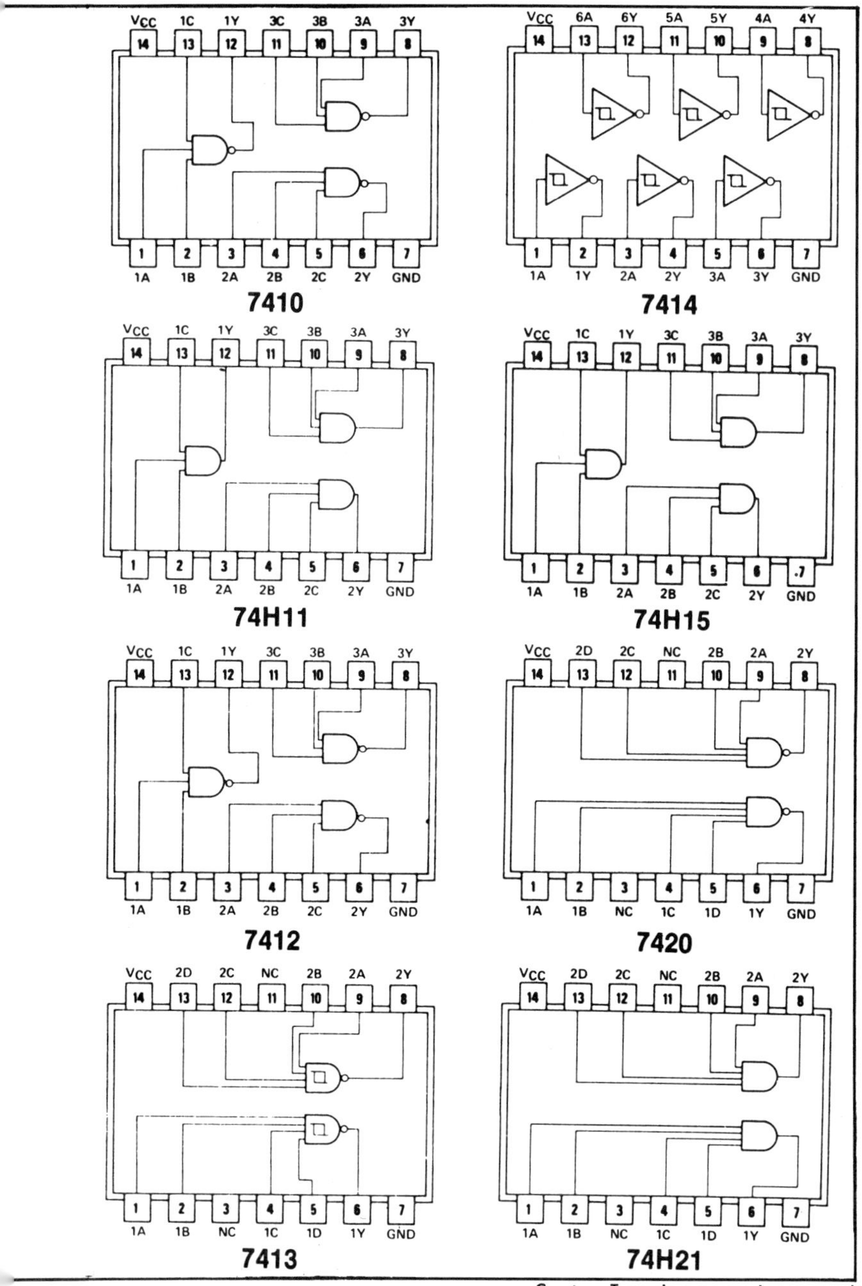

Courtesy Texas Instruments Incorporated

configurations.

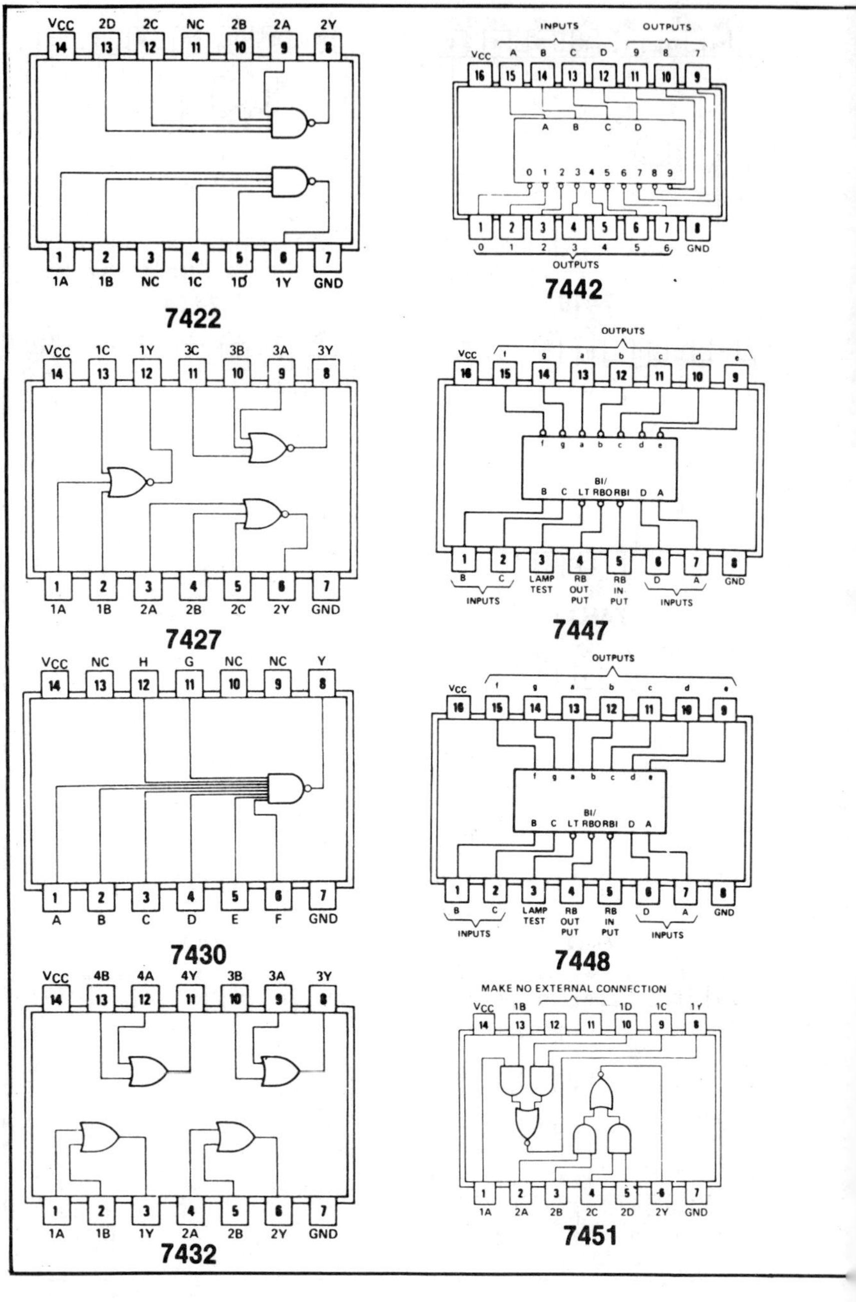

7422
Vcc 2D 2C NC 2B 2A 2Y
1A 1B NC 1C 1D 1Y GND
7427
Vcc 1C 1Y 3C 3B 3A 3Y
1A 1B 2A 2B 2C 2Y GND
7430
Vcc NC H G NC NC Y
A B C D E F GND
7432
Vcc 4B 4A 4Y 3B 3A 3Y
1A 1B 1Y 2A 2B 2Y GND
7442
INPUTS
OUTPUTS
Vcc A B C D 9 8 7
0 1 2 3 4 5 6 GND
OUTPUTS
7447
OUTPUTS
Vcc f g a b c d e
BI/
B C LT RBO RBI D A
B C INPUTS
LAMP TEST
RB OUT PUT
RB IN PUT
D A INPUTS
GND
7448
OUTPUTS
Vcc f g a b c d e
BI/
B C LT RBO RBI D A
B C INPUTS
LAMP TEST
RB OUT PUT
RB IN PUT
D A INPUTS
GND
7451
MAKE NO EXTERNAL CONNECTION
Vcc 1B 1D 1C 1Y
1A 2A 2B 2C 2D 2Y GND

Fig. 1-22. Pin

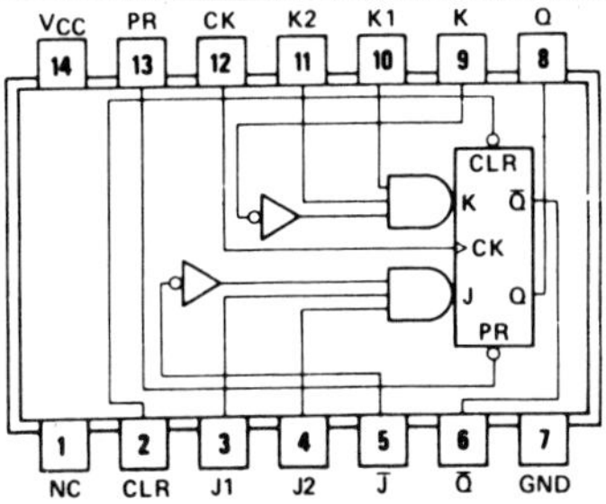

7470

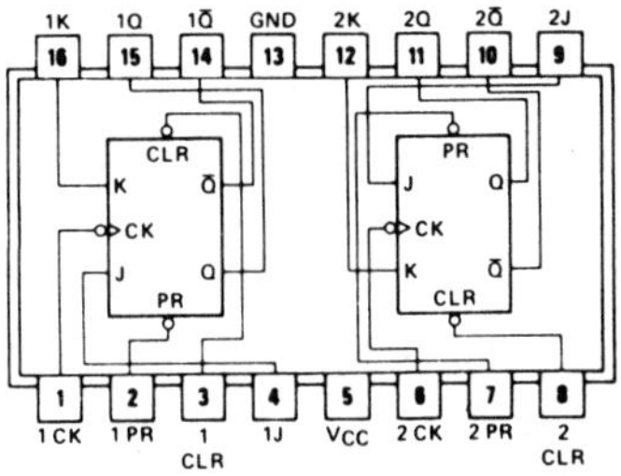

7476

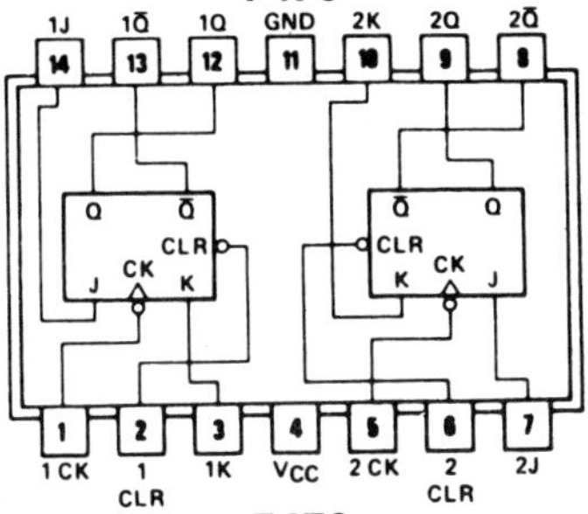

7473

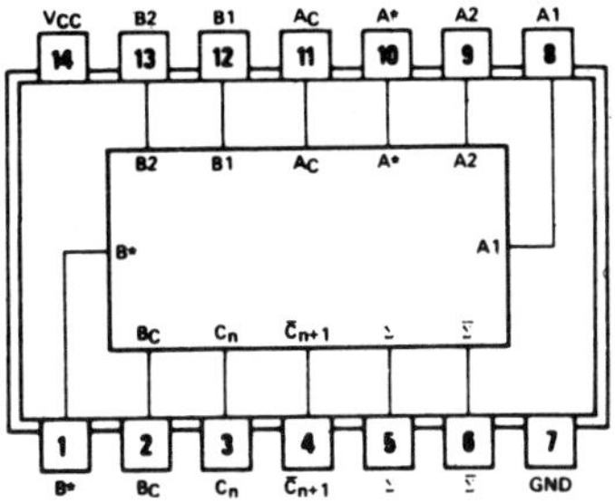

7480

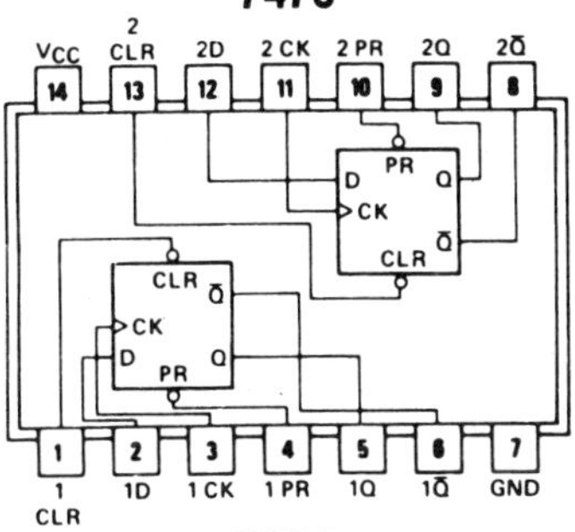

7474

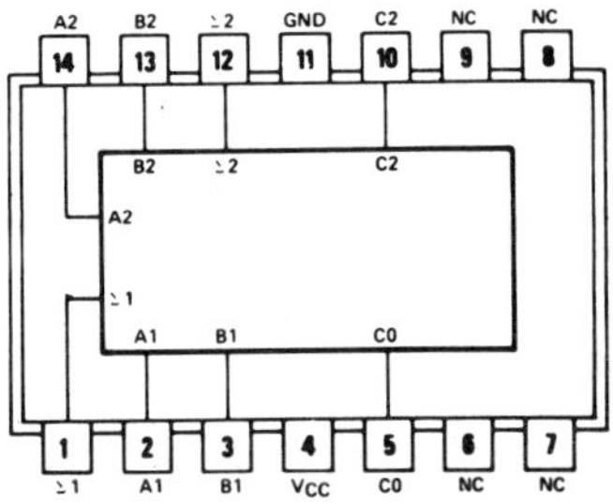

7482

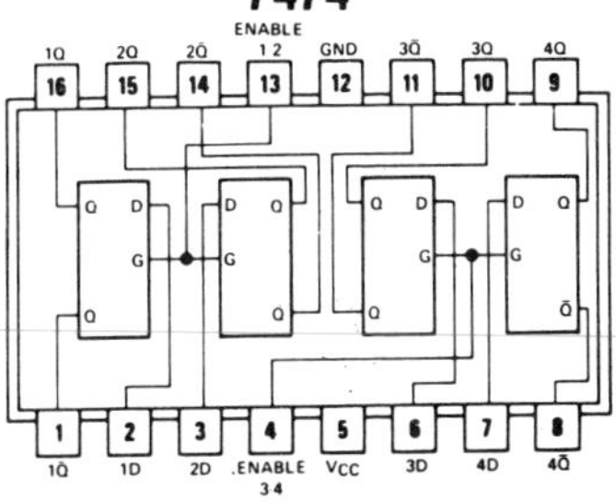

7475

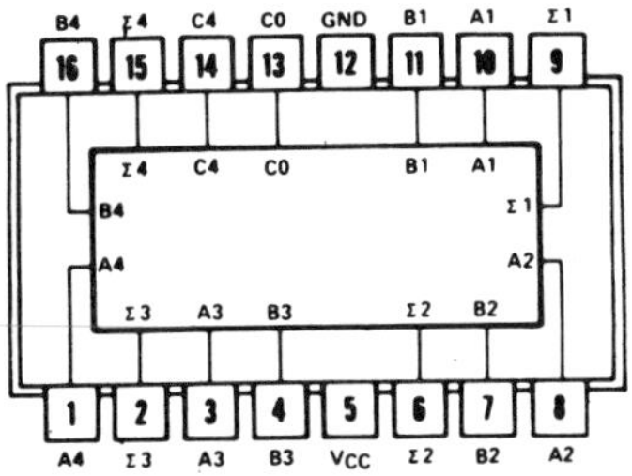

7483

Courtesy Texas Instruments Incorporated

configurations.

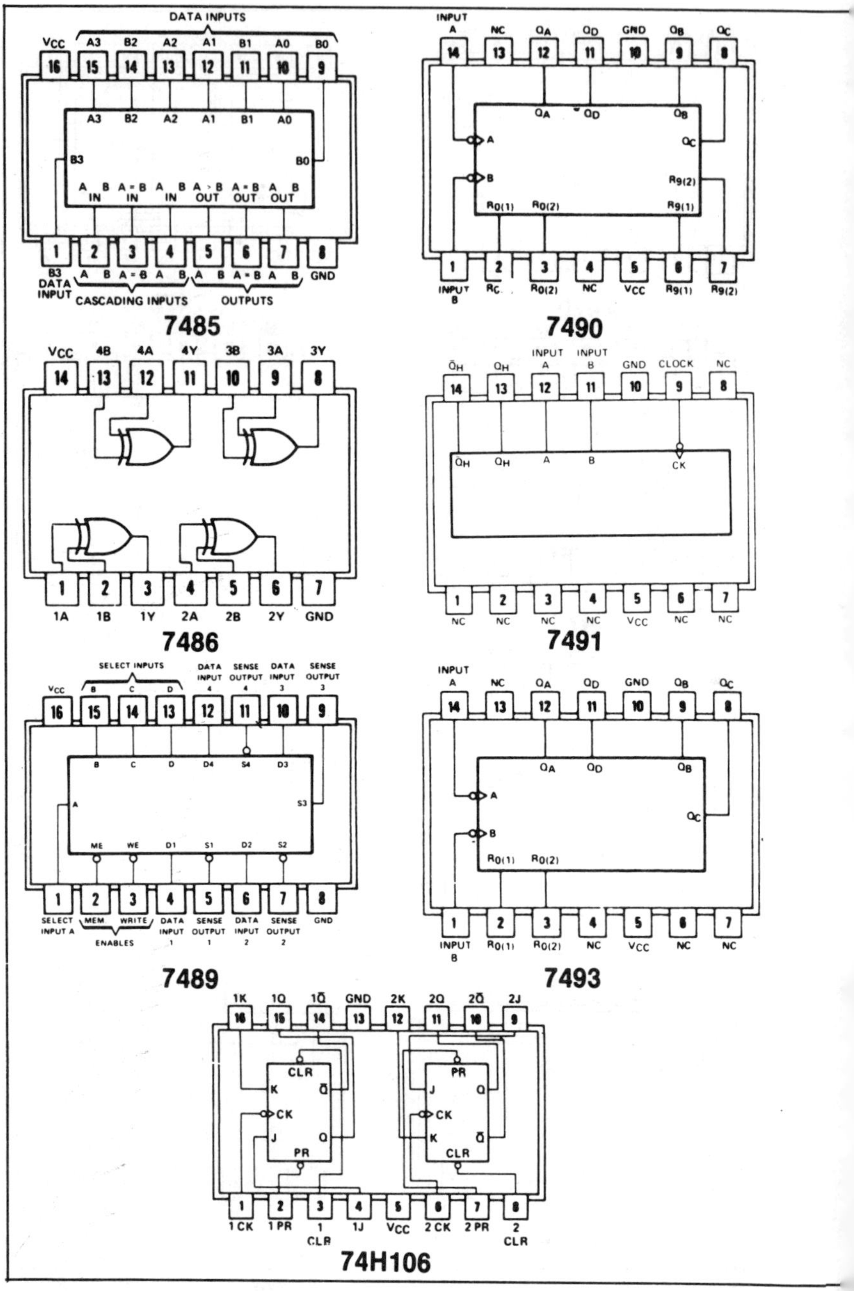

DATA INPUTS
VCC A3 B2 A2 A1 B1 A0 B0
16 15 14 13 12 11 10 9
A3 B2 A2 A1 B1 A0
B3 B0
A B A=B A B A>B A=B A B
IN IN IN OUT OUT OUT
1 2 3 4 5 6 7 8
B3 DATA INPUT
A B A=B A B A B A=B A B GND
CASCADING INPUTS OUTPUTS
7485
INPUT A NC QA QD GND QB QC
14 13 12 11 10 9 8
QA QD QB
A QC
B R9(2)
R0(1) R0(2) R9(1)
1 2 3 4 5 6 7
INPUT B NC R0(2) NC VCC R9(1) R9(2)
7490
VCC 4B 4A 4Y 3B 3A 3Y
14 13 12 11 10 9 8
1 2 3 4 5 6 7
1A 1B 1Y 2A 2B 2Y GND
7486
Q̄H QH INPUT A INPUT B GND CLOCK NC
14 13 12 11 10 9 8
Q̄H QH A B CK
1 2 3 4 5 6 7
NC NC NC NC VCC NC NC
7491
SELECT INPUTS
VCC B C D
DATA INPUT 4 SENSE OUTPUT 4 DATA INPUT 3 SENSE OUTPUT 3
16 15 14 13 12 11 10 9
B C D D4 S4 D3
A S3
ME WE D1 S1 D2 S2
1 2 3 4 5 6 7 8
SELECT INPUT A MEM WRITE ENABLES DATA INPUT 1 SENSE OUTPUT 1 DATA INPUT 2 SENSE OUTPUT 2 GND
7489
INPUT A NC QA QD GND QB QC
14 13 12 11 10 9 8
QA QD QB
A QC
B
R0(1) R0(2)
1 2 3 4 5 6 7
INPUT B R0(1) R0(2) NC VCC NC NC
7493
1K 1Q 1Q̄ GND 2K 2Q 2Q̄ 2J
16 15 14 13 12 11 10 9
CLR K Q̄ CK J Q PR
PR J Q CK K Q̄ CLR
1 2 3 4 5 6 7 8
1 CK 1 PR 1 CLR 1J VCC 2 CK 2 PR 2 CLR
74H106

Fig. 1-23. Pin

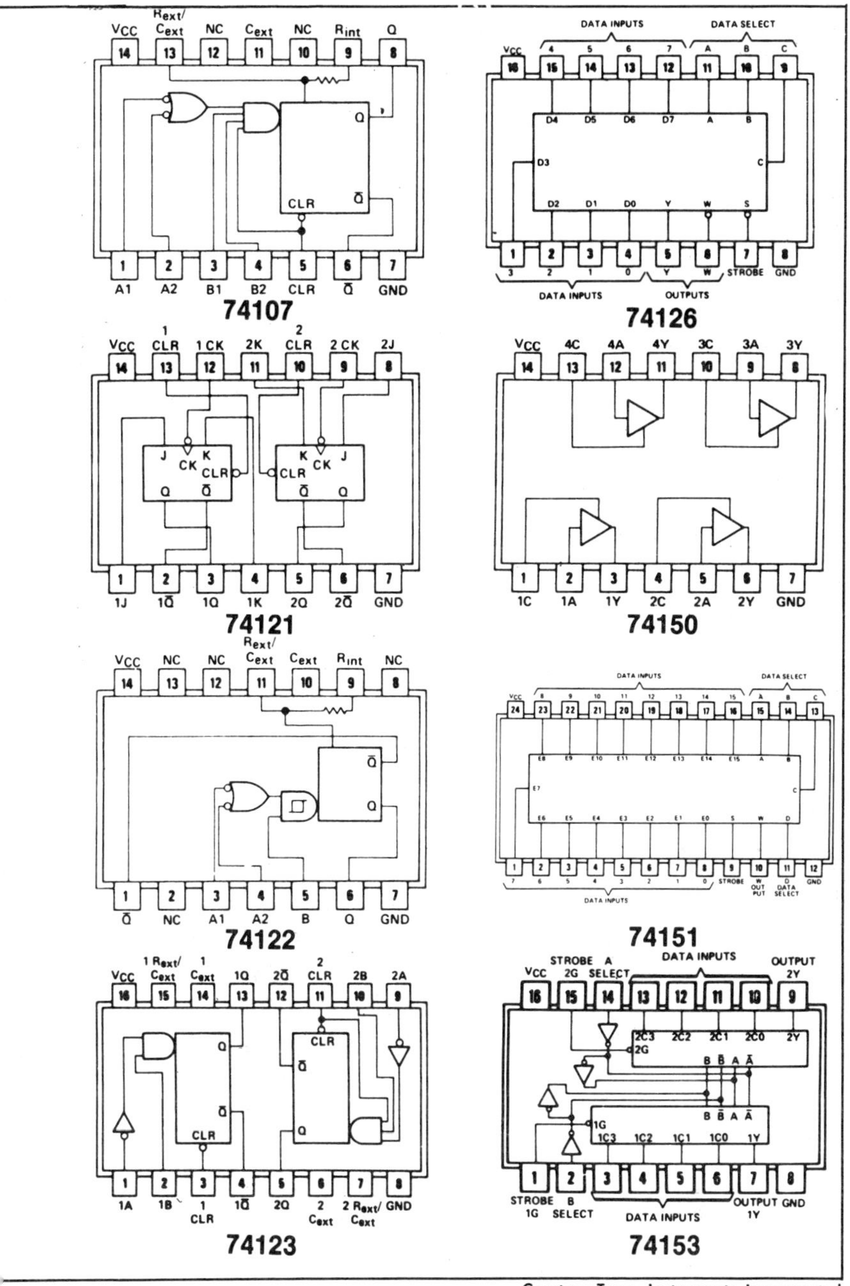

Courtesy Texas Instruments Incorporated

configurations.

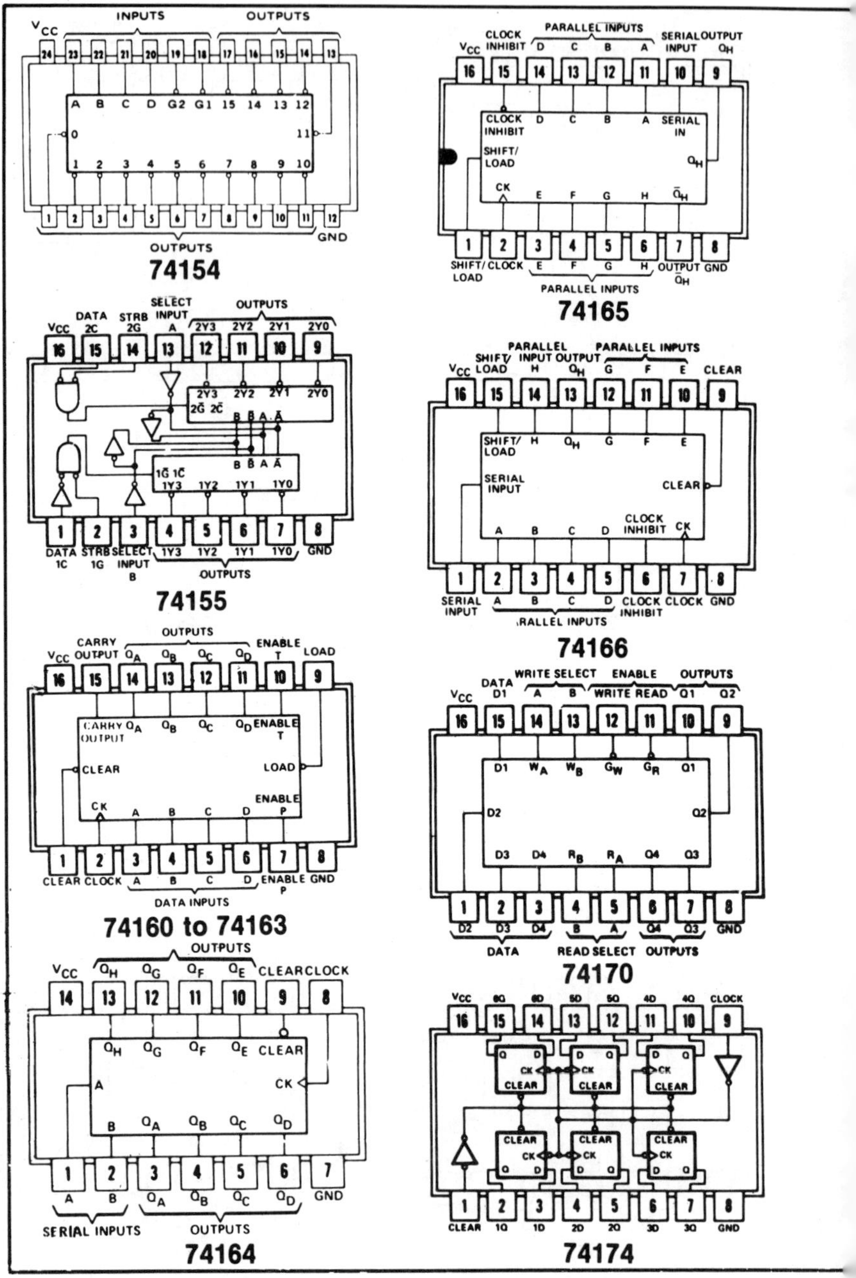

INPUTS
OUTPUTS
VCC
GND
74154
74165
SHIFT/LOAD
CLOCK
CLOCK INHIBIT
SERIAL INPUT
OUTPUT QH
PARALLEL INPUTS
74155
DATA 2C
STRB 2G
SELECT INPUT A
DATA 1C
STRB 1G
SELECT INPUT B
74166
CLEAR
74160 to 74163
CARRY OUTPUT
ENABLE T
LOAD
ENABLE P
DATA INPUTS
74170
WRITE SELECT
ENABLE
WRITE READ
DATA
READ SELECT
74164
SERIAL INPUTS
CLEAR CLOCK
74174

Fig. 1-24. Pin

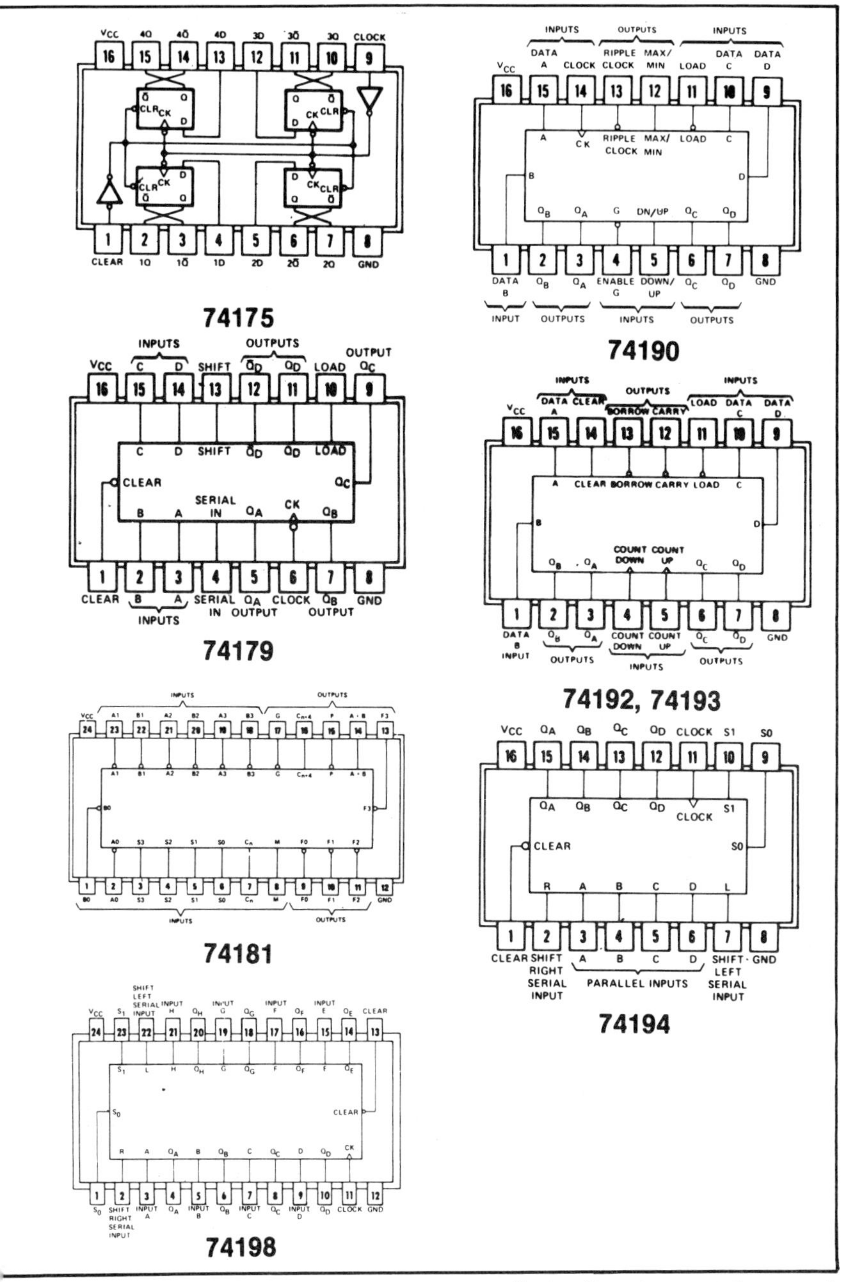

Courtesy Texas Instruments Incorporated

configurations.

If you have not already done so, please read the "Rules for Setting Up Experiments" section provided earlier in this chapter. In Fig. 1-25, we show the types of hardware that you will generally require for this book. The complete breadboarding system is portable and the experiments can be performed anywhere.

EXPERIMENT NO. 1

Purpose

The purpose of this experiment is to explore the operating characteristics of the LR-25 Breadboarding Station Outboard® module.

Schematic Diagram of Outboard (Fig. 1-26)

Step 1

Carefully study the diagram of the LR-25 Outboard module. Note that it contains two sets of four lamp monitors, one set of four logic switches, two debounced pulsers, a clock (with provision for a timing capacitor), and a diode-protected power input. It also contains a socket for a Hewlett-Packard numeric or hexadecimal display. You will find that this breadboarding station will permit you to perform almost any experiment in *Logic & Memory Experiments Using TTL Integrated Circuits, Books 1 and 2.*

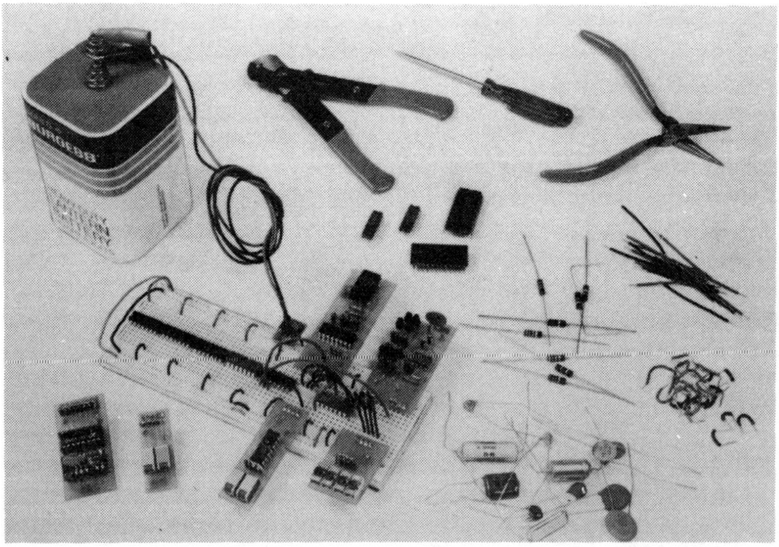

Fig. 1-25. A complete breadboarding system requires little hardware: a socket, a battery, some chips, wire, resistors, capacitor, and a few tools will suffice.

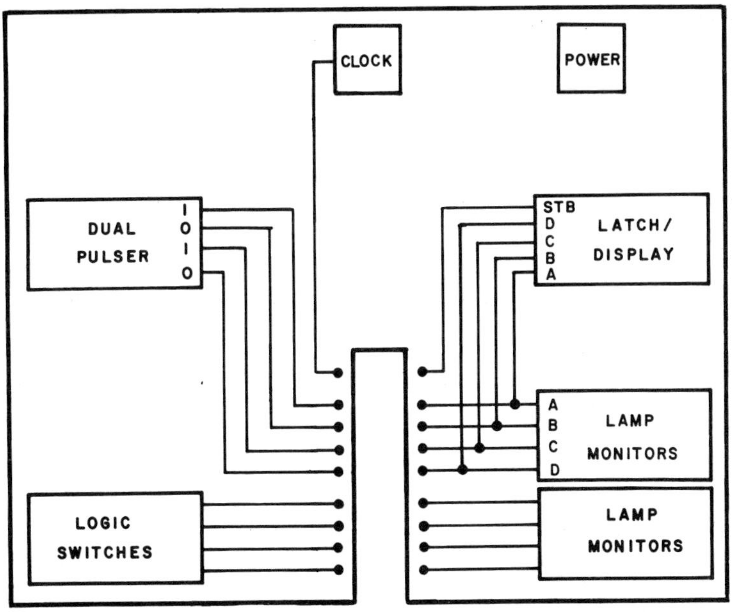

Fig. 1-26.

Step 2

Insert the breadboarding station on one end of an SK-10 or SK-50 breadboarding socket. Make certain that the two power pins, +5 Volts and GROUND, are connected to the appropriate outer bus strips. These pins apply power to the solderless terminals at the edges of the socket.

Step 3

Apply power to the breadboarding station. The power LED should light. To test the operation of the station, we encourage you to perform the early experiments in this chapter, as listed next.

Step 4

Do Experiments Nos. 2 and 3 with the aid of the lamp monitors on the breadboarding station module.

Step 5

Do Experiments Nos. 4 and 5 with the aid of the logic switches, lamp monitors, and latch/display on the breadboarding station. The latch/display replaces the seven-segment display in Experiment No. 5.

Step 6

Do Experiments Nos. 6 and 7 with the aid of the breadboarding station module.

Step 7

Do Experiment No. 8 with the breadboarding station module.

Step 8

Do Experiment No. 9 with the breadboarding station module. You will require a 0.33-microfarad capacitor to maintain a clock rate of approximately 0.7 Hz.

Step 9

Do Experiment No. 10 with the breadboarding station module.

EXPERIMENT NO. 2

Purpose

The purpose of this experiment is to demonstrate the logic 0 and logic 1 states with the use of lamp monitors. The experiment also gives you experience in wiring the SK-10 breadboarding socket.

Schematic Diagram of Circuit (Fig. 1-27)

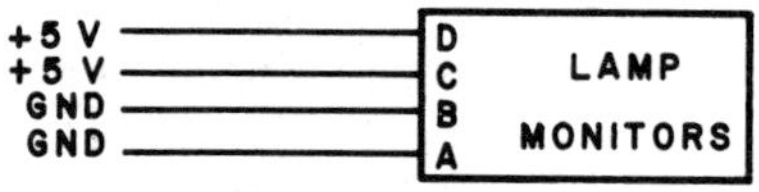

Fig. 1-27.

Step 1

Carefully study the SK-10 breadboarding socket shown in Fig. 1-28. This universal breadboard contains 840 noncorrosive spring contacts, each having an expected contact life of 10,000 contacts. Note that there are eight sets of twenty-five electrically connected solderless terminals on the outside edges of the socket, and 128 sets of five electrically connected solderless terminals in the center of the socket. A groove down the center of the breadboarding socket separates the 128 sets into two sets of 64. Turn the SK-10 socket over. Note the paper-tape backing. Never remove this backing because you may short the interconnections if they touch a metal surface.

Step 2

Carefully study the breadboard set-ups shown in Figs. 1-29 and 1-30. As shown in Fig. 1-30, gently insert the LR-1 Power Outboard

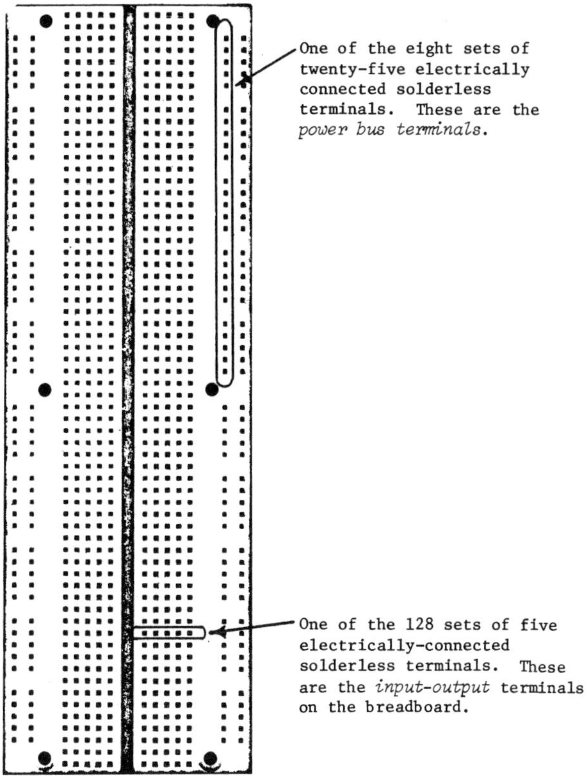

Courtesy E & L Instruments, Inc.

Fig. 1-28. The SK-10 socket.

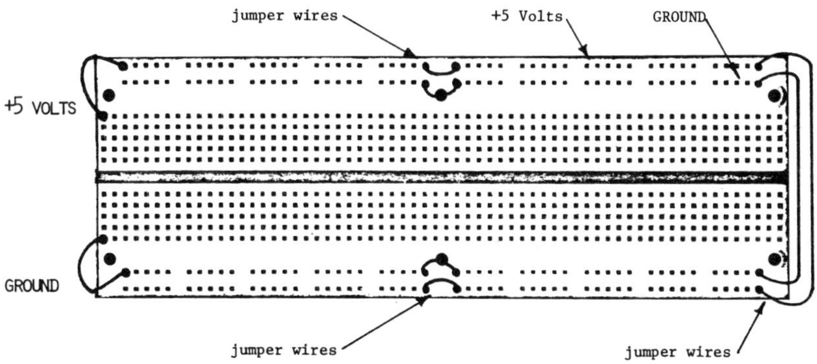

Courtesy E & L Instruments, Inc.

Fig. 1-29. The SK-10 socket ready to attach to a battery or power supply.

module so that it straddles four sets of twenty-five electrically connected solderless terminals on one edge of the SK-10 socket. Connect the four additional jumper wires as shown in Fig. 1-30. With the aid of these jumper wires, there are now 100 solderless terminals at GROUND potential (0 volts) and 100 solderless terminals at ap-

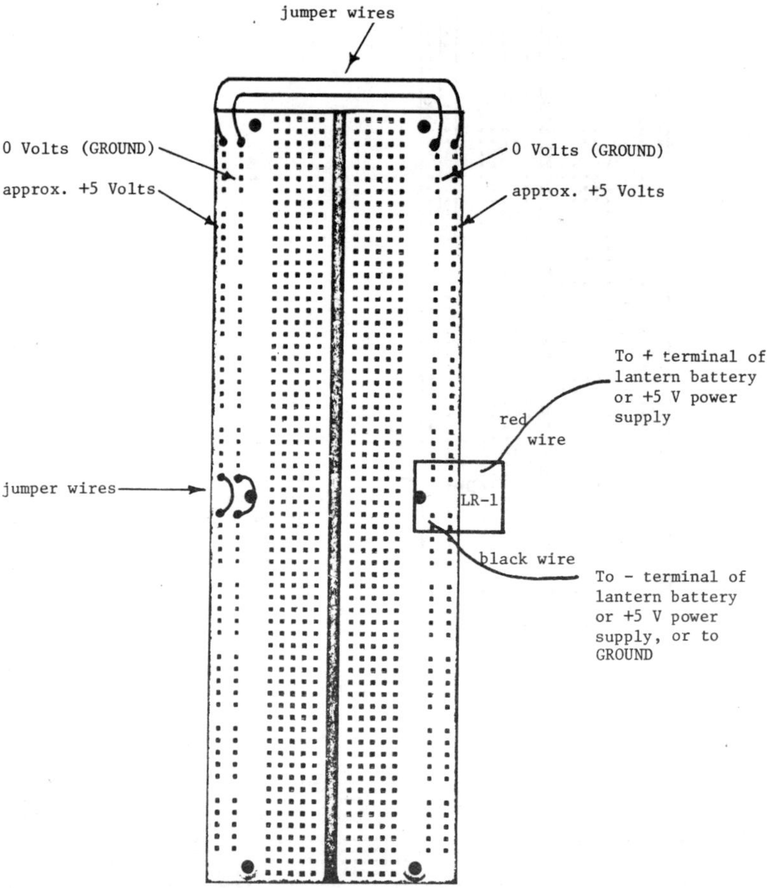

Courtesy E & L Instruments, Inc.

Fig. 1-30. The SK-10 socket with the LR-1 power module properly attached.

proximately +5 volts. The higher potential should never be less than +4.75 volts nor greater than +6.0 volts, the typical operating range for the 7400-series integrated-circuit chips. Do not connect the *alligator clips* that are attached to the LR-1 power module to any power source yet.

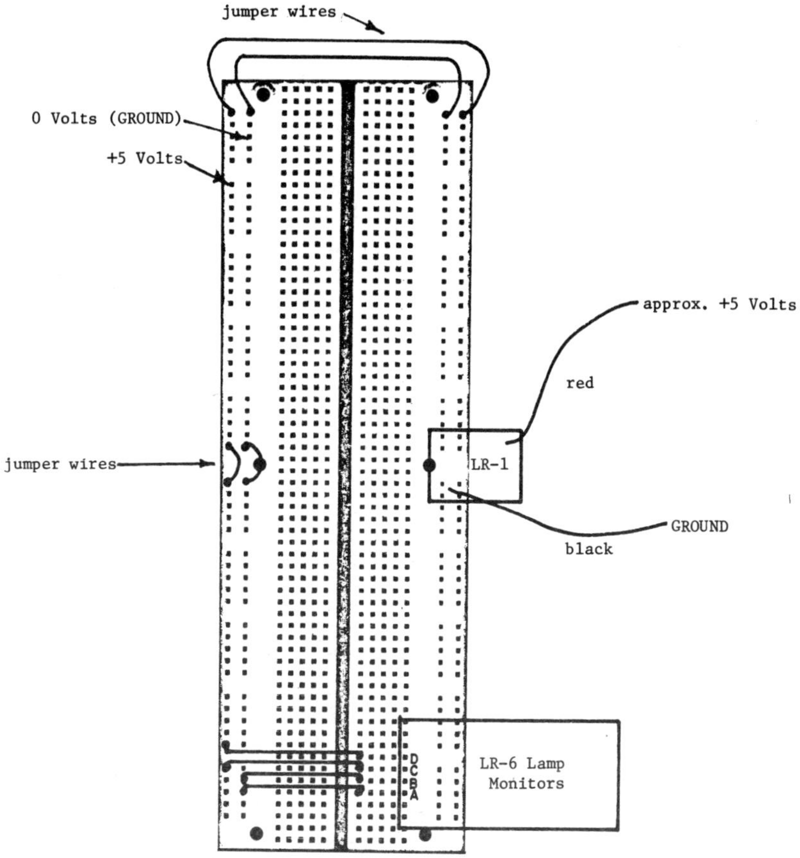

Courtesy E & L Instruments, Inc.

Fig. 1-31. Wiring diagram for Experiment No. 2.

Step 3

Study the set-up shown in Fig. 1-31. Note that we have now added another Outboard module, this time the LR-6 lamp monitor module. We have located the module in this position for a reason, which will become apparent in a later experiment. Gently insert the LR-6 module into the SK-10 socket.

Step 4

Refer again to Fig. 1-31 and connect the four wires to the four inputs of the LR-6 lamp monitor module as shown. The wired circuit should correspond exactly to the schematic diagram shown on the preceding page.

Step 5

Connect the two alligator clips to the battery and write down in the following space how many lamp monitors are lit.

If you have wired this circuit correctly, the lamp monitors that we have defined as C and D in Fig. 1-31 should be lit. Lamp monitors A and B in Fig. 1-31 should be unlit. If you do not observe this result, disconnect the alligator clips from the battery and repeat the experiment.

Step 6

For the LR-6 lamp monitor module, we shall *usually* define a lit lamp as an "ON" or "logic 1" state and an unlit lamp as an "OFF" or "logic 0" state. In Steps 4 and 5 above, which lamp monitors are in the logic 1 state?

In Steps 4 and 5 above, which lamp monitors are in a logic 0 state?

If you have answered these two questions correctly, you should conclude that lamp monitors A and B are both in an "OFF" or "logic 0" state, whereas lamp monitors C and D are both in an "ON" or "logic 1" state. With the use of a simple +6-volt lantern battery or +5-volt power supply, only two different logic states are possible: a "logic 0" state and a "logic 1" state. This is an important result. In fact, *the majority of simple digital electronic circuits are based upon two states, a "logic 0" state and a "logic 1" state, which respectively correspond to voltages that differ by 5 volts, the lower potential usually, but not always, being the "logic 0" state.*

Step 7

Carefully study the "schematic diagram of circuit" given for this experiment and draw a schematic diagram in which all four lamp monitors, A through D, exist in a "logic 1" state. Use the space below for your answer. Wire this circuit on the breadboard and demonstrate that it works.

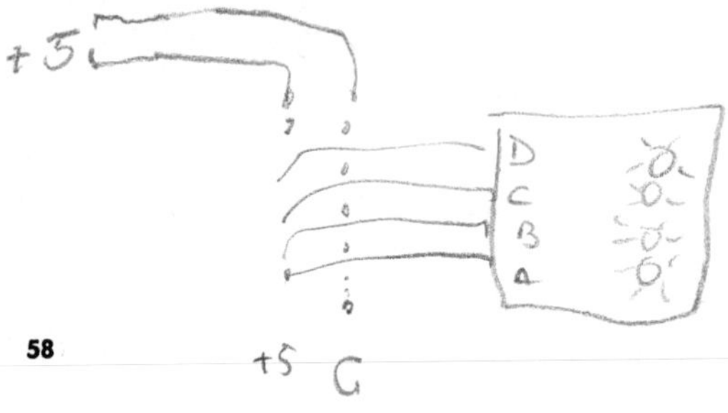

Questions

1. Using a lamp monitor, how would you determine whether an electrical terminal, integrated-circuit pin, wire on a resistor, etc., is at a logic 0 or logic 1 state?

2. In Fig. 1-30, how many different logic states are possible once the LR-1 power module is properly connected to the two terminals of a lantern battery? What are these logic states?

3. How many additional wires can you easily connect to the "A" input of the LR-6 lamp monitor module shown in Fig. 1-31?

EXPERIMENT NO. 3

Purpose

The purpose of this experiment is to demonstrate what happens when you incorrectly connect the LR-1 power module terminals to a lantern battery.

Schematic Diagram of Circuit (Fig. 1-32)

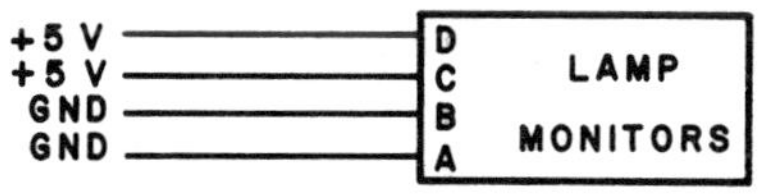

Fig. 1-32.

Step 1

You should successfully complete Experiment No. 2 before you undertake this one. The schematic diagram of the circuit and the breadboarding set-up shown in Fig. 1-31 apply equally well to this experiment.

Step 2

Connect the black wire and alligator clip to the (+6) terminal on the lantern battery or (+5) terminal on the power supply. Connect the red wire and alligator clip to the (−) terminal on the battery or power supply. Indicate below whether lamp monitors C and D on the LR-6 Lamp Monitor Outboard® module are lit or unlit.

NONE

Step 3

Carefully inspect the LR-1 power module. Do you observe any device that is lit on the module? NO

Step 4

Reverse the black wire and red wire connections to the lantern battery or power supply so that the wires are connected correctly. Carefully inspect the LR-1 power module once more. Do you observe any device that is lit on the module?

Yes

Are lamp monitors C and D on the LR-6 monitor module now lit as well?

Yes

If you have performed Steps 2 and 3 correctly, you should observe that no device on the LR-1 power module is lit and that lamp monitors C and D on the LR-6 lamp monitor module are unlit. If you have performed Step 4 correctly, you should observe that there is a small device on the LR-1 power module that emits a red light when the wires on the module are properly connected to a power source. Further, lamp monitors C and D on the LR-6 lamp monitor module are lit as well. With these observations in mind, what do you believe is the function of the lit device, which we call a *light-emitting diode* (*LED*), on the LR-1 power module? is lit when polarity + voltage is correct

The function of the LED on the LR-1 power module is to provide an indication when a sufficient and correct voltage is applied to the breadboard. If the alligator clips are reversed, the LED will remain unlit. If the voltage at the alligator clips is too low, the LED will be dimly lit.

Questions

1. How can you determine whether or not the LR-1 power module is properly connected to a power supply?

Check lamp

2. If the LR-1 power module is improperly connected, are you able to observe a "logic 1" state at any solderless terminal on the SK-10 breadboard?

EXPERIMENT NO. 4

Purpose

The purpose of this experiment is to demonstrate the operation of the LR-2 Logic Switch Outboard module.

Schematic Diagram of Circuit (Fig. 1-33)

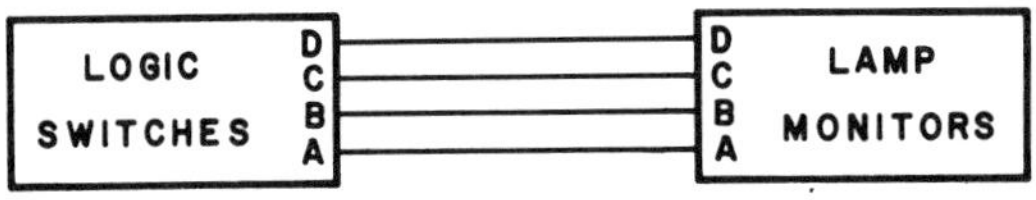

Fig. 1-33.

Step 1

Disconnect the alligator clips from the battery or power supply but leave the LR-1 and L-6 modules in place. Gently plug in the LR-2 Logic Switch module in the location shown in Fig. 1-34 and make the wire connections shown between the LR-2 and LR-6 modules. It is convenient to place the display modules on one side of the SK-10 socket and the source modules on the other side.

Step 2

Note the manner in which we have labeled the input pins on the LR-6 module and the output pins on the LR-2 module in Fig. 1-34. We would again like to emphasize the fact that the labeling of these two Outboard modules is completely arbitrary, and can be ABCD, DCBA, 1248, 8421, 0123, 3210, or anything else. What is important is that you decide what the designations should be for your own breadboarding system.

Step 3

Connect the two alligator clips to the battery or power supply. Compare your final circuit with the schematic in Fig. 1-33. You should observe a close correspondence between the two.

Flip the four logic switches so that they are all in the "ON" position. Are the four lamp monitors lit or unlit?

Flip the four logic switches so that they are all in the "OFF" position. Are the four lamp monitors lit or unlit now?

If you have performed this experiment correctly, you should observe four lit lamp monitors when the logic switches are "ON," and four unlit lamp monitors when the logic switches are "OFF."

Step 4

For the LR-2 Logic Switch module, we shall define a switch in the "ON" position as a "logic 1" state and a switch in the "OFF" position

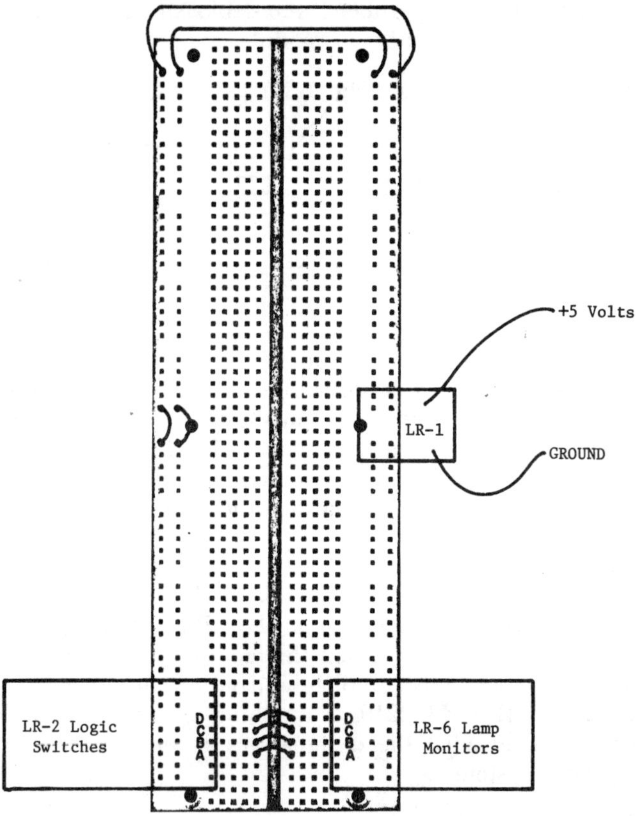

Courtesy E & L Instruments, Inc.

Fig. 1-34. Wiring diagram for Experiment No. 4.

as a "logic 0" state. We can write the results obtained in Step 3 in the following manner:

	Logic States of Logic Switches D C B A
All switches in the "ON" position: All switches in the "OFF" position:	1 1 1 1 0 0 0 0
	Logic States of Lamp Monitors D C B A
All lamp monitors lit: All lamp monitors unlit:	1 1 1 1 0 0 0 0

We can conclude that:

- *A logic switch in the "OFF" or "logic 0" state produces a lamp monitor in the unlit or "logic 0" state.*
- *A logic switch in the "ON" or "logic 1" state produces a lamp monitor in the lit or "logic 1" state.*

Therefore, using these two rules, we can write the following data:

Inputs (Logic states of logic switches)				Outputs (Observed logic states of lamp monitors)			
D	C	B	A	D	C	B	A
1	1	1	1	1	1	1	1
0	0	0	0	0	0	0	0

Step 5

Vary the settings of the four logic switches between the "OFF" and "ON" positions and write down the resulting logic states of the four lamp monitors that you observe by actually performing the experiment. Refer again to the schematic diagram and to Fig. 1-34 to ensure that you have wired up the circuit correctly. Remember again that:

Logic switch in "OFF" position = "logic 0" state
Logic switch in "ON" position = "logic 1" state
Lamp monitor in unlit position = "logic 0" state
Lamp monitor in lit position = "logic 1" state

Place your answers in the following table.

Inputs (Logic states of logic switches)				Outputs (Observed logic states of lamp monitors)			
D	C	B	A	D	C	B	A
0	0	0	0	0	0	0	0
0	0	0	1	0	0	0	1
0	0	1	0	0	0	1	0
0	0	1	1	0	0	1	1
0	1	0	0	0	1	0	0
0	1	0	1	0	1	0	1
0	1	1	0	0	1	1	0
0	1	1	1	0	1	1	1
1	0	0	0	1	0	0	0
1	0	0	1	1	0	0	1
1	0	1	0	1	0	1	0
1	0	1	1	1	0	1	1
1	1	0	0	1	1	0	0
1	1	0	1	1	1	0	1
1	1	1	0	1	1	1	0
1	1	1	1	1	1	1	1

Questions

1. How can you determine whether or not a logic switch on the LR-2 logic switch is in the "OFF" or "ON" position? Provide two different ways whereby you can make this determination.

if the corresponding lamp is lit the switch is on

look at the switch position

2. If the LR-1 power module is improperly connected, are you able to observe a "logic 1" state with a logic switch on the LR-2 logic switch module in the "ON" position? Why or why not?

3. What similarities between the logic switch and lamp monitor columns in the table (Step 5) do you observe? Why do these similarities exist?

EXPERIMENT NO. 5

Purpose

The purpose of this experiment is to demonstrate the operation of the LR-4 Seven-Segment LED Display Outboard module. A *display* is a device that provides a visual presentation of a single or group of digital electronic signals.

Schematic Diagram of Circuit (Fig. 1-35)

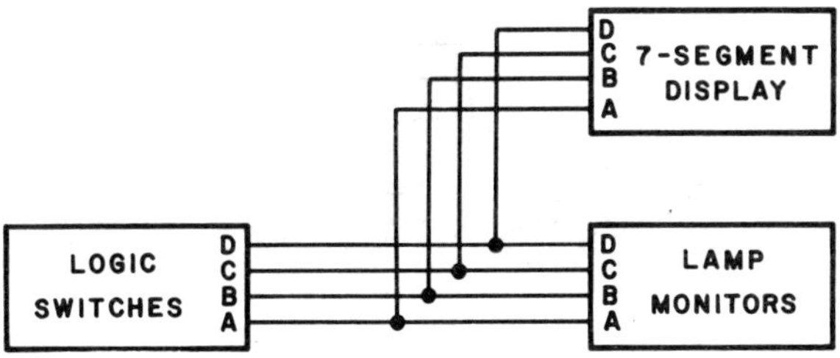

Fig. 1-35.

Step 1

Carefully study the schematic diagram (Fig. 1-35) and the experimental set-up shown in Fig. 1-36. Note that when two wires cross each other in space but are not connected, we use the following symbol:

On the other hand, when two wires are connected to each other, the following symbols apply:

If they do not cross.   If they cross.

Disconnect the alligator clips from the power source and gently insert the LR-4 seven-segment LED display module in the place shown in Fig. 1-36. Do not disconnect the wiring from the preceding experiment, but rather add the four additional wire connections shown in the figure. Apply power to the breadboard.

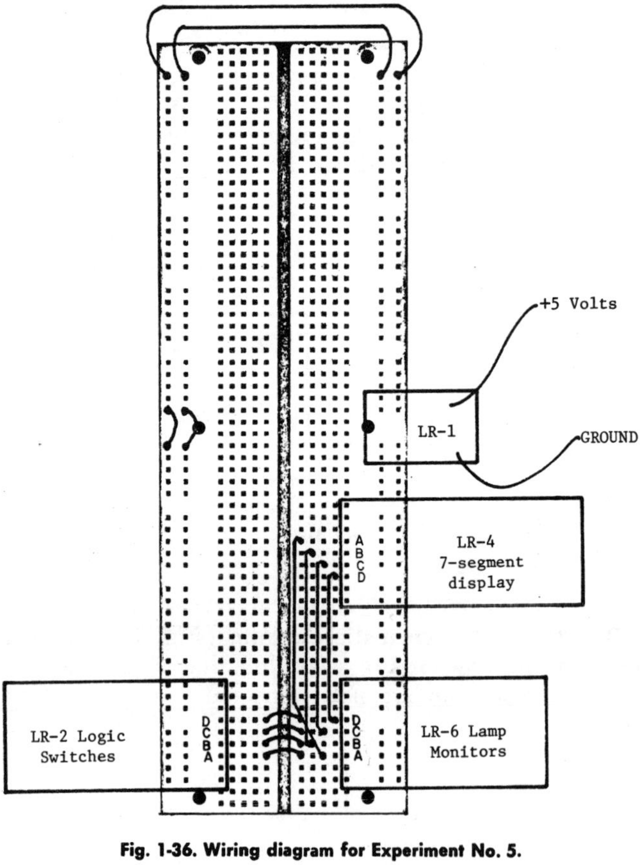

Fig. 1-36. Wiring diagram for Experiment No. 5.

Step 2

Flip the four logic switches so that they are all in the "OFF" position. What reading do you observe on the LR-4 seven-segment display module?

Flip the four logic switches so that they are all in the "ON" position. What reading, if any, do you now observe on the seven-segment display?

If you have performed these two experiments correctly, you should have observed a readout of decimal zero when all four logic switches were in the "OFF" position, and no readout whatsoever (a blank display) when all four logic switches were in the "ON" position. If you do not observe these results, check your wiring and repeat these experiments.

The labeling of the LR-4 seven-segment display module is not arbitrary; 1 represents the 2^0 bit, 2 the 2^1 bit, 4 the 2^2 bit, and 8 the 2^3 bit in a 4-bit binary word. In the DCBA notation, A represents the 2^0 bit, B the 2^1 bit, C the 2^2 bit, and D the 2^3 bit.

Step 3

Place logic switches B, C, and D in the "OFF" or "logic 0" state and logic switch A in the "ON" or "logic 1" state and write down the readout on the seven-segment LED display in the space below:

If you did this experiment correctly, you should observe a decimal one on the display.

Step 4

Place logic switches A, C, and D in the "OFF" position and logic switch B in the "ON" position and write down the readout on the seven-segment LED display in the space below.

If you did this experiment correctly, you now should observe a decimal two on the display.

Step 5

Place logic switches A, B, and D in the "OFF" position and logic switch C in the "ON" position and write down the readout that you observe on the seven-segment LED display.

If you did this experiment correctly, you should observe a decimal four on the display.

Step 6

Place logic switches A, B, and C in the "OFF" position and logic switch D in the "ON" position and write down the readout that you observe on the seven-segment LED display.

If you did this experiment correctly, you should observe a decimal eight on the display.

Step 7

Vary the settings of the four logic switches between the "OFF" and "ON" positions and write, in the following table, the decimal numbers or unusual symbols that you observe on the LR-4 seven-segment LED display module.

Questions

1. Did you expect to see the decimal numbers 10, 11, 12, 13, 14, and 15? Why or why not? NO

   not enough segments

2. If the LR-1 Power module is improperly connected, are you able to observe any decimal numerals or other symbols on the LR-4 seven-segment LED display module? Why or why not?

   no

   improper polarity

Inputs (Logic states of logic switches)				Output (Observed number or symbol on seven-segment LED display)
D	C	B	A	
0	0	0	0	0
0	0	0	1	1
0	0	1	0	2
0	0	1	1	3
0	1	0	0	4
0	1	0	1	5
0	1	1	0	6
0	1	1	1	7
1	0	0	0	8
1	0	0	1	9
1	0	1	0	⊏
1	0	1	1	⊐
1	1	0	0	⊔
1	1	0	1	⊆
1	1	1	0	⊨
1	1	1	1	Blank display

EXPERIMENT NO. 6

Purpose

The purpose of this experiment is to demonstrate the operation of the LR-4 Seven-Segment LED Display Outboard module as a zero/one LED display.

Schematic Diagram of Circuit (Fig. 1-37)

Step 1

The circuit in this experiment is somewhat similar to the circuit of the preceding experiment. Leave the LR-2, LR-4, and LR-6

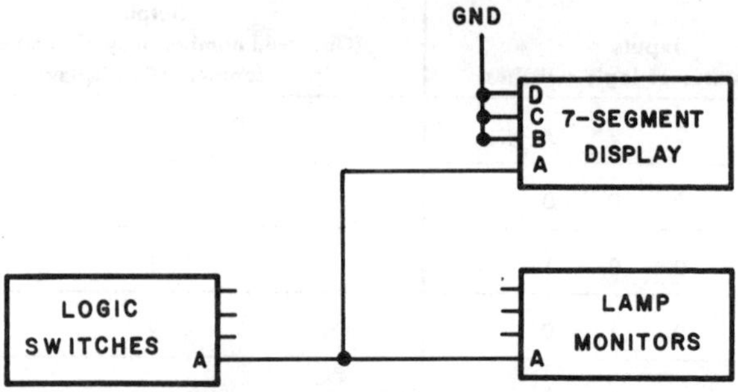

Fig. 1-37.

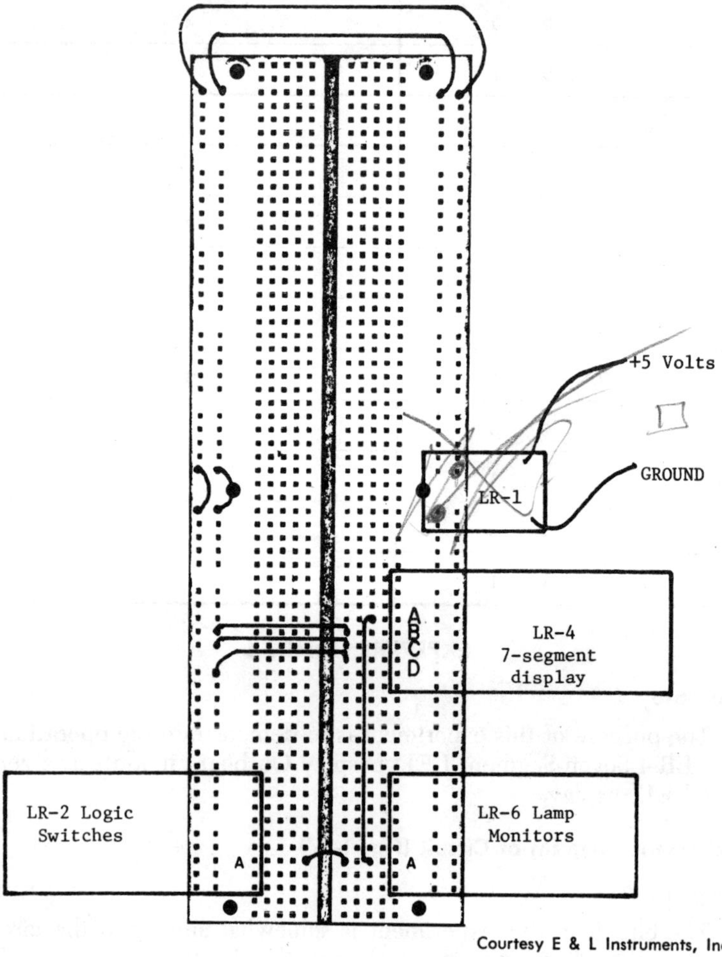

Courtesy E & L Instruments, Inc.

Fig. 1-38. Wiring diagram for Experiment No. 6.

modules in place, disconnect the alligator clips to the power source, and re-wire the modules as shown in the schematic (Fig. 1-37). Attempt to wire the circuit directly from the diagram, then check with Fig. 1-38 to determine whether you have wired the circuit correctly. (Any lamp monitor and logic switch will suffice.)

Step 2

Connect the alligator clips to the power source and flip logic switch A in the "OFF" and "ON" positions. In the space provided below, write down what you observe for both lamp monitor A and the seven-segment LED display that is acting as a "zero/one" display.

Input	Outputs	
Logic Switch A	**Lamp Monitor A**	**Zero/One Display**
"OFF" = "logic 0" state "ON" = "logic 1" state	off on	0 1

If you have performed this experiment correctly, you should observe a decimal zero when logic switch A is "OFF" and a decimal one on the zero/one display when logic switch A is "ON."

Questions

1. Input pins B, C, and D on an LR-4 seven-segment LED display module all at a logic 0 state, as shown in the schematic diagram. What reading would you observe on the display if pin A were at +5 volts? At ground potential?

   at +5V 0/1 = 1

   at 0V = 0

2. Input pins A, C, and D on a seven-segment LED display are all at a logic 0 state. What reading would you observe on the display if pin B were at +5 volts? At ground potential?

   at +5 = 2

   0V = 0

3. Input pins A, B, and D on a seven-segment display are all at a logic 0 state. What reading would you observe on the display if pin C were at +5 volts? At ground potential?

4. Input pins A, B, and C on a seven-segment LED display are all at a logic 0 state. What reading would you observe on the display if pin D were at +5 volts? At ground potential?

EXPERIMENT NO. 7

Purpose

The purpose of this experiment is to demonstrate the operation of the LR-7 Debounced Dual Pulser Outboard module. A *pulser* is a logic switch that generates a *clock pulse*. A *clock pulse* is a complete logic cycle, i.e., a transition from logic 0 to logic 1 and back to logic 0 (a *positive* clock pulse), or a transition from logic 1 to logic 0 and back to logic 1 (a *negative* clock pulse).

Schematic Diagrams of Circuits (Figs. 1-39 and 1-40)

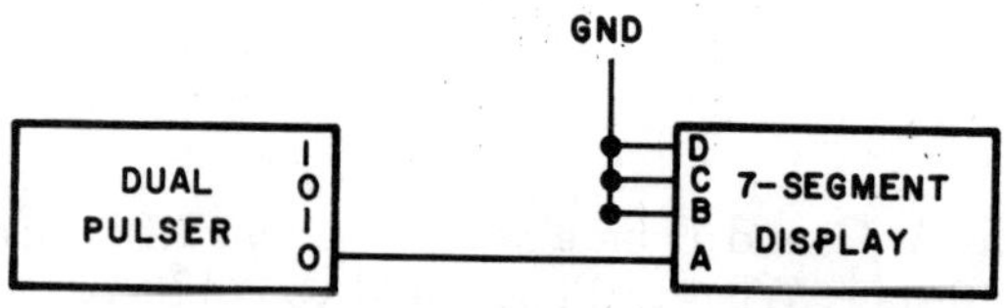

Courtesy E & L Instruments, Inc.

Fig. 1-39.

Step 1

The above circuit requires a new Outboard module, the LR-7 debounced dual pulser module. Disconnect the alligator clips from the power source and remove all wires from the preceding experiment except the jumper wires that tie together the 100 ground solderless terminals and the 100 +5 Volts solderless terminals. Leave the LR-1, LR-2, LR-4, and LR-6 Outboards in place. Add the LR-7 module to the SK-10 socket immediately opposite the LR-4 module as shown in Fig. 1-41. Make the four wire connections shown.

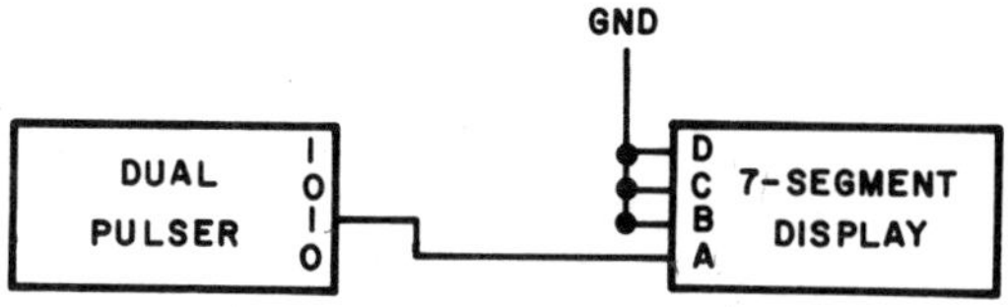

Fig. 1-40.

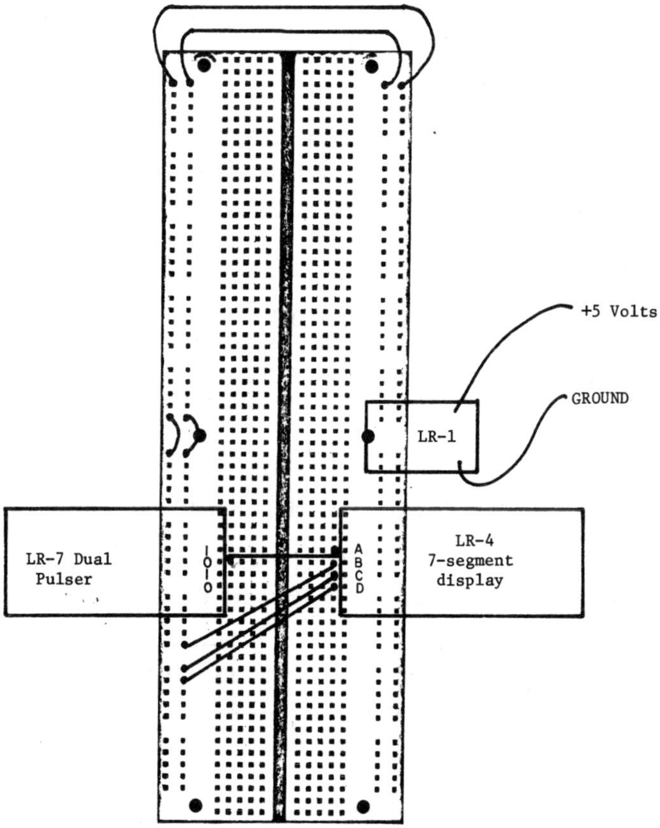

Fig. 1-41. Wiring diagram for Experiment No. 7 (Step 1). For Step 6, the wire between the "1" output of the LR-7 dual pulser module and A on the LR-4 seven-segment LED display module.

Step 2

Study the schematic diagram in Fig. 1-39. Note that we have connected the "0" output from one of the two pulsers on the LR-7 module to the A input of the seven-segment display. The B, C, and D inputs of the display are grounded, which means that the display can only show either decimal 0 or decimal 1, *i.e.,* it is operating as a "zero/one" display. You will find that the zero/one display is very handy, especially in subsequent chapters.

Step 3

Remember that to operate the pulser correctly, you must push the little plastic button in, not down, if you have an older Outboard module. A fingernail, screwdriver, or a small blunt instrument can be used.

Step 4

Apply power to the breadboard by connecting the pair of alligator clips to the power source. What decimal number do you observe on the seven-segment LED display?

You should observe a decimal 0. If you do not, check the wire connection between the pulser and the seven-segment display. Push the pulser button in and hold it in. Now what decimal number do you observe?

You should observe a decimal 1. If you do not have a seven-segment LED display, use a lamp monitor instead. When the pulser button is pushed in, the lamp monitor should be lit. When a lamp monitor is used, the schematic diagram is like that shown in Fig. 1-42.

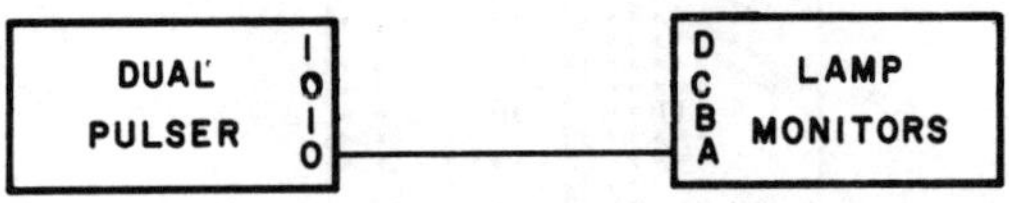

Fig. 1-42.

Step 5

Repeat Step 4 several times, then write what you observed in the space provided.

Input Pulser	Output Seven-Segment Display	or	Lamp Monitor
released			
pressed			

In the next space provided, explain how you would generate a positive clock pulse, *i.e.*, a transition from logic 0 to logic 1 and back to logic 0.

Step 6

Observe the subtle difference in the top and bottom schematic diagrams and wire the bottom circuit. This is a rather trivial change and you can make it with the power source still connected to the breadboarding socket. No integrated circuits are involved, so there is little possibility that any damage will result.

Step 7

Press and release the pulser button several times. You should observe that the display is at decimal 0 when the pulser is pressed, and at decimal 1 when the pulser is released. Write down your observations in the space below.

Input Pulser	Output Seven-Segment Display	or	Lamp Monitor
released			
pressed			

In the space below, explain how you would generate a negative clock pulse, i.e., a transition from logic 1 to logic 0 and back to logic 1.

Step 8

Draw a schematic diagram in which the "1" output from a pulser is connected to lamp monitor B and the "0" output from the same pulser is connected to lamp monitor A.

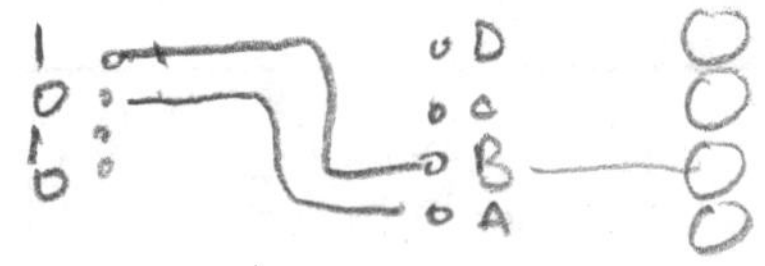

Wire this circuit on the breadboard and demonstrate to yourself the following results:

Input Pulser	Outputs Lamp Monitor A	 Lamp Monitor B
released	unlit	lit
pressed	lit	unlit

Recall that a lit lamp monitor corresponds to a logic 1 state and that an unlit lamp monitor corresponds to a logic 0 state. Based upon the results from the preceding table, write the proper logic states for the two lamp monitors, A and B, in the following space.

Input Pulser	Outputs Logic State of Lamp Monitor A	 Logic State of Lamp Monitor B
released	0	1
pressed	1	0

We assign the word, *complement,* to the two outputs from the pulser. The "0" output is the complement of the "1" output, and the "1" output is the complement of the "0" output. In other words, a logic 0 state is the complement of a logic 1 state, and vice versa.

Step 9

How many pulsers are there on the LR-7 Dual Pulser module?

2

Do both pulsers behave in an identical manner? If you are not sure of the answer to this question, try the other pulser in the above experiments.

Questions

1. What is the function of a pulser? to change from one logic state to its compliment and back quickly

2. If the LR-1 Power module is improperly connected, are you able to observe a logic 1 output from the LR-7 Dual Pulser module? Why or why not? no same as before

3. A seven-segment LED display with inputs B, C, and D all connected to ground is called a zero/one display. A schematic diagram for a circuit in which the "0" output from a pulser is connected to such a zero/one display can be written as

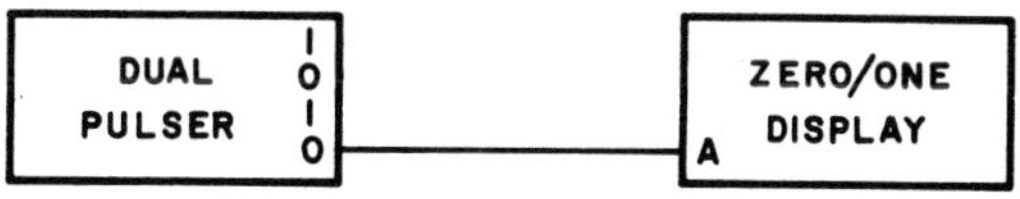

Fig. 1-43.

Do such a zero/one display and a lamp monitor behave similarly? If your answer is yes, explain why? yes all off on the lm = 0 on the led and 1 = 1

4. Write the complement of the words given in the following table.

Word	Complement of Word
ON	off 0
NO GO	go 1
OFF	on 1
START	stop 0
STOP	start 1

EXPERIMENT NO. 8

Purpose

The purpose of this experiment is to demonstrate the operation of an integrated circuit, in this case the *7490 decade counter.*

Pin Configuration of Integrated-Circuit Chip (Fig. 1-44)

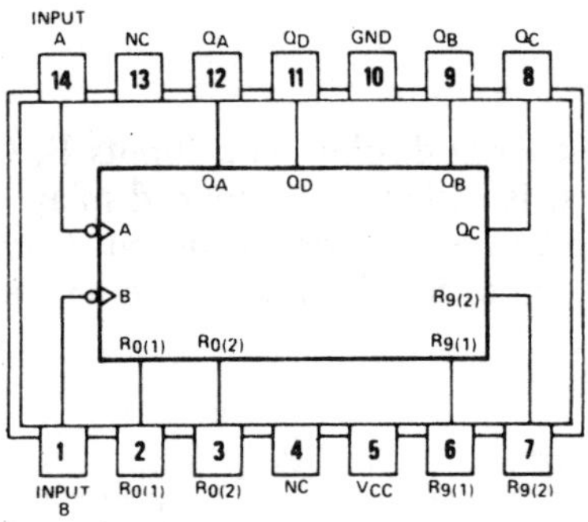

7490

Courtesy Texas Instruments Incorporated

Fig. 1-44.

Schematic Diagram of Circuit (Fig. 1-45)

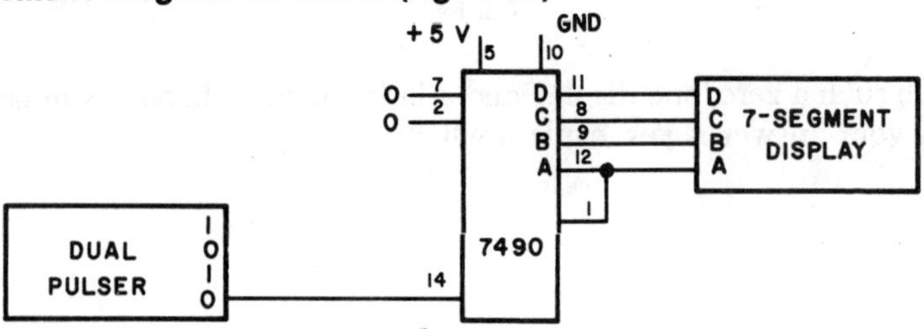

Fig. 1-45.

Step 1

This is the first experiment in which you wire a digital circuit that includes an integrated-circuit "chip." Pick out the 7490 chip from among your collection of chips and study it carefully. Hold the chip such that the pins face away from you. You are now looking down on the top of the chip. Note the small semicircular notch or dot in the plastic. This notch or dot is used to identify the pin-1/pin-14 end of the integrated-circuit chip, as shown in Fig. 1-46. With the aid of this illustration, it should be easy for you to pick out any of the fourteen pins on the chip.

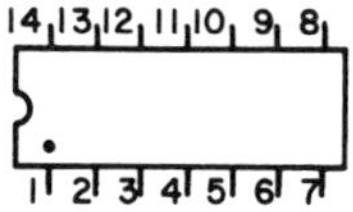

Fig. 1-46.

In the digital electronics area, manufacturers of integrated-circuit chips provide either a *pin configuration* or else a *schematic representation* (also known as a *block diagram*) that indicates pin function as a function of pin number. The pin configuration for the 7490 chip is shown in Fig. 1-44. We prefer the use of schematic representations, in which inputs, outputs, and power connections are grouped together by function. For the 7490 chip, such a representation is shown in Fig. 1-47. This figure is easier to understand than the pin configuration even though both figures provide the same basic information. If we compare the two figures, we observe the following:

Pin No.	Pin Configuration Diagram	Schematic Representation
1	INPUT B	Clock input BD
2	$R_0(1)$	Reset input
3	$R_0(2)$	Reset input
4	No connection (NC)	—
5	V_{cc}	+5 Volts
6	$R_9(1)$	Reset input
7	$R_9(2)$	Reset input
8	Q_C	Output C
9	Q_B	Output B
10	GND	GND
11	Q_D	Output D
12	Q_A	Output A
13	No connetion (NC)	—
14	INPUT A	Clock input A

We can thus summarize the characteristics of the 7490 chip in the following manner:

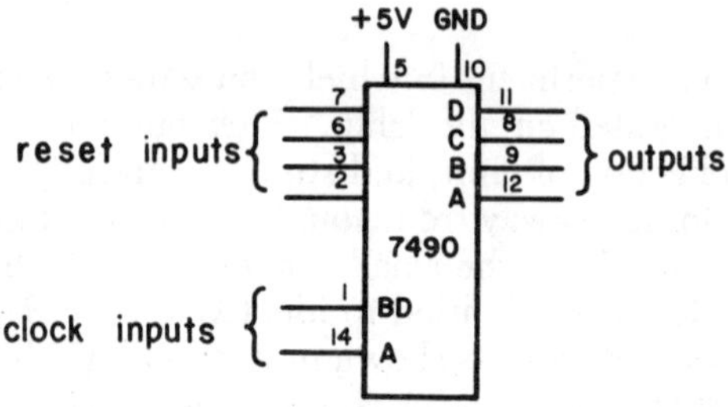

Fig. 1-47.

- Pins 5 and 10 are the power connections to the chip. Pin 5 must be at +5 volts and pin 10 at ground potential for the chip to function properly.
- Pins 12, 9, 8, and 11 are the four outputs, A, B, C, and D, from the chip.
- Pins 14 and 1 are the "clock inputs" to the chip. One or the other of these inputs is used in conjunction with a pulser, clock, or other device that can produce a series of clock pulses.
- Pins 2, 3, 6, and 7 are "reset inputs." When pins 2 and 3 are both at a logic 1 state, the counter will give the following output: A = 0, B = 0, C = 0, and D = 0. When pins 6 and 7 are both at logic 1, the counter will give the following output: A = 1, B = 0, C = 0, and D = 1. Pin 7 will usually be set to logic 0. Pin 2 will be also set to logic 0 except in about five circuits in this book. With pins 2 and 7 at logic 0, the counter will be able to count.

The above information should make the transition to wiring an integrated circuit a little bit easier for you.

Step 2

In order to perform this experiment, you will require the following:

1. Wire of various colors and lengths. We recommend the use of short 1-inch or 1⅛-inch lengths of black (GND) or red (+5V) wire for connecting pins 2, 7, 10, and 5 to either GND or +5 Volts. Never insert a bent wire into a solderless breadboarding terminal. Although you will always make a connection, you will gradually damage the solderless terminals if you always use bent wires.
2. A single 7490 decade counter integrated circuit.
3. An SK-10 breadboard with an LR-1 power module or the LR-25 breadboarding station.

4. A power source.
5. The LR-4 seven-segment LED display module.
6. The LR-7 Debounced Dual Pulser module or the LR-25 breadboarding station.

We would also like to emphasize some of the hints that we have previously given:

- You will find a pair of pliers to be quite handy to (a) straighten bent wire, (b) straighten bent or crooked integrated-circuit chip pins, and (c) insert and remove hookup wire in crowded places on the SK-10 breadboard.
- Use a small screwdriver to remove an integrated-circuit chip from the SK-10 breadboard. Insert the screwdriver into the groove and pry the chip up slowly.
- *Always disconnect the power to the breadboard when you are making connections to integrated-circuit chips. In our experience, we have observed that students who fail to heed this suggestion have a tendency to burn out integrated-circuit chips.*
- *Always check your wiring before you apply power to the breadboard. In our experience, this is a second common fault made by students who make incorrect wire connections, are in a hurry to apply power to the breadboard, and are then surprised to find out that their integrated-circuit chip burns up. A correctly wired integrated-circuit chip will almost never burn out.*
- Pay particular attention to how you insert an integrated-circuit chip into the solderless terminals on the SK-10 socket. It takes some practice to do it correctly. Make sure that pins on the integrated-circuit chip are vertical. The manufacturer of the SK-10 breadboard has provided the following recommendations:

  As has occurred only too often, when first using the SK-10 solderless breadboarding socket, the student or designer has difficulty in inserting integrated-circuit chips into the socket. It is quite strongly suggested that the following procedure be used to avoid insertion problems:

  "One row of the integrated-circuit chip's contact pins is placed into the proper holes of the socket. DO NOT push the integrated-circuit chip in, however! Slant the integrated-circuit chip at an angle of approximately 30° with the breadboard, keeping the one row of contact pins engaged in their holes. With a single outward pressure (in the direction of the engaged terminals), 'roll' the second set of terminals into the proper holes. Only after this second set of terminals is properly engaged should the integrated-circuit chip be pushed in by pressing directly down on the top of the chip."

Step 3

Taking a closer look at the schematic diagram of the circuit for this experiment, you should observe that

1. Pin 11 on the 7490 chip is connected to input 8 (D) on the seven-segment display.
2. Pin 8 on the 7490 chip is connected to input 4 (C) on the display.
3. Pin 9 on the 7490 chip is connected to input 2 (B) on the display.
4. Pin 12 on the 7490 chip is connected both to input 1 (A) on the display and also to pin 1 on the 7490 chip.
5. Pin 5 on the 7490 chip is connected to +5 Volts on the SK-10 socket.
6. Pins 2, 7, and 10 are all connected to GROUND on the SK-10 socket.
7. Pin 14 on the 7490 chip is connected to the "0" output from the pulser.
8. All other pins—3, 4, 6, and 13—are left unconnected!

Step 4

Wire the circuit as shown in the schematic diagram and described in Step 3. Try to wire the circuit directly from the diagram.

Step 5

Check your wired circuit. Make certain that it is correct.

Step 6

Apply power to the breadboard and observe what happens. You may observe a decimal 0, a decimal 8, a blank reading, or some other numeral or symbol on the display. To "reset" the 7490 counter and display to decimal 0, disconnect the wire between pin 2 on the 7490 counter and GROUND. You should observe that the display immediately goes to a decimal 0.

If you do not observe a decimal zero, disconnect the power source and search for mistakes in your wiring of the circuit. Make certain that the LED on the LR-1 Power module is lit, indicating that you have properly connected the LR-1 module to the power source.

If you do observe a decimal 0 on the display, *reconnect pin 2 on the 7490 chip to GROUND.*

Step 7

Now press and release the pulser *once*. The reading on the seven-segment LED display should change to a decimal one.

Press and release the pulser once more. The reading on the display should change to a decimal two.

Press and release the pulser a third time. The reading on the display should change to a decimal three.

If you observe the predicted results, continue with the experiment. If nothing happens when you press and release the pulser button, you may have an error in your circuit. Finally, if you don't observe a simple 0, 1, 2, 3 sequence, but rather a nonsequential group of four numbers such as 0, 4, 9, 1, either your pulser is not *debounced* or else you have incorrectly wired outputs A, B, C, and D to the display. We shall not discuss the concept of a *debounced* pulser at this time. However, if your pulser is not debounced, you will be unable to continue this experiment. Stop here, answer what questions you can, save your circuit, and proceed to Experiment No. 9.

Step 8

Repeatedly press and release the pulser button. You should observe that you can count on the seven-segment LED display from 0 to 9 and back to 0 again. This is what we mean by a *decade counter,* which is defined in the following way:

> "A decade counter is a logic device that has 10 stable states and may be cycled through these states by the application of 10 clock or pulse inputs. A decade counter usually counts in a binary sequence from state 0 through state 9 and then cycles back to 0. Sometimes referred to as a divide-by-10 counter."[1]

Step 9

Does the number on the seven-segment LED display change when you press or when you release the pulser?

If you have wired up the pulser correctly, the number should change when you *release* the pulser. If it does not, please check the output pin on the pulser to which you have wired pin 14 on the 7490 chip.

Step 10

Repeatedly press and release the pulser button until you cycle from 0 to 9. In the following space indicate how many different segments must be lit for every different number. Remember that the LR-4 module is a *seven*-segment LED display.

Step 11

Rather than counting the number of lit segments, let us instead determine which of the seven segments is lit for each different deci-

Decimal number	Number of lit segments
0	6
1	2
2	5
3	5
4	4
5	5
6	5
7	3
8	7
9	5

mal numeral. Fig. 1-48 shows a lettering scheme for the seven-segment display in which the letters *a* through *g* are used to denote the seven individual segments of the display.

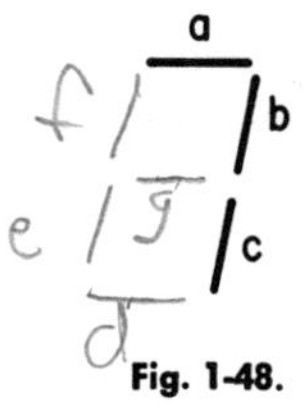

Fig. 1-48.

Cycle the display until a decimal one appears. Which two segments are lit?

The correct answer to this question is segment *b* and segment *c*.

Continue to cycle the display reading until a decimal eight appears. Now which segments are lit?

The correct answer here is: All seven segments are lit!

Step 12

Let us be more systematic in the way we determine which segments are lit for a given decimal number on the display. Let us, in fact, determine which segments are lit for each of the different possible decimal numbers.

In the following table, write in the columns beneath the letters *a* through *g* whether that segment is lit or unlit for each of the ten different decimal numbers. We have helped somewhat by providing answers for decimal one and decimal eight.

	Segments						
Decimal number	**a**	**b**	**c**	**d**	**e**	**f**	**g**
0	lit	lit	Lit	lit	Lit	lit	Unlit
1	unlit	lit	lit	unlit	unlit	unlit	unlit
2	lit	Lit	Unlit	Lit	Lit	Unlit	Lit
3	Lit	Lit	Lit	Lit	Unlit	Unlit	Lit
4							
5	lit	unlit	lit	lit	unlit	lit	lit
6							
7							
8	lit	lit	lit	lit	lit	lit	lit
9	lit	Lit	lit	Unlit	unlit	lit	lit

Step 13

Save the experimental set-up. It will be used in Experiment No. 9.

Questions

1. In your own words, define the term, "seven-segment display."

a display with 7 seperate units in it that can indicate different symbols

2. When does the reading on the seven-segment display change in the preceding experiments, when the pulser is pushed in or when it is released?

released

3. Can you suggest any reasons why you are unable to observe any of the unusual symbols that you have previously observed in Step

7 of Experiment No. 5? You may not be able to answer this question correctly, but take a stab at it.

4. If the LR-1 Power module is improperly connected, are you able to observe a count sequence using the 7490 integrated-circuit chip?

5. What does the 7490 decade counter integrated-circuit chip do? Describe the specific function that you have observed in this experiment.

6. In your own words, define the term, "clock pulse." Give at least one example where a clock pulse is required for a digital electronic circuit to operate correctly.

7. The table in Step 12 is called a "truth table." Explain, if you can, what is the object or purpose of a truth table. You may not be able to answer this question, but take a stab at it.

EXPERIMENT NO. 9

Purpose

The purpose of this experiment is to demonstrate the operation of the LR-5 Clock Outboard module, which will be employed to "clock" a 7490 decade counter integrated-circuit chip.

Pin Configuration of Integrated-Circuit Chip (Fig. 1-49)

Schematic Diagram of Circuit (Fig. 1-50)

Step 1

Remove the pulser, logic switch, and lamp monitor modules as well as all connections between them and the LR-4 module or the 7490 chip. Leave all connections between the 7490 chip, the LR-4

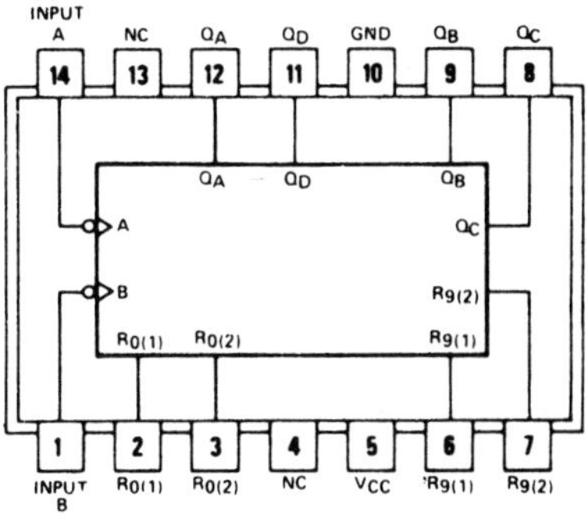

Fig. 1-49.

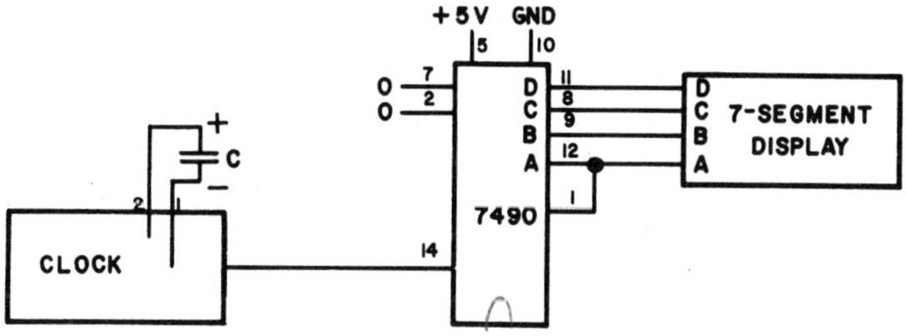

Fig. 1-50.

module, +5 Volts, and GROUND intact. Gently insert the LR-5 Clock module in the location shown in Fig. 1-51. Run a wire between pin 14 on the 7490 chip and the CK pin on the LR-5 module; the CK pin on the LR-5 module is the second from the left, as viewed from the top with the output pins facing you.

Step 2

The LR-5 Clock Outboard module is supplied with 0.03 μF, 0.33 μF, and 3.3 μF capacitors (μF is the abbreviation for microfarad, a commonly used unit for capacitance) that are inserted into the two pins near the back of the module. The 0.03-μF and 0.33-μF capacitors should have the numbers 0.03 and 0.33 (or 334) written somewhere on the surface of the "disc." The 3.3-μF capacitor will be an "electrolytic" capacitor, in which case it will have a positive terminal, which must be connected to a point that is at higher potential than the point to which the negative terminal is connected. In the schematic diagram, we identify the positive ("+") and negative ("−") ends on the capacitor as well as the pins on the 555 Timer integrated-circuit chip to which the capacitor is connected. Identify the pin that is connected directly to GROUND; the negative end of the electrolytic capacitor is connected at this point. By following the correct procedure for connecting an electrolytic capacitor to the

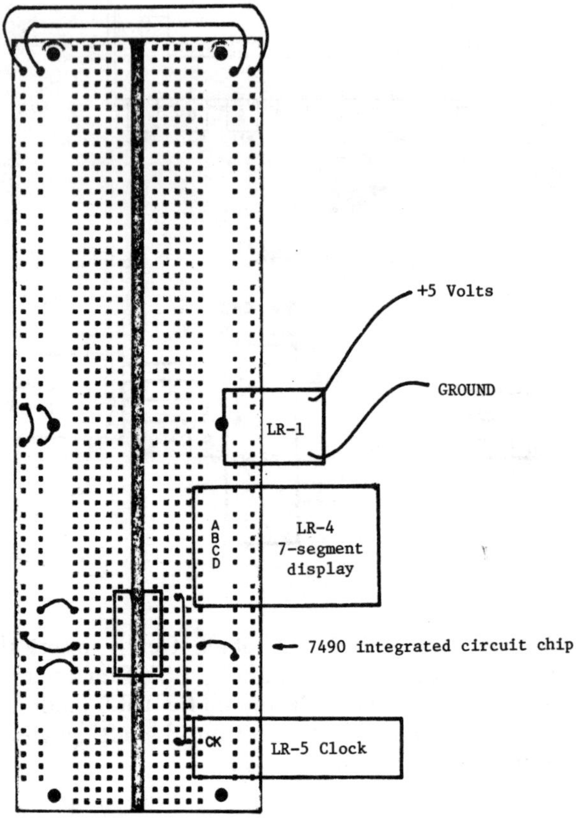

Courtesy E & L Instruments, Inc.

Fig. 1-51. Partial wiring diagram for Experiment No. 9.

LR-5 module, you will not damage the capacitor. This special procedure does not apply to Mylar or disc capacitors, either end of which can be connected to the GROUND pin on the LR-5 module.

Step 3

Fig. 1-51 shows one way in which the circuit for this experiment can be constructed. Note that only ten wire connections are required.

Step 4

Insert a 0.33-μF capacitor in the two socket pins on the LR-5 Clock module. If you forget to do this, your integrated-circuit clock will operate at too high a frequency for you to time it with a wristwatch.

Step 5

Apply power to the breadboard and observe that the seven-segment LED display is now counting from 0 to 9. If you do not observe this result, disconnect the power from the breadboard and carefully check your circuit. Are you certain that pin 14 of the 7490 chip is connected to the CK output on the LR-5 Clock module?

Step 6

The decimal reading on the seven-segment LED display will change every time there is a clock pulse from the LR-5 Clock module. Obtain a watch with a second hand and determine how many clock pulses occur in a period of one minute. Write your answer in the spaces provided.

For a .33 μF capacitor, there are 55 clock pulses each minute.

Step 7

Now change the value of the timing capacitor inserted into the two socket pins of the Clock Outboard module. For each capacitor employed, determine the number of clock pulses per *second.* Fill in the following table.

Timing capacitor, μF	*Number of clock pulses per second*
.33	1 per sec
3.3 μF	10 per min
.033	10 per Sec

If you use an electrolytic capacitor, make certain that the negative capacitor terminal is connected to the ground socket pin on the Clock Outboard module.

When we performed these measurements, we obtained the following results:

Timing Capacitance, μF	*Frequency, Counts/Sec*	*1/Frequency, Seconds/Count*
0.10	6.97	0.144
0.10	7.35	0.136
0.33	1.23	0.81
1.0	0.34	2.9
2.2	0.140	7.2
3.3	0.106	9.4

When the reciprocal frequency is plotted against the timing capacitance, a linear relationship is obtained, as shown in Fig. 1-52. Note that the tolerance of the capacitors was no better than ±20%.

Step 8

Calculate the reciprocal frequency as a function of timing capacitance and plot your results on Fig. 1-53. Is your curve linear? What is the slope of the curve?

Timing Capacitance, μF	*Frequency, Counts/Sec*	*1/Frequency, Seconds/Count*

Four our measurements, we calculated a slope of 3.08 Hz^{-1} μF^{-1}. Thus, for relatively large capacitance values, the relationship between frequency and timing capacitance is approximately

$$\nu = 0.31/C$$

where the frequency, ν, is in hertz and the timing capacitance is in microfarads.

Questions

1. Define the term, "clock."

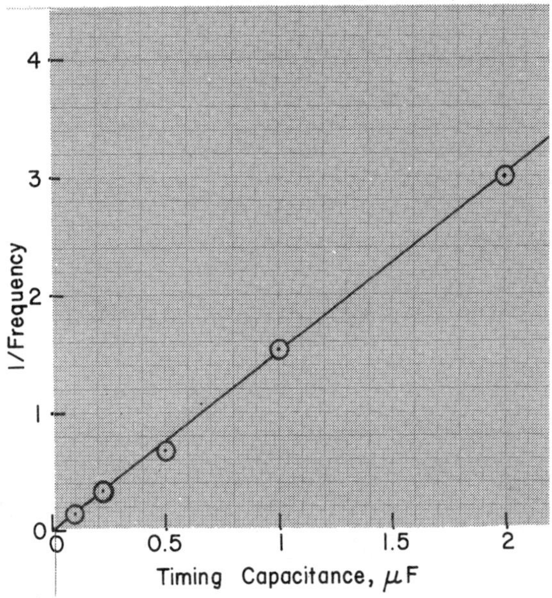

Fig. 1-52.

2. Can you suggest several reasons why we need a "clock" in digital electronic circuits? This question may not be easy to answer, but try your best.

3. Both the pulser and the Clock Outboard module act as "clocks." In what way are they different from each other? What advantages do you detect for the use of a pulser as a "clock?"

4. What advantages does a repetitive clock such as the LR-5 Clock module have over a nonrepetitive clock such as the pulser?

5. Could you have obtained capacitance values for unknown capacitors within the range of 0.10 and 1.0 microfarad in Step 8?

Fig. 1-53.

If the answer is yes, explain how you would actually calculate the unknown capacitances.

6. Why is it useful to make a graph of experimental data? In your answer, please explain what advantages that a graph has over a table. Take a careful look at Step 7 when you give your answer to this question.

7. Did you have any difficulties in measuring the number of counts/minute for any of the capacitors that you studied? We

encountered difficulties both with the 0.10- and the 3.3-microfarad capacitors. Explain the problems that you had and why your curve may not be quite as accurate as the one that we have provided in Step 7.

8. Did you use a longer time interval than 1 minute to measure the number of counts/minute for the 3.3-microfarad capacitor? Why or why not?

9. Did you use a shorter time interval than 1 mintue to measure the number of counts/minute for the 0.10-microfarad capacitor? Why or why not?

10. How accurately did you know the capacitance values for the capacitors that you employed in Step 7? Was the stated value accurate to 2%, 5%, or 10%? Our capacitors were probably similar to yours, and we certainly obtained some scatter in our experimental data. We tried to eliminate the scatter by making several measurements for each capacitor and by developing a technique whereby we could start or stop the 7490 decade counter with an error of no more than ± 1 second. We shall discuss this technique in Chapter 2, Experiment No. 1.

11. According to the data given in Step 7, how many counts/minute would you observe for a 0.35-microfarad capacitor? Show your calculations.

12. Using the experimental relationship, $\nu = 0.31/C$, given in Step 8, estimate the counting frequency for the following capacitors:

Capacitor	*Counting Frequency, Hz*
0.01 μF	
0.005 μF	
0.002 μF	
0.001 μF	

EXPERIMENT NO. 10

Purpose

The purpose of this experiment is to demonstrate the operation of the LR-26 Latch/Display Outboard module, or the equivalent function on the LR-25 breadboarding system Outboard module.

Schematic Diagram of Circuit (Fig. 1-54)

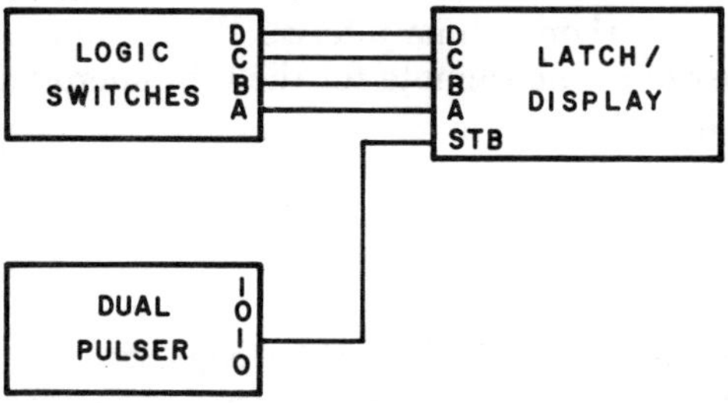

Fig. 1-54.

Step 1

For this experiment, you will only require the four logic switches, a pulser, and the LR-26 Latch/Display Outboard module, or the equivalent LR-25 breadboarding system, which contains all of these functions. If the 7490 integrated-circuit chip is still wired, we recommend that you remove it. You will wire the chip to the Latch/Display Outboard module in a subsequent experiment.

Step 2

Wire the circuit shown in the schematic diagram.

Step 3

Perhaps the most important wire connection in this experiment is the one from the "1" output of the pulser to the strobe, STB, input to the latch/display. When this input to the display is at logic 1, you will observe that nothing happens. When this input is at logic 0, the latch/display will follow the data applied at the input pins A, B, C, and D. You will now see how this is done.

Step 4

Set the four logic switches all to logic 0. Press and release the pulser. What do you observe on the latch/display?

0

We observed that the latch/display went to 0.

Step 5

Now set the four logic switches all to logic 1. Notice that the display reading has not changed; it is still at 0. This observation demonstrates that the display does not change unless the STB input is at logic 0. Press and release the pulser. What do you now observe?

F

Your observation will depend upon the type of Hewlett-Packard numeric display you have. If it is the 5082-7340 hexadecimal display, then you should observe the letter "F." If it is one of the two numeric displays, then you should observe a blank display. Which type of display do you have? 5082-7340

Step 6

Set the logic switches to D = 0, C = 1, B = 0, and A = 1. Press and release the pulser. What do you observe on the latch/display?

5

We observed the numeral 5.

Step 7

Set the logic switches to D = 1, C = 0, B = 0, and A = 1. Press and release the pulser. What do you observe now?

9

We observed the numeral 9.

Step 8

Change the wire connection between STB and the pulser to the "0" output of the pulser. Now you will be able to observe changes in the logic switch settings as soon as you make them. Vary the settings of the four logic switches between the "0" and "1" positions and write down in the table on the following page the numerals, letters, or symbols that you observe. Compare your observations with the predicted values for the two types of Hewlett-Packard displays.

Inputs				Output		
(Logic States of Logic Switches)				Observed Numeral, Letter or Symbol	Hewlett-Packard Hexadecimal Display	Hewlett-Packard Numeric Display
D	C	B	A			
0	0	0	0	0	0	0
0	0	0	1	1	1	1
0	0	1	0	2	2	2
0	0	1	1	3	3	3
0	1	0	0	4	4	4
0	1	0	1	5	5	5
0	1	1	0	6	6	6
0	1	1	1	7	7	7
1	0	0	0	8	8	8
1	0	0	1	9	9	9
1	0	1	0	A	A	all dots lit
1	0	1	1	B	B	blank
1	1	0	0	C	C	blank
1	1	0	1	D	D	—
1	1	1	0	E	E	blank
1	1	1	1	F	F	blank

What type of Hewlett-Packard 5082-7300 series indicator do you have, the hexadecimal or the numeric indicator?

Hexi

Questions

1. If you would study the LR-4 Seven-Segment Display module, you would see an integrated circuit chip on the module. In case you are curious, this is the 7447 decoder/driver chip discussed in Chapter 5. On the LR-26 Latch/Display module, there is no integrated circuit chip. Why not? Where is the decoder/driver? Can you see it?

EXPERIMENT NO. 11

Purpose

The purpose of this experiment is to demonstrate the operation of the LR-27 Octal Latch/Display Outboard module. Octal code is discussed in Chapter 5.

Schematic Diagram of Circuit (Fig. 1-55)

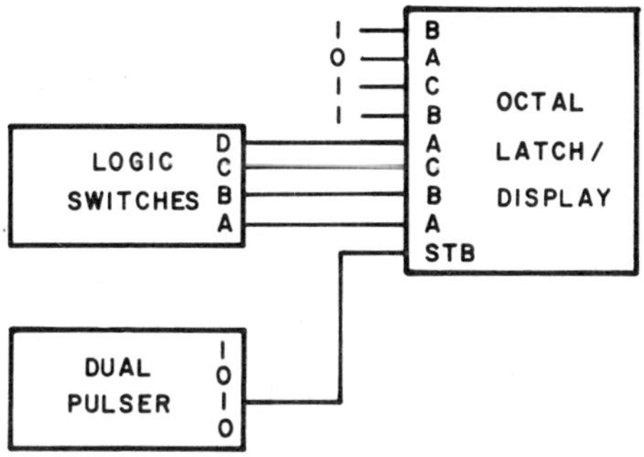

Fig. 1-55.

Step 1

Wire the circuit shown in the schematic diagram. Make certain that the power is disconnected. Keep in mind that "0" represents GROUND potential and "1" represents +5 volts. This experiment is somewhat similar to Experiment No. 10.

Step 2

Set the four logic switches all to logic 0. Apply power to the breadboarding socket. What reading do you observe on the LR-27 octal latch/display?

We observed a reading of 260_8. The little "8" represents octal code. We shall explain this observation in a moment.

Step 3

Set the logic switches to D = 1, C = 0, B = 0, and A = 1. When you did this in Experiment No. 10 and then pressed and released the pulser, you observed a 9 on the latch/display. Press and release the pulser and write the number that you observe on the display in the space here.

We observed a reading of 271_8.

Step 4

Change the wire connection between STB and the pulser to the "0" output of the pulser. Now vary the settings of the four logic switches between the "0" and "1" positions and write down in the following table the octal numbers that you observe on the display.

Inputs Logic States of Logic Switches	Output Observed Octal Number
D C B A	
0 0 0 0	260_8
0 0 0 1	
0 0 1 0	
0 0 1 1	
0 1 0 0	
0 1 0 1	
0 1 1 0	
0 1 1 1	
1 0 0 0	
1 0 0 1	271_8

You need not go any higher than this in the table.

Step 5

There is no D input in octal code. Only A, B, and C inputs are allowed, and the largest 3-bit binary number is A = 1, B = 1, and

C = 1. What you have done in this experiment, *as a consequence of the wiring of the LR-27 Octal Latch/Display Outboard module*, is to convert an 8-bit binary number into three octal digits, where an octal digit can have any value ranging from 0 to 7. Octal refers to eight states, and note that there are indeed eight states in the digits 0 through 7. This conversion is done from the least-significant bit to the most-significant bit, in two groups of three and one group of two. For further details about octal code, we refer you to Chapter 5.

In Experiment No. 10, when DCBA = 1001, you observed a 9 on the display. Here, you observe a 1 on the least-significant digit. Can you guess why? If so, write your suggestion in the following space.

Questions

1. What, in your opinion, is the difference between octal and hexadecimal code?

2. It is possible to convert an 8-bit binary number into a three-digit octal code. Explain how this is done.

3. It is also possible to convert an 8-bit binary number into a two-hexadecimal-digit code. Explain how this is done. This is a difficult question, and you may not be able to answer it on the basis of the experience that you have had so far in digital electronics.

4. Which code do you think you would prefer, octal or hexadecimal? Why?

5. The answers in the table in Step 4 of this experiment are the 8-bit ASCII code representations for the numerals 0 through 9. What is ASCII code? If you do not know, you might look at Chapter 5.

EXPERIMENT NO. 12

Purpose

The purpose of this experiment is to demonstrate the use of the LR-26 Outboard module as a display for a 7490 counter that is clocked by the LR-5 Clock Outboard module.

Pin Configuration of Integrated-Circuit Chip (Fig. 1-56)

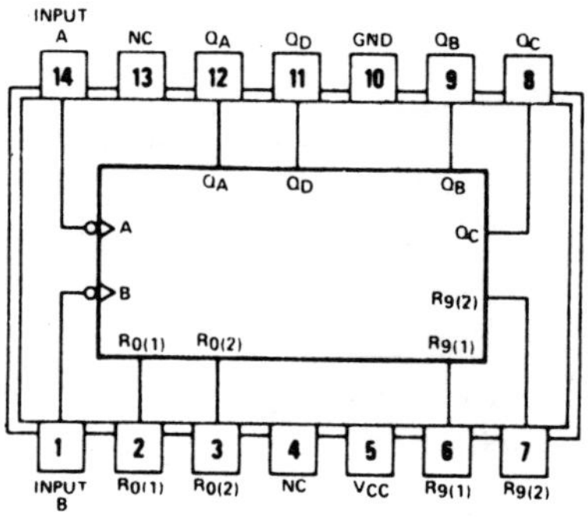

7490

Fig. 1-56.

Schematic Diagram of Circuit (Fig. 1-57)

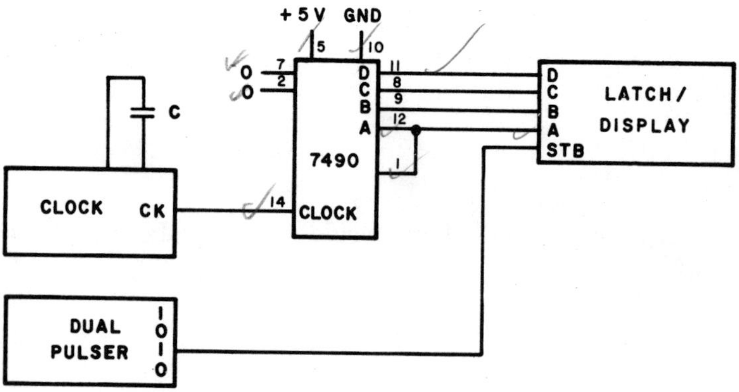

Fig. 1-57.

Step 1

Wire the circuit shown in the schematic diagram. Make certain that the power is disconnected. Note the similarity between this circuit and the one given in Experiment No. 9.

Step 2

Choose a capacitor in the range of 0.1 to 1.0 μF for your clock. We recommend a value of 0.33 μF, which will give you a clock rate of approximately 1.0 Hz.

Step 3

Apply power to the circuit and observe whether or not anything happens.

We observed that the reading on the latch/display did not change. Why not?

Step 4

Press the pulser in and hold it in. What do you now observe on the latch/display?

We observed sequential counting on the display.

Step 5

Change the connection between the STB input to the display and the pulser to the "0" output of the pulser. What do you now observe?

Our circuit now displayed sequential counting.

Step 6

Press the pulser in and hold it in. What happens now?

We did not observe any more counting on the latch/display. Why not?

Questions

1. What, in your opinion, is the purpose of the STB input to the latch/display? Explain its function.

2. We call the STB input the "strobe" or "enable" input. Can you guess what these two adjectives mean? If so, give your suggestion in the space below.

3. Why were you not able to see either a blank display, a minus sign, or all dots lit if you had a numeric indicator, or the letters A through F if you had a hexademical indicator?

4. Which display do you prefer, the one on the LR-4 Outboard module or the one on the LR-26 Outboard module? Why?

WHAT HAVE WE ACCOMPLISHED IN THIS CHAPTER?

Review of Objectives

It is appropriate to review our initial objectives. We stated in the Introduction that at the end of this chapter you would be able to do the following:

1. Set up any experiment given in subsequent chapters by following the steps, instructions, and/or schematic diagrams provided.

   In the experiments that you have performed in this chapter, you have had a sample of the steps, instructions, and schematic diagrams that are given to help you out. If you have successfully performed the experiments given, you can look forward to many more experiments given in subsequent chapters with the knowledge that you generally will be able to do many, if not most, of them.

2. Demonstrate the operation of the principal OUTBOARD® modules and associated equipment, including
   battery or power supply
   breadboard
   logic switches
   lamp monitors
   pulsers
   clock
   seven-segment LED display
   integrated-circuit chips
   capacitors

   You have used each of the above modules in one or more experiments in this chapter. You have also read a description of the modules in the introductory material in this chapter. You have responded to a variety of questions concerning the operation of the various modules. Thus, you should now have reasonable familiarity with the above auxiliary functions and components. You may be lacking in familiarity with integrated circuits, but this deficiency will be quickly corrected in the next two chapters.
3. Identify the schematic symbols *employed in this book* for the following auxiliary functions:
   battery or power supply
   logic switches
   lamp monitors
   pulsers
   clock
   seven-segment LED display
   as well as for the following components:
   integrated-circuit chips
   capacitors
   resistors

   In the introductory material prior to the experiments in this chapter, you were provided with the appropriate schematic symbols for all of the auxiliary functions and components listed above. You have encountered most of these functions and components in schematic diagrams throughout the experiments in the chapter.
4. Make wire connections between the above auxiliary functions and components using a breadboard.

   You have done this in the group of twelve experiments provided in this chapter. The SK-10 socket has served as a breadboard.

5. Handle the above auxiliary functions, components and breadboard with sufficient care that none will be destroyed during the entire series of experiments in this book.

   The authors cannot be sure that you have succeeded in this objective. Nevertheless, a large number of helpful hints and suggestions were provided. We hope that the more important hints were heeded.

6. Recognize that the majority of simple digital electronic circuits are based on two states, a "logic 0" state and a "logic 1" state, which usually correspond to voltages of approximately 0 volt and +5 volts with TTL integrated circuits.

   This point was made in Experiment No. 2, Step 6.

2

Gating a Digital Signal

INTRODUCTION

One of the most important concepts that you must learn in digital electronics is the concept of a *gate*. While there are many ways to approach this concept, we believe that one of the best is to observe a gate perform a useful function, such as the gating of a train of clock pulses. This particular application of a gate is widely used in digital electronics and can be easily wired using the knowledge you have gained in Chapter 1.

OBJECTIVES

At the completion of this chapter, you will be able to do the following:

1. Demonstrate, using appropriate integrated-circuit chips and auxiliary functions, how you can "enable" and "disable" a clock input to a 7490 decade counter integrated-circuit chip.
2. Compare the gating signals to and the outputs from the following four different types of 2-input gates:

   2-input AND gate
   2-input NAND gate
   2-input OR gate
   2-input NOR gate

   when the "input data" to each of these gates is a train of clock pulses.

3. Pass a test in which you are asked to define, in your own words and without the aid of this or any other text on digital electronics, the following terms:

to inhibit	OR gate
to gate	NAND gate
to enable	NOR gate
to disable	positive logic
to strobe	negative logic
gate (noun)	clock
AND gate	gate, gating, and gated (adjectives)

4. Write the correct symbols for the following logic devices:

   2-input AND, NAND, OR, and NOR gates
   3-input AND, NAND, and NOR gates
   4-input AND and NAND gates
   inverter
   buffer
   driver

DEFINITIONS

amplifier—A device that draws power from a source other than the input signal and that produces as an output an enlarged reproduction of the essential features of its input.[1]

AND gate—A binary circuit with two or more inputs and a single output, in which the output is logic 1 only when all inputs are logic 1, and the output is logic 0 if any one of the inputs is logic 0.[1]

binary—A numbering system using a base number, or radix, of 2. There are two digits (0 and 1) in the binary system.[1]

binary code—A code in which each code element is one of two distinct kinds or values, e.g., the presence or absence of a pulse.[1]

binary counter—(1) A counter that produces one output pulse for every two, four, eight, sixteen, etc., input pulses. (2) A counting circuit, each stage of which has two, four, eight, sixteen, etc., states.

binary digit—A character that represents one of the two digits in the number system that has a radix of two. Also called a bit. Either of the digits, 0 or 1, that may be used to represent the binary conditions on or off.[1]

binary number system—A number using two symbols (usually denoted 0 and 1) and having 2 as its base, just as the decimal system uses 10 symbols (0, 1, 2, 3, 4, 5, 6, 7, 8, and 9) and has a base of 10. Also called binary notation.[1]

binary numeral—The binary representation of a number, e.g., 101_2.[1]

binary signal—A voltage or current that carries information in the form of changes between two different states that are a discrete interval apart.[1] These two states are usually called the logic 0 and logic 1 states.

clock input—That flip-flop input terminal whose condition or change of conditions controls the admission of data through the synchronous inputs of the flip-flop and thereby controls the output state of the flip-flop. The clock signal permits data signals to enter the flip-flop and, after entry, directs the flip-flop to change state accordingly.[1] Many MSI chips have flip-flops inside, and consequently require clock pulses for their operation. Such clock pulses enter at the clock input pins of the chips.

disconnect—To break an electric circuit. A device or group of devices that remove electrical continuity from between two conductors of a circuit.[1]

(to) disable—To prevent the passage of binary signals by application of the proper signal to the disable terminal of the device.[1]

driver—An element coupled to the output stage of a circuit to increase the power- or current-handling capability, or fan-out, of the stage.[1]

(to) enable—To permit a circuit to be activated by the removal of a suppression signal.[1]

fan-in—The number of inputs that can be connected to a logic circuit.[1] A fan-in of one is what a single input on a 7400 NAND gate chip requires to operate correctly.

fan-out—The number of parallel loads within a given logic family that can be driven from one output mode of a logic circuit.[1]

gate (noun)—A circuit having two or more inputs and one output, the output depending upon the combination of logic signals at the inputs. There are four gates, called AND, OR, NAND, and NOR.[1]

gate (noun)—A signal used to trigger the passage of other signals through a circuit. This signal is called a gating signal.[1]

gate (adjective)—Pertaining to a gate (noun).

(to) gate (verb)—To control the passage of a digital signal, such as a pulse or train of clock pulses.

gated (adjective) (past participle)—(1) The past participle of the verb to gate. (2) Pertaining to a device that requires a gating signal for its operation. This term does not apply to the operation of simple gates. (The term, gated gate is redundant.)

gating (gerund)—The action of controlling the passage of a digital signal, such as a pulse or train of clock pulses.

gating (adjective)—Pertaining to a device, signal, etc., that modifies the operation of a gate circuit.

inverter—A circuit that has one input and one output. Its function is to invert the input. When the input is high, the output is low, and vice versa. The inverter is sometimes called a NOT circuit, since it produces the reverse of the input. A device or circuit that complements a Boolean function.[1] When the input is a logic 0, the output is a logic 1, and vice versa.

(to) inhibit—To prevent an action from taking place or data from being accepted by applying an appropriate signal (generally a logic 1 in positive logic) to the proper input.[1]

NAND gate—A combination of a NOT function and an AND function in a binary circuit that has two or more inputs and one output. The output is logic 0 only if all inputs are logic 1; it is logic 1 if any input is logic 0. With logic of the opposite polarity ("negative logic"), this type of gate becomes a NOR gate.[1]

NOR gate—An OR gate followed by an inverter to form a binary circuit in which the output is logic 0 if any of the inputs is logic 1 and is logic 1 only if all the inputs are logic 0. With the opposite logic polarity ("negative logic"), this type of gate is a NAND gate.[1]

NOT circuit—A binary circuit with a single output that is always the opposite of the single input. Also called an inverter circuit.[1]

negative logic—A form of logic in which the more positive voltage level represents logic 0 and the more negative level represents logic 1.[1] In many cases, the more negative level is also known as GROUND, but it is at a logic 1 state in a negative logic system.[1]

OR gate—Also called an OR circuit. A gate that performs the function of logical "inclusive or." It produces an output whenever any one (or more) of its inputs is energized. For a 2-input OR gate, when both inputs are at logic 0, the output is at logic 0. When either or both inputs are at logic 1, the output is at logic 1.[1]

positive logic—A form of logic in which the more positive logic level represents 1 and the more negative level represents 0.[1]

switch—A mechanical or electrical device that completes or breaks the path of the current or sends it over a different path.[1]

to strobe—To gate a digital input signal into a counter, register, memory, etc.[1]

train of clock pulses—A sequence of clock pulses.

transmit—To send information from one location to another.[1]

7400, 7402, 7408, and 7432—Popular 7400-series 2-input logic gates.

7490 and 7493—Popular 7400-series counters.

WHAT IS A DIGITAL SIGNAL?

Digital signals are discrete or discontinuous signals whose various states are discrete intervals apart (Rudolf F. Graf, *Modern Dictionary of Electronics*, Howard W. Sams & Co., Inc.). The typical digital signal in digital electronics is a *binary signal*, which is defined as a voltage or current that carries information in the form of changes between two different states that are a discrete interval apart. One of these states is called the "logic 0" state, and the other, the "logic 1" state. For voltage signals, the "logic 0" state is typically at GROUND potential, whereas the "logic 1" state varies between +3 and +5 volts. For current signals, the "logic 0" state is typically 0 milliamperes, whereas the "logic 1" state is 20 milliamperes.

WHAT OPERATIONS CAN WE PERFORM ON A SINGLE DIGITAL SIGNAL?

Consider Fig. 2-1, in which we indicate that a digital device converts an input digital signal into an output digital signal. For purposes of argument, we might consider the input digital signal to be

Fig. 2-1. Diagram of a simple digital device.

a train of clock pulses, which is represented by the symbol shown in Fig. 2-2. The question then becomes: What happens to this train of clock pulses as it passes through different digital devices? We can answer this question in tabular form by listing

Fig. 2-2. A train of clock pulses.

- The operation that the digital device performs on the input digital signal, in this case, a train of clock pulses.
- The digital device that is used to perform this operation.
- The schematic symbol for this digital device.

On the following pages are listed the different operations that are commonly performed on a *single digital signal.* This list summarizes many of the important gating operations that you may encounter in digital electronics.

Operation	*Digital Device*	*Digital Device Symbol*
SINGLE OUTPUT DEVICES		
Transmit the signal unchanged	wire	
Invert the signal	inverter	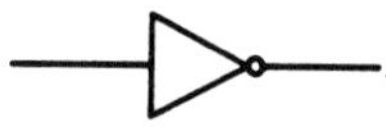

Operation	Digital Device	Digital Device Symbol
Increase the signal voltage level	amplifier, driver	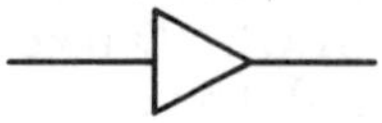
Increase the signal current level	buffer	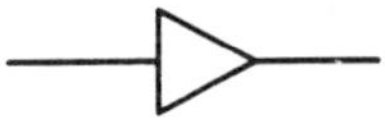
Transmit the signal, *or* disconnect the signal from the output line	single-pole single-throw switch	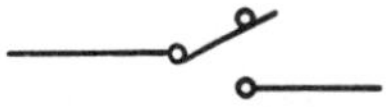
Transmit *or* block the signal	AND gate	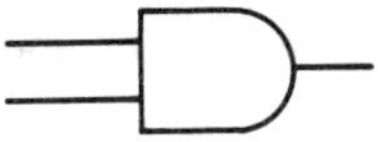
Invert *or* block the signal	NAND gate	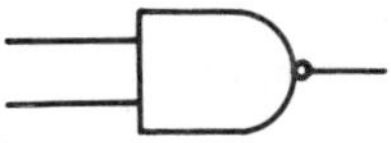
Transmit *or* block the signal	OR gate	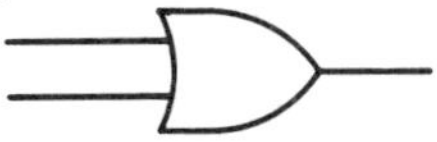
Invert *or* block the signal	NOR gate	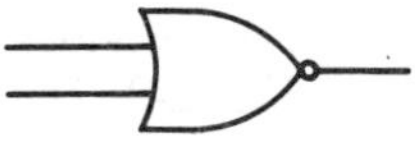
Transmit *or* block the signal, *or* disconnect the signal from the output line	three-state AND gate	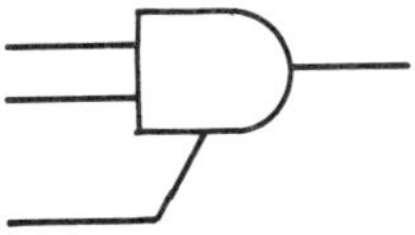
Invert *or* block the signal, *or* disconnect the signal from the output line	three-state NAND gate	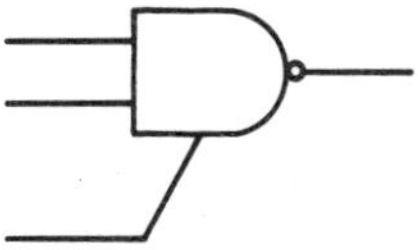
Transmit *or* block the signal, *or* disconnect the signal from the output line	three-state OR gate	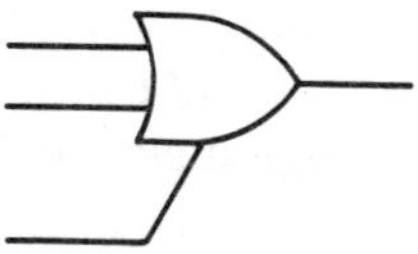
Invert *or* block the signal, *or* disconnect the signal from the output line	three-state NOR gate	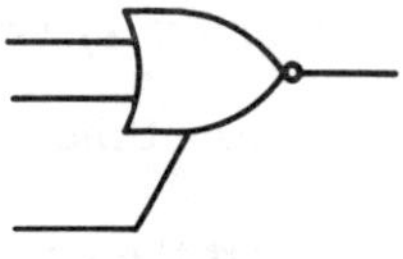

Operation	Digital Device	Digital Device Symbol
Re-route the signal	single-pole double-throw switch	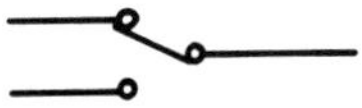
Re-route the signal	wafer switch	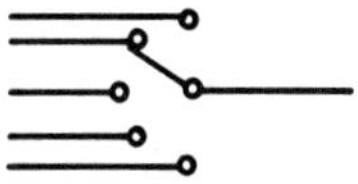
Re-route *or* block the signal	multiplexer	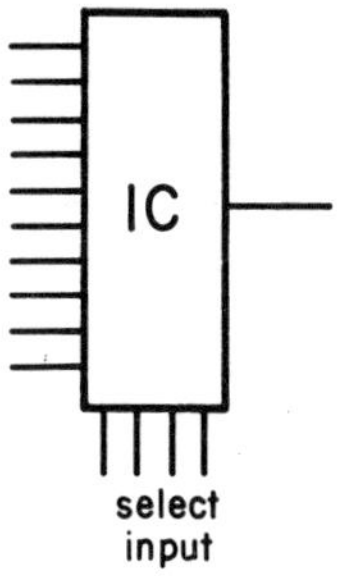

SINGLE INPUT DEVICES

Operation	Digital Device	Digital Device Symbol
Re-route the signal	single-pole double-throw switch	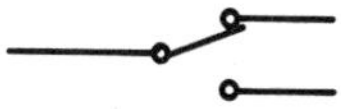
Re-route the signal	wafer switch	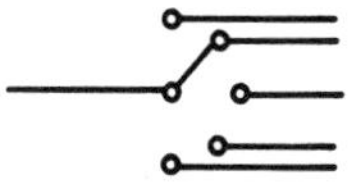
Re-route the signal	demultiplexer	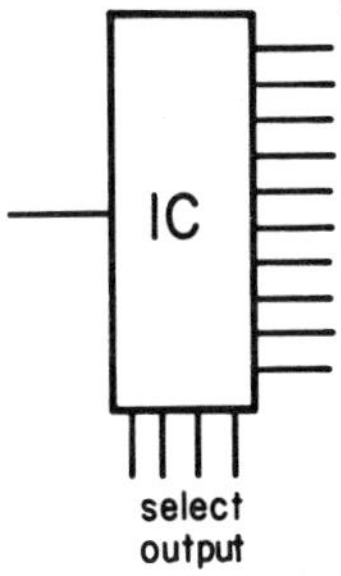

THE NOUN: GATE

The term, *gate*, has a variety of meanings, so it is appropriate for us to carefully state various possible definitions. The noun, *gate*, can have several meanings, all of which are given in the *Modern Dictionary of Electronics*.

1. A circuit having two or more inputs and one output, the output depending upon the combination of logic signals at the inputs. There are four gates, called AND, OR, NAND, and NOR. The definitions assume positive logic is used. In computer work, a gate is often called an AND circuit.
2. A signal used to trigger the passage of other signals through a circuit. This signal is called a *gating signal,* where the word *gating* is used as an adjective.
3. One of the electrodes of a field-effect transistor (FET).
4. A circuit in which one signal (usually a square wave or a pulse) switches another signal on or off.

All of these situations are commonly encountered in digital electronics. When reading an article on digital electronics, you may occasionally have to do a bit of sleuthing in order to determine in which sense the word gate is being used. In this book, we shall most commonly use definition 1 above.

THE VERBS: TO GATE AND TO STROBE

The verb, *to gate,* generally has only a single meaning: *To control the passage of a digital signal, such as a pulse or train of clock pulses.* The use of this verb is very common in digital electronics, e.g., "The information was gated into a memory," or "We then gate the counter . . . ." When we speak of "gating a clock," we are using the verb, to gate, in its gerund form. The gerund, *gating,* can generally have several meanings:

1. Selecting those portions of a wave that exist during certain intervals or which have certain magnitudes.
2. Applying a rectangular voltage to the grid or cathode of a cathode-ray tube, to sensitize it during the sweep time only.
3. Applying a specific waveform to perform electronic switching.
4. Controlling the passage of a digital signal, such as a pulse or train of clock signals.

You commonly encounter definition number 4 in digital electronics. A related term is the verb, *to strobe,* which generally means: *To gate a digital input signal into a counter, register, memory, etc.* This term is frequently encountered in the literature of computer interfacing.

THE ADJECTIVES: GATE, GATED, AND GATING

Finally, there are the three adjectives, *gate, gated,* and *gating.* These adjectives are employed in the following senses:

gate circuit
gate pulse
gate signal

gated buffer
gated driver
gated flip-flop

gating circuit
gating pulse
gating signal

They are best defined individually. (NOTE: We are aware of the fact that you may not know or understand many of the terms used in the definitions above and below, e.g., field-effect transistor (FET), electrode, wave, grid, rectangular voltage, cathode, cathode-ray tube, sweep time, waveform, coincidence, delay circuit, low-impedance, inverting driver, pulse differentiation, fan-out, stage, flip-flop, etc. It is not necessary to know any of these terms in order to proceed with this chapter. Please continue.)

buffer—A noninverting digital circuit element that may be used to handle a large fan-out.[1]

driver—An element coupled to the output stage of a circuit to increase the power- or current-handling capability, or fan-out, of the stage; for example, a clock driver is used to supply the current necessary for a clock line.[1]

fan-in—The number of inputs that can be connected to a logic circuit.[1] (This term is related to fan-out.)

fan-out—The number of parallel loads within a given logic family, in our case TTL, that can be driven from one output mode of a logic circuit.[1]

gate circuit—A circuit that passes a signal only when a gating pulse is present.[1]

gate pulse—A pulse that enables a gate circuit to pass a signal. The gate pulse generally has a longer duration than the signal to ensure time coincidence.[1]

gate signal—That signal generated by some form of delay circuit.[1] This term can have other meanings as well.

gated buffer—A low-impedance inverting driver circuit that may be used as a line driver for pulse differentiation or in multivibrators.[1] In general, a buffer that is gated.

gated driver—In general, a driver that is gated.

gated flip-flop—A flip-flop that has a steering circuit that prevents both flip-flop outputs from becoming "0" at the same time. Alternating current input pulses must be used to prevent the flip-flop from oscillating. This situation can arise if both input lines are made high simultaneously.[1]

flip-flop—A two-stage multivibrator circuit having two stable states. A trigger signal changes the circuit from one state to the other, and the next trigger signal changes it back to the first state. For counting and scaling purposes, a flip-flop can be used to deliver one output pulse for each two input pulses.[1]

gating circuit—A circuit that operates as a selective switch and allows conduction only during selected time intervals, or when the signal magnitude is within specified limits.[1]

gating pulse—A pulse that modifies the operation of a gate circuit.[1]

gating signal—A signal that modifies the operation of a gate circuit.[1]

trigger—A pulse that starts an action.[1]

The above definitions should be sufficient to illustrate the uses of the three adjectives given. Return to these definitions as you proceed through the book.

SWITCH VS. GATE: WHAT IS THE DIFFERENCE?

A *switch* can be defined as *a mechanical or electrical device that completes or breaks the path of the current or sends it over a different path.*[1] In *Digital Electronics for Scientists,*[4] the authors discuss the following differences between a switch and a gate:

- An open switch prevents passage of a signal, whereas an open gate allows a signal to pass.
- A gate is essentially a one-way device, whereas a switch is a two-way device.
- When a gate is closed, a signal cannot pass.
- When a switch is closed, a signal can pass in either direction.
- When a gate is open, a signal can pass in only one direction.
- When a switch is open, a signal cannot pass in either direction.

GATE SYMBOLS

The symbolic representation of gates has become standard throughout the digital electronics industry. We shall use these standard symbols in most experiments in this book. For purposes of discussion in this chapter, we shall also use several more general representations of a gate, as shown in Figs. 2-3 through 2-5. All of these schematic symbols emphasize one of the several important functions of logic gates: the gating of digital data from the input of the gate to the output of the gate.

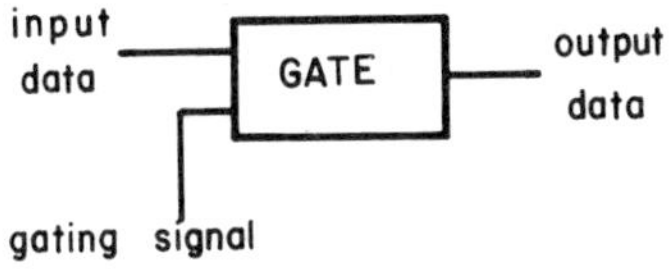

Fig. 2-3. A 2-input gate symbol.

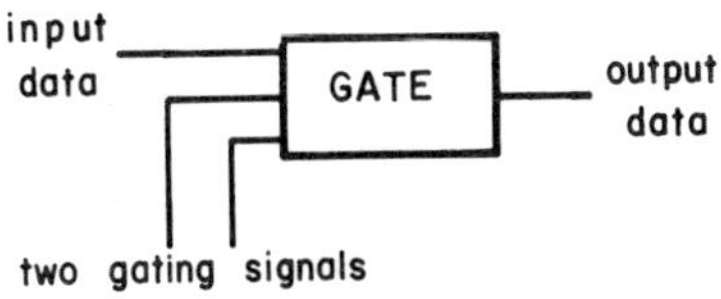

Fig. 2-4. A 3-input gate symbol.

Fig. 2-5. A multiple-input gate symbol.

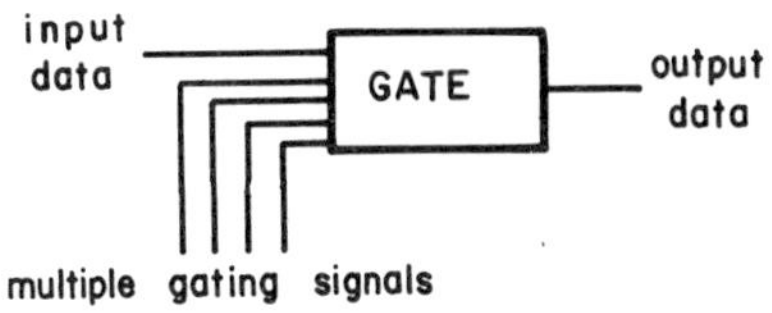

Fig. 2-3 is a general representation of a 2-input gate. One of the inputs can be considered as the "data input" and the other input is for the gating signal, which causes the data input either to be blocked by the gate or transmitted through the gate to the output line. Fig. 2-4 is a general representation of a 3-input gate. It still has a single "data input," but in addition there are two separate gating signal inputs. Depending on the nature of the gate and the logic states of the two gating signals, the data can either be blocked or transmitted through the gate to the output line. Fig. 2-5 shows a general representation of a multiple-input gate. In general, only the AND and NAND gates are readily available with multiple inputs. However, the use of multiple-input gates in medium-scale integration (MSI) is quite widespread.

Gates are also widely used to make logical decisions on digital signals that are not specifically considered as either "data input" or a gating signal. In such cases, it is far more common to use the schematic symbols shown in Figs. 2-6 through 2-8.

Apart from the general representation for a gate, which was given above, there are also specific representations for the four common gates, AND, NAND, OR, and NOR. In this book, we shall use two symbols for such gates: (1) A symbol that emphasizes the fact that we are

Fig. 2-6. A 2-input logic gate symbol in which A and B are inputs and Q is the output.

Fig. 2-7. A 3-input logic gate symbol in which A, B, and C are inputs and Q is the output.

Fig. 2-8. A 4-input logic gate symbol in which A, B, C, and D are inputs and Q is the output.

"gating" input data through the gate, and (2) A symbol that reflects the fact that the gate is being used as a logic gate, in which there is usually no specific distinction between "input data" and "gating signals." The two kinds of symbol representations are compared in Figs. 2-9 through 2-14.

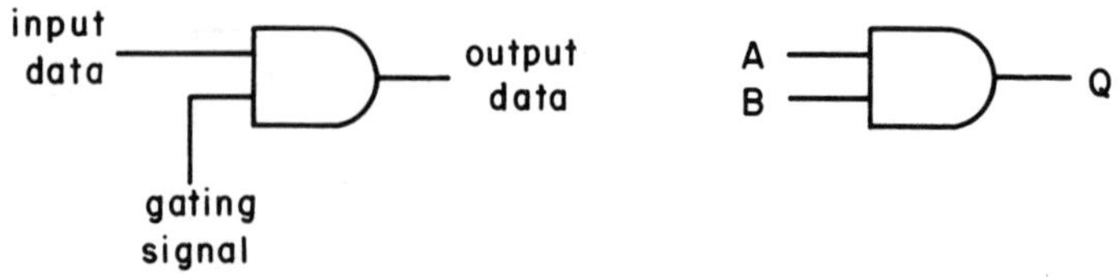

Fig. 2-9. Two kinds of symbols for a 2-input AND gate.

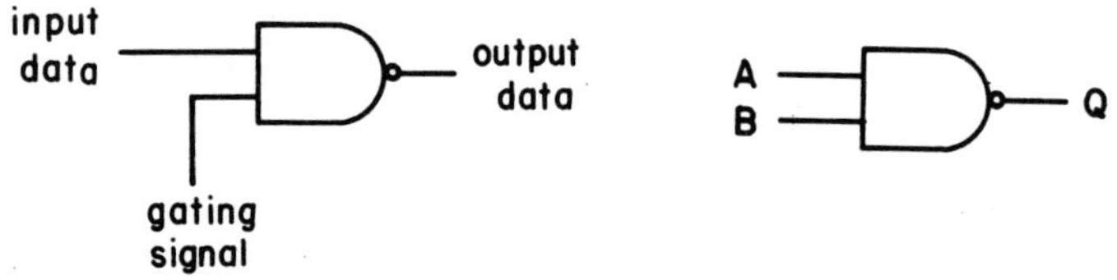

Fig. 2-10. Two kinds of symbols for a 2-input NAND gate.

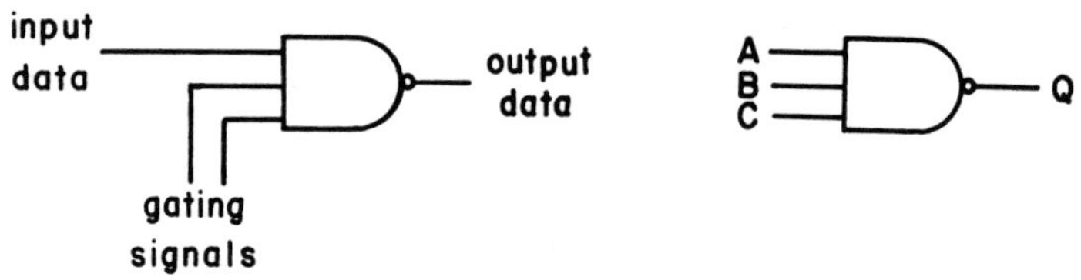

Fig. 2-11. Two kinds of symbols for a 3-input NAND gate.

In addition to the above gates, there are also the schematic symbols for an inverter, a buffer, a driver, and an amplifier. These symbols are shown in Figs. 2-15 and 2-16.

Inverters, buffers, and drivers are not gates, but they do perform useful operations on digital signals. Inverters can be created from some of the gates. This will be discussed shortly.

MANIPULATING A TRAIN OF CLOCK PULSES

A *clock pulse* has been previously defined as a complete logic cycle, i.e., a transition from logic 0 to logic 1 and back to logic 0, which is represented by the pulse symbol, as shown in (Fig. 2-17),

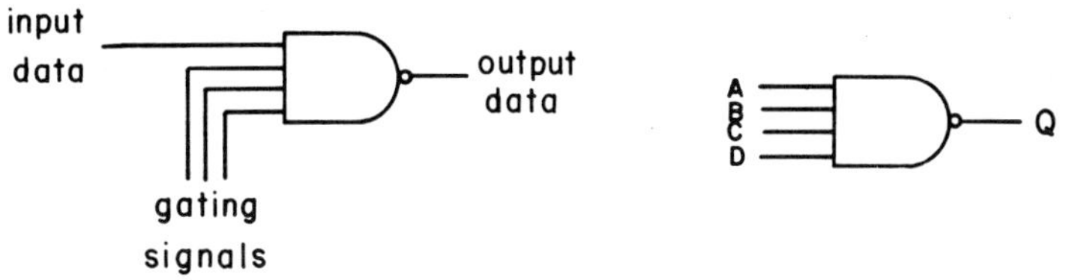

Fig. 2-12. Two kinds of symbols for a 4-input NAND gate.

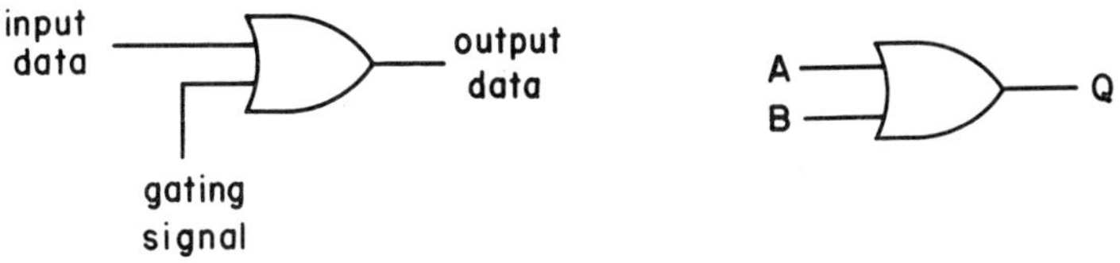

Fig. 2-13. Two kinds of symbols for a 2-input OR gate.

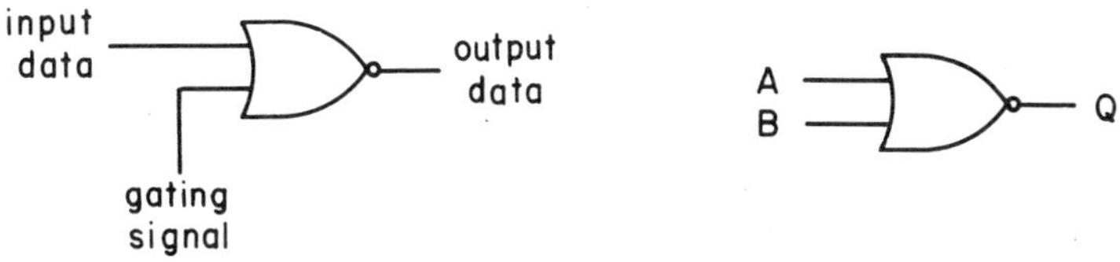

Fig. 2-14. Two kinds of symbols for a 2-input NOR gate.

or a transition from logic 1 to logic 0 and back to logic 1, which is represented by a somewhat similar model (Fig. 2-18)

A *train of clock pulses* simply is a sequence of clock pulses, which can be represented by the symbols (Fig. 2-19)
for a normal and an inverted train of clock pulses, respectively. The normal and inverted trains are *complements* of each other.

A train of clock pulses can be

- Transmitted unchanged
- Inverted
- Blocked
- Re-routed
- Disconnected from the output line
- Amplified

or some combination of these operations, such as

- Transmitted or blocked
- Inverted or blocked
- Transmitted and amplified
- Inverted and amplified
- Transmitted or disconnected from the output line
- Transmitted or blocked or disconnected from the output line

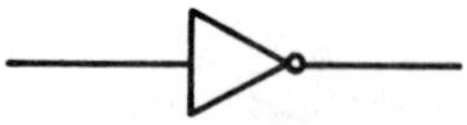

Fig. 2-15. Symbol for an inverter.

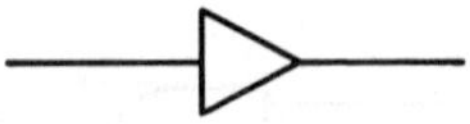

Fig. 2-16. Symbol for a buffer, driver, or an amplifier and representing devices used to increase fan-out, or drive, of digital logic gates.

- Inverted or blocked or disconnected from the output line
- Re-routed or blocked
- Inverted and re-routed or blocked

The simplest operation that can be performed on a train of clock pulses is to transmit them unchanged. This is accomplished using a wire. The diagram for this operation is shown in Fig. 2-20. The next simplest operation on a train of clock pulses is to invert them; this is accomplished using an inverter. The diagram for this operation is (Fig. 2-21),

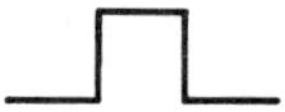

Fig. 2-17. Diagram of a positive clock pulse.

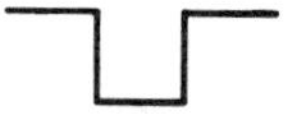

Fig. 2-18. Diagram of a negative clock pulse.

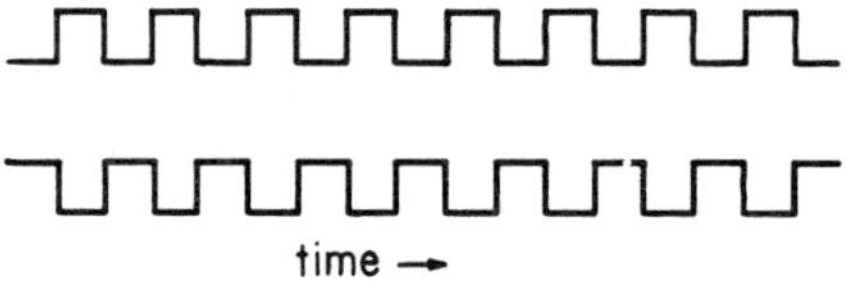

Fig. 2-19. Trains of positive and negative clock pulses.

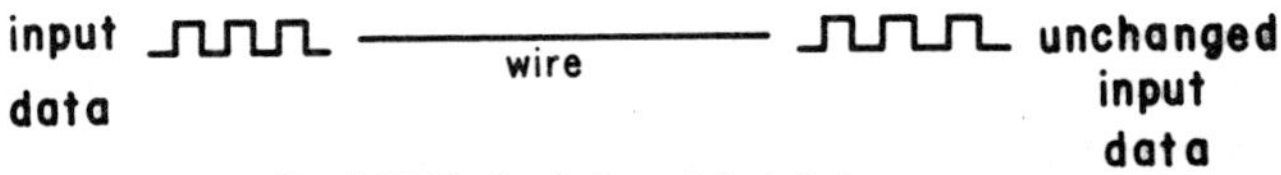

Fig. 2-20. Transmission of digital data.

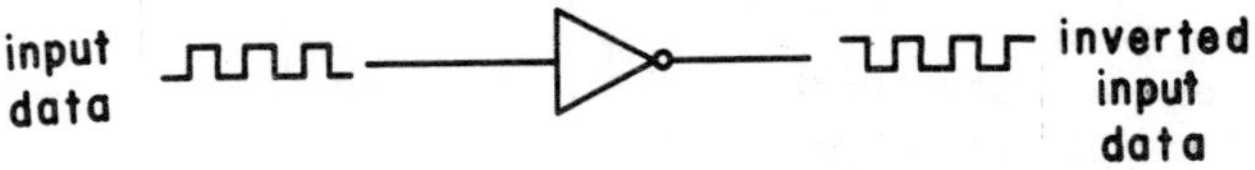

Fig. 2-21. Inversion of digital data.

A third simple operation that you can perform on a train of clock pulses is to amplify them using a buffer, driver, or amplifier (Fig. 2-22),

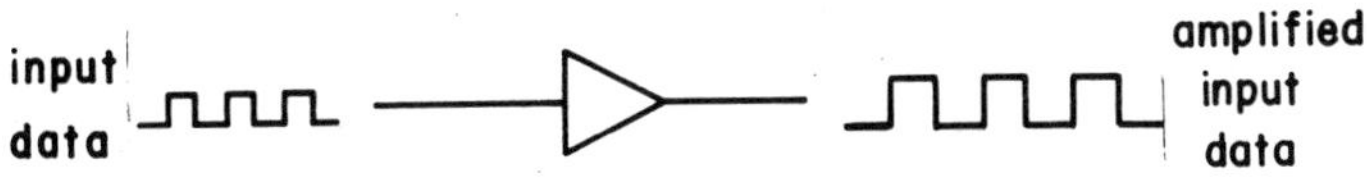

Fig. 2-22. Buffering of digital data (signal current is amplified).

Beyond these three operations, you encounter the concept of "gating a train of clock pulses," an operation which will be described in the following section.

GATING A TRAIN OF CLOCK PULSES

When you "gate a train of clock pulses," you either block the train of clock pulses or transmit them to the output line in the normal or inverted form. Four simple gates are commonly used for this purpose: the AND gate, the NAND gate, the OR gate, and the NOR gate. These gates will now be discussed individually.

The AND Gate

When the gating signal to an AND gate is at a "logic 0" state, input data, such as a train of clock pulses, cannot pass through the gate. This situation is represented in Fig. 2-23. Note that the output remains at a *quiescent* level of logic 0. On the other hand, when the gating signal to the AND gate is at a "logic 1" state, input data is directly transmitted to the output without inversion (Fig. 2-24).

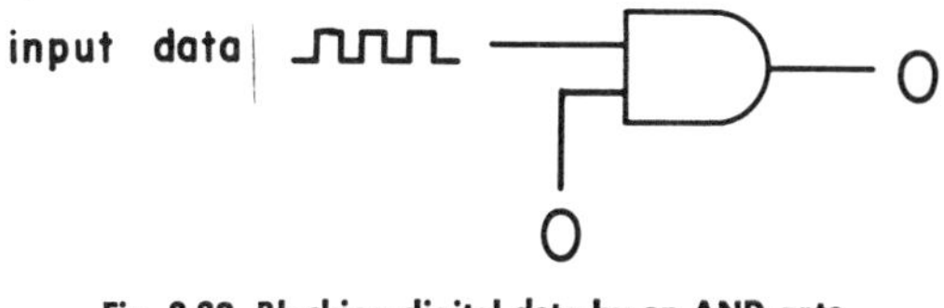

Fig. 2-23. Blocking digital data by an AND gate.

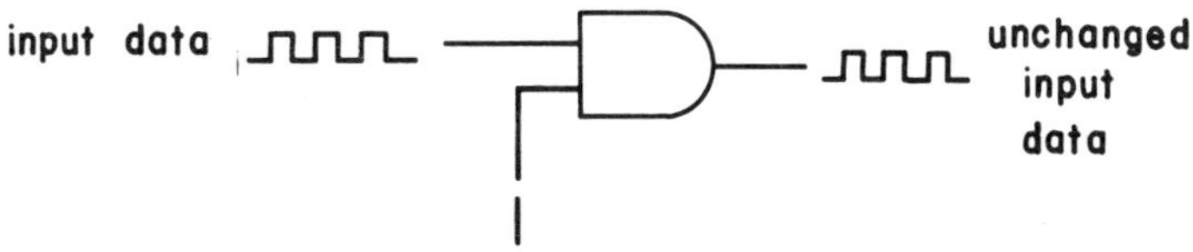

Fig. 2-24. Transmission of digital data by an AND gate.

The NAND Gate

The NAND gate, which is also known as a "negated" AND gate, is similar to the AND gate. When the gating signal is at a "logic 0" state, input data is blocked, as in Fig. 2-25. Note the quiescent output, which is now logic 1. On the other hand, when the gating signal is at a "logic 1" state, inverted input data is available at the output of

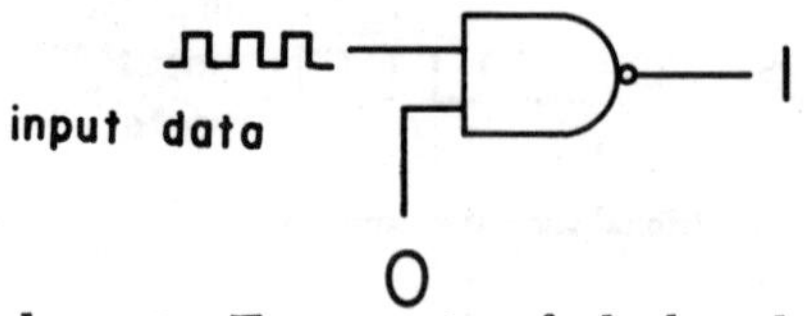

Fig. 2-25. Blocking digital data by a NAND gate.

the gate. For a train of clock pulses, this operation can be represented by the diagram in Fig. 2-26.

The OR Gate

For the OR gate, input data is blocked through the use of a "logic 1" state at the gating signal input (Fig. 2-27). The use of a "logic 0" gating signal causes input data to be directly transmitted to the output without inversion (Fig. 2-28).

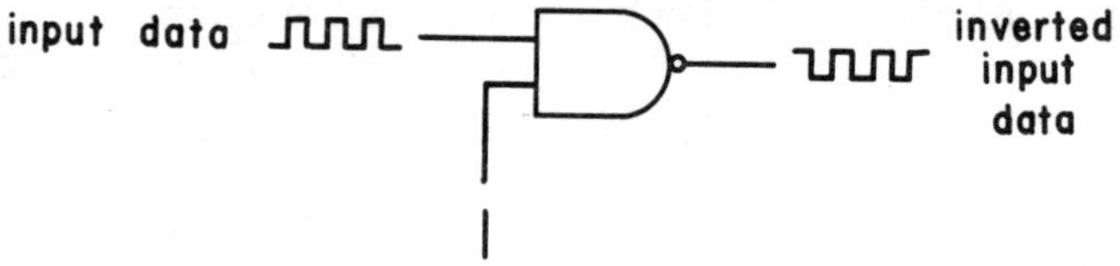

Fig. 2-26. Transmission of digital data by a NAND gate.

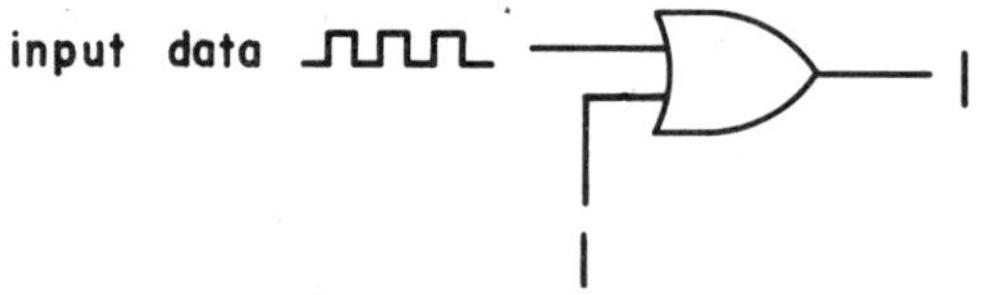

Fig. 2-27. Blocking digital data by an OR gate.

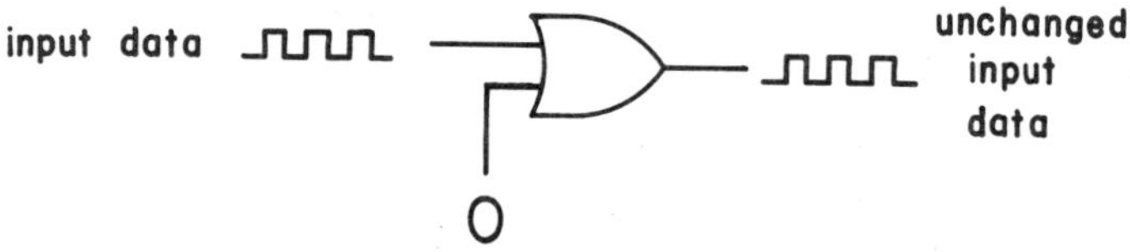

Fig. 2-28. Transmission of digital data by an OR gate.

The NOR Gate

The NOR gate, or "negated" OR gate, bears the same relationship to the OR gate as the NAND gate bears to the AND gate. Like the OR gate in Fig. 2-29, the use of a "logic 1" gating signal completely blocks input data from appearing at the output of the gate. Finally, the use of a "logic 0" gating signal causes inverted input data to be available at the output, as shown in Fig. 2-30.

Comparison of Gating Behavior

In the above sections, we were concerned about whether the gate blocked the data or not, whether the unblocked data was normal or

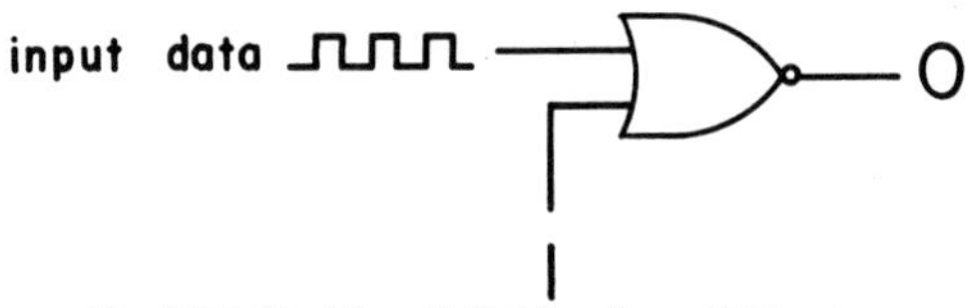

Fig. 2-29. Blocking digital data by a NOR gate.

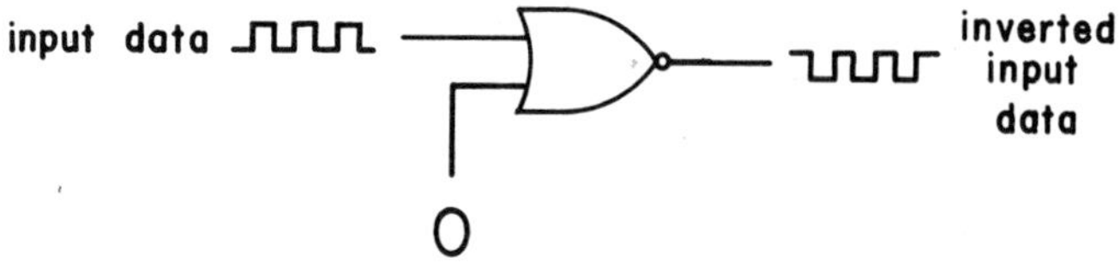

Fig. 2-30. Transmission and inversion of digital data by a NOR gate.

inverted, and what was the output logic state from the gate when the input data was blocked. The important results that we obtained can be summarized by the following data. One set gives the logic state of the gating signal required to open each of the four gates, and the other gives both the logic state of the gating signal as well as the gate output when the input data is blocked.

This data shows the logic state of the gating signal required to enable each of the four gates. The nature of the output data signal is also listed.

Gate	*Logic State of Gating Signal*	*Nature of Output Data*
AND	1	unchanged input data
NAND	1	inverted input data
OR	0	unchanged input data
NOR	0	inverted input data

This data shows the logic state of the gating signal, and the nature of the output signal, when each of the four gates is blocked, i.e., disabled.

Gate	*Logic State of Gating Signal*	*Logic State of Quiescent Output*
AND	0	0
NAND	0	1
OR	1	1
NOR	1	0

OTHER GATING TERMINOLOGY: TO INHIBIT OR TO ENABLE/DISABLE A GATE

When we have described the characteristics of a gate, we have used the words "open," "close," "block," and "unblock." Three other verbs that you will frequently encounter are to *inhibit, to enable,* and *to disable* a gate. These three verbs mean the following:

to inhibit—To prevent an action from taking place, to prevent data from being accepted by applying an appropriate signal to the proper input.[1]

to enable—To permit a circuit to be activated by the removal of a suppression signal.[1]

to disable—To prevent the passage of binary signals by application of the proper signal to the disable terminal of the device.[1]

In the context of our previous discussion, it should be clear that

TO OPEN A GATE = TO ENABLE A GATE
= TO UNBLOCK A GATE
= TO STROBE A GATE
TO CLOSE A GATE = TO DISABLE A GATE
= TO INHIBIT A GATE
= TO BLOCK A GATE

Open, enable, strobe, and unblock are essentially synonyms, as are close, disable, inhibit, and block. Since the terms, *enable, disable,* and *inhibit* have fewer connotations than open, close, block, and unblock, professionals in the area of digital electronics prefer to use the former group of terms, even though many individuals may not initially know what they mean. In digital electronics, the terms enable, disable, and inhibit simply are more precise than open, close, block, etc. We do not agree with the use of unnecessary jargon, but, frankly, we do like the sound of words such as strobe, enable, disable, inhibit, and the like. They are crisp and quite precise as to their meanings. In digital instrumentation, these terms are frequently encountered on the output pins or output panel. One day even your spouse might say, "Honey, please enable the mileage meter in the car and see how well we are conserving gas." Unlikely? Perhaps, but not impossible.

POSITIVE VS. NEGATIVE LOGIC

The terms, *positive logic* and *negative logic,* can be defined as follows:

positive logic—A form of logic in which the more positive logic level (in our case, +5 Volts) represents logic 1 and the more negative logic level (in our case, GROUND) represents logic 0.

negative logic—A form of logic in which the more positive logic level (in our case +5 Volts) represents logic 0 and the more negative logic level (in our case, GROUND) represents logic 1.

For example, when the two inputs to a 2-input positive NAND gate are each at +5 volts, the output is at ground potential. We refer you to Book 2, Chapter 7 for a more detailed discussion of positive and negative logic and an explanation why it is useful to distinguish between them.

REPRESENTING INTEGRATED CIRCUITS: A 2-INPUT AND GATE

Now that you have read the general background that applies to simple logic gates, you are ready to perform a variety of experiments that demonstrate how you can "gate" a clock. Before you start the experiments, however, we would like to discuss what we believe is an extremely important topic: What is the most effective way to represent the wiring to an integrated-circuit chip? In Chapter 1, in Fig. 1-51, we gave the actual physical location of auxiliary functions and integrated circuits plus part or all of the wiring necessary on the SK-10 breadboard. As we move to more complicated digital electronic circuits, e.g., circuits that may require four or five different auxiliary functions and four or five integrated-circuit chips, it will become progressively more difficult to provide illustrations such as those given in Fig. 1-51. We have observed that any wiring diagram that shows the actual physical location of the pins, components, functions, and wires *becomes too complicated to read when more than twenty wires are required.* Clearly, we must develop a better way to help you wire digital circuits. The sooner we succeed in this task, the easier it will be for you to wire complex circuits.

Perhaps the most important problem that we should tackle right now is: How can we provide a diagram or symbolic representation that would make it easy for you to wire an integrated circuit chip? Let us answer this question by considering how we might represent the wiring to a 7408 *quad 2-input positive* AND *gate,* the "pin configuration of which is shown in Fig. 2-31. The number, 7408, is the IC number assigned to any 2-input positive AND gate that has the same pin configuration and electrical specifications of the original

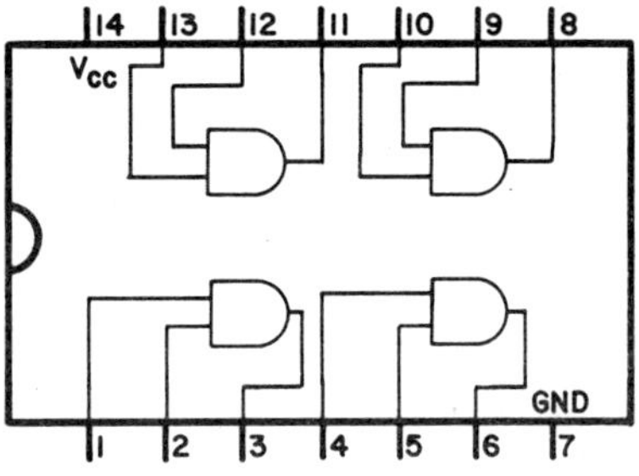

Fig. 2-31. Pin configuration of the 7408 quad 2-input positive AND gate integrated circuit.

7408 IC chip originally manufactured by Texas Instruments Incorporated. This particular chip is available from at least five different manufacturers, including Signetics, National Semiconductor, Sprague, Stewart Warner Corporation, and TI. The term, *quad,* means that there are four separate 2-input positive AND gates on the chip. The adjective, *positive,* means that there are four positive logic AND gates on the IC chip. Finally, it is clear that there are two inputs to each of the four AND gates, hence the description, *2-input.*

Let us now assume that we wish to wire a single 2-input positive AND gate, in which the two inputs are A and B and the output is Q. The appropriate symbol is shown in Fig. 2-32; this is what you would

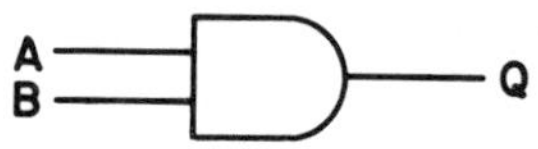

Fig. 2-32. Simple AND gate without pin numbers or IC identification, as is frequently the case.

be given in a schematic diagram. What would be the most effective way for us to show you how to wire this simple digital circuit? We might show you the exact placement of components and wires on a breadboarding socket, as is done in Fig. 2-33. This would be fine for such a simple circuit, but it has the serious fault that it does not force you to become independent in the wiring of digital circuits.

A second approach is shown in Fig. 2-34, in which we draw to scale the 7408 integrated-circuit chip and show you the pin numbers and connections to make. Note that we have numbered pins 1, 2, 3, 7, and 14; these are the pins that must be connected on your breadboard. In our opinion, this approach is not appreciably better than the first one described above. With a complicated circuit containing four or five integrated circuits and many wires, you may have some trouble understanding what to do.

A third approach, and one that we like much better, is shown in Fig. 2-35. This diagram will be used with MSI chips, such as counters, decoders, registers, memories, etc. This representation uses the same principle that we have previously discussed in Step 3 of Experiment No. 7 in Chapter 1. The pins are no longer arranged in sequential order; instead, they have been rearranged to make the schematic diagram easier to read! *This representation does not have a semicircular notch in it, so it does not represent the pin configuration of the actual 7408 integrated-circuit chip.* Note that the two inputs are on the left, the output is on the right, and the two power connections—GROUND and +5 Volts—are on the top (see Fig. 2-36). We shall use this type of representation throughout this book, especially for medium-scale integrated circuits, such as counters, shift registers, decoders, multiplexers, and the like.

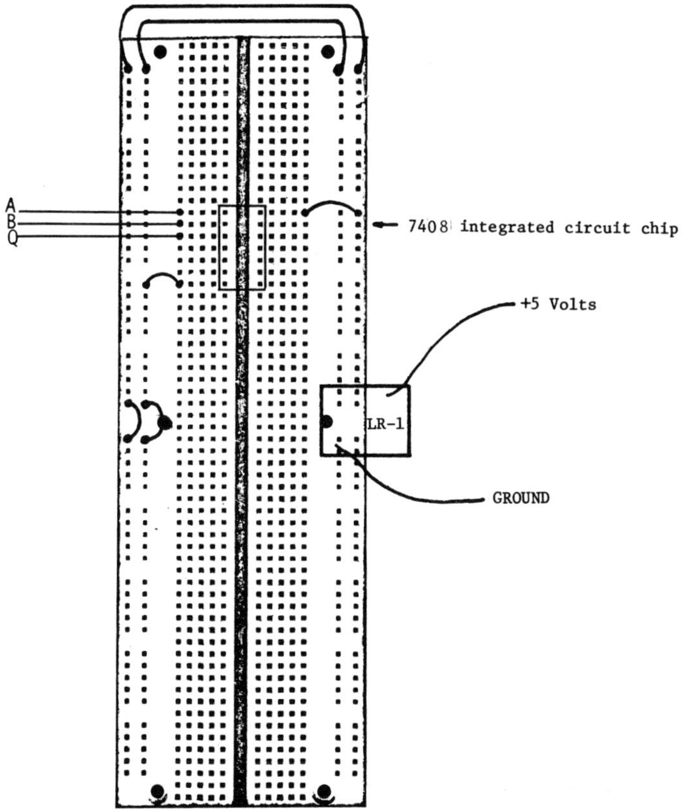

Fig. 2-33. Wiring diagram for a single 2-input positive AND gate in the 7408 IC chip.

To finish our discussion, we must point out that there is still a more common way to represent the pin configuration for a simple gate integrated circuit. This is shown in Fig. 2-37. Note that we have

Fig. 2-34. One approach showing how to wire the AND gate circuit of Fig. 2-32.

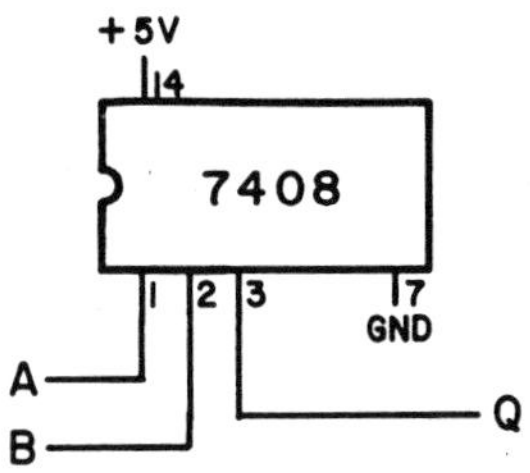

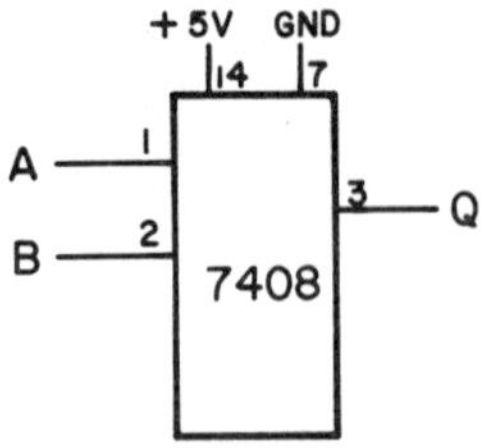

Fig. 2-35. Another way to wire the AND gate circuit shown in Fig. 2-32.

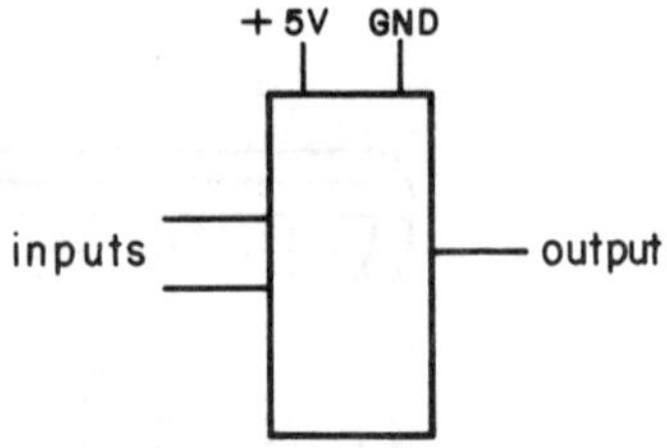

Fig. 2-36. A general representation similar to that given in Fig. 2-34.

used the correct symbol for a 2-input AND gate, and have modified it by providing the pin numbers required for the two inputs and the single output. These are the numbers 1 and 2, and 3, respectively. Note that we have not indicated the two power connections at pins 7 and 14 of the 7408 chip. *In such a representation, it is assumed that you will make these two power connections; otherwise, the IC will not operate!* The quantity, "¼ 7408" means that we are only using one-fourth of the 7408 chip. Since the chip contains four AND gates, one-fourth of the chip means that we are using a single AND gate. This type of representation is common in the literature of digital electronics; most schematic diagrams represent simple gates in this manner.

We shall make one minor change to this widely used method of representing simple gates and their actual pin connections: We shall leave off the ¼, ⅓, ⅙, or ½ before the number of the integrated circuit. Thus, our final representation (the one that we shall use in this book) is shown in Fig. 2-38 for a single 2-input AND gate on the 7408 integrated-circuit chip. Note the absence of the power connections.

Remember, in order to properly connect the gate in Fig. 2-38, you must make certain that the power connections are properly made at pins 7 and 14 on the integrated circuit. Failure to make such connections will give you an inoperative gate.

The above discussion is extremely important, and we request that you re-read it if you do not understand the line of reasoning used. Diagrams such as Fig. 2-33 are neither necessary nor practical for each experiment. By learning how to wire digital circuits from a

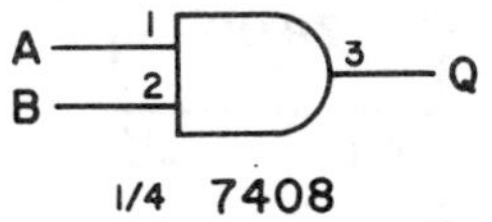

Fig. 2-37. A common way to represent simple gates in schematic diagrams.

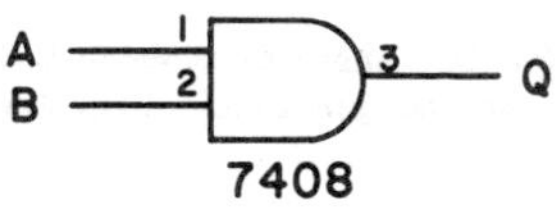

Fig. 2-38. A simple gate used in this book.

more general circuit representation, you will gain the ability to wire circuits on any type of breadboard using any kind of hardware.

PIN CONFIGURATIONS FOR SIMPLE 2-INPUT GATE INTEGRATED CIRCUITS

In this chapter, you will perform experiments using four very simple gate integrated circuits:

- The 7408 quad 2-input positive AND gate
- The 7400 quad 2-input positive NAND gate
- The 7432 quad 2-input positive OR gate
- The 7402 quad 2-input positive NOR gate

It is appropriate to comment on them briefly.

The 7408 Quad 2-Input Positive AND Gate Integrated Circuit

The "pin configuration" or "pin assignment" for the 7408 quad 2-input positive AND gate IC is shown in Fig. 2-39. This chip has been

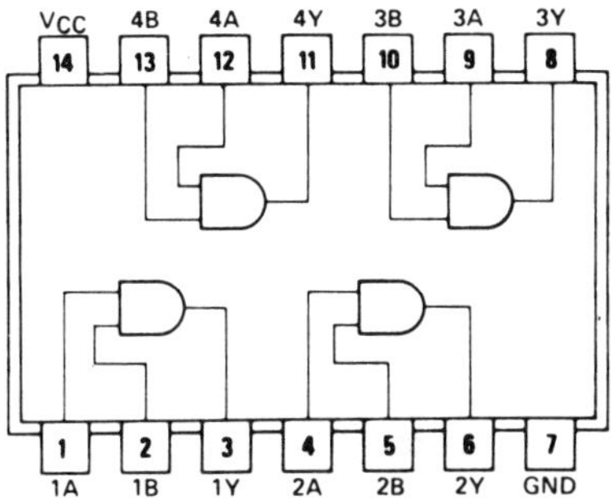

Fig. 2-39. Pin configuration of the 7408 quad AND gate IC.

described previously. Observe that there are four *independent* 2-input AND gates, which can be schematically represented by the four diagrams provided in Fig. 2-40.

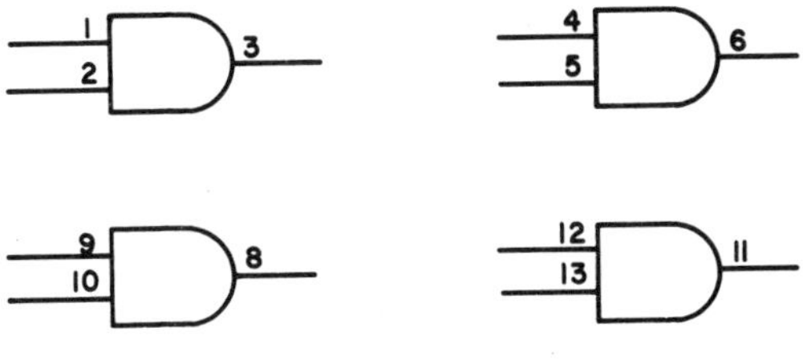

Fig. 2-40. Diagrams of the four independent 2-input AND gates of the 7408 IC.

The 7400 Quad 2-Input Positive NAND Gate Integrated Circuit

The 7400 quad 2-input positive NAND gate IC is very similar in appearance to the 7408 IC chip. Like the 7408 chip, the power is applied at pin 7 (GROUND) and pin 14 (+5 Volts). The pin configuration is shown in Fig. 2-41, and the four *independent* 2-input NAND gates can be represented as shown in Fig. 2-42.

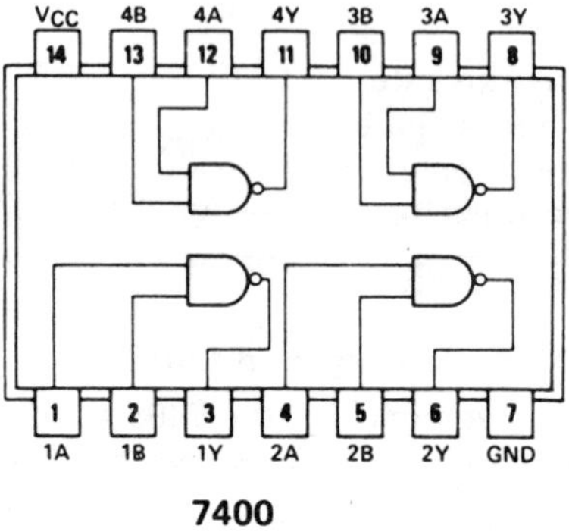

Fig. 2-41. Pin configuration of the 7400 quad 2-input positive NAND gate IC.

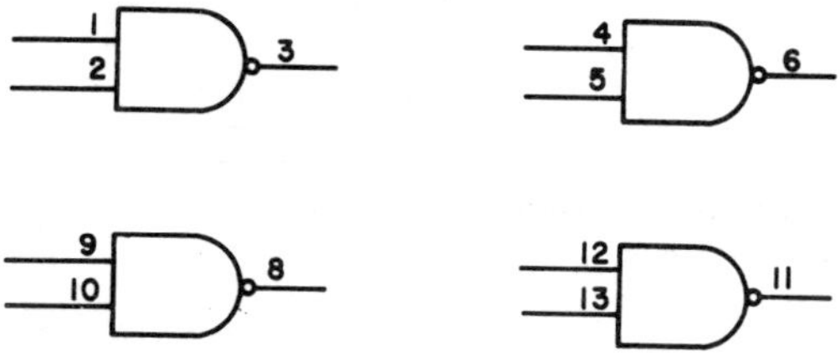

Fig. 2-42. Diagrams of the four independent 2-input NAND gates of the 7400 IC.

The 7432 Quad 2-Input Positive OR Gate Integrated Circuit

The 7432 quad 2-input OR gate IC is, as you would guess, quite similar to the 7400 and 7408 chips. The gate inputs, gate outputs, and power connections are all located in the same place, as can be seen on the pin configuration schematic in Fig. 2-43.

The four *independent* 2-input OR gates can be represented as shown in Fig. 2-44

The 7402 Quad 2-Input Positive NOR Gate Integrated Circuit

Finally, there is the 7402 quad 2-input positive NOR gate IC. This IC chip is slightly different from the other three; though the power connections are in the same place (pins 7 and 14), the gate inputs and outputs are slightly different. You will want to carefully study the pin configuration diagram provided in Fig. 2-45.

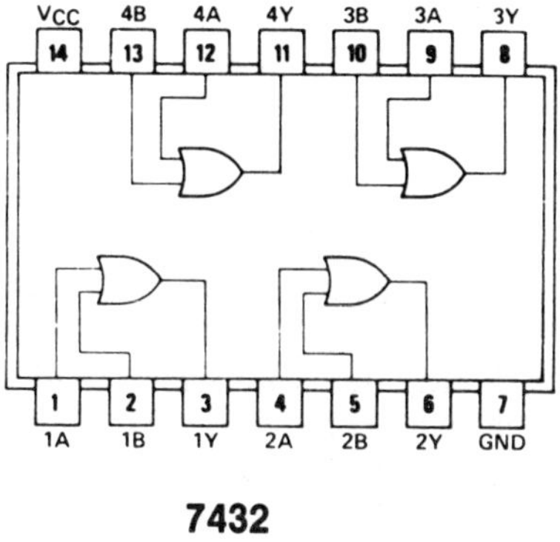

Fig. 2-43. Pin configuration of the 7432 quad 2-input positive OR gate IC.

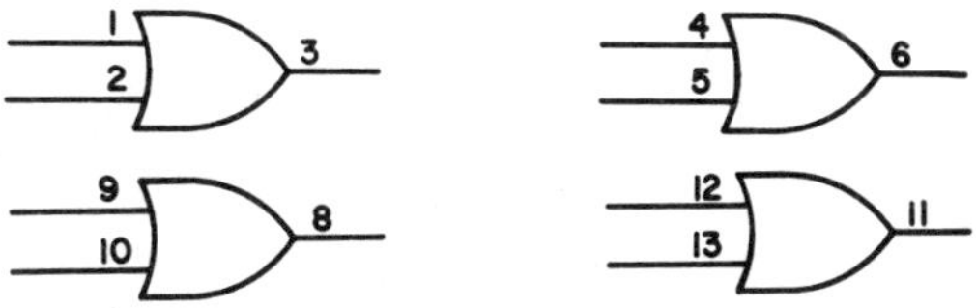

Fig. 2-44. Diagrams of the four independent 2-input OR gates of the 7432 IC.

The four *independent* 2-input NOR gates can be represented as shown in Fig. 2-46. Compare these pin numbers to the numbers given for the AND, NAND, and OR gates in Figs. 2-40, 2-42, and 2-44.

INTRODUCTION TO THE EXPERIMENTS

The preceding section completes our background discussion that leads to experiments in which you learn how to "gate" a digital signal, a train of clock pulses. This is not all that we shall have to

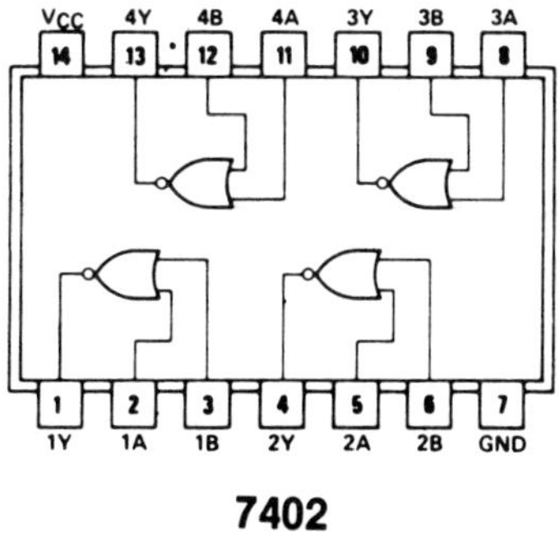

Fig. 2-45. Pin configuration of the 7402 quad 2-input NOR gate IC.

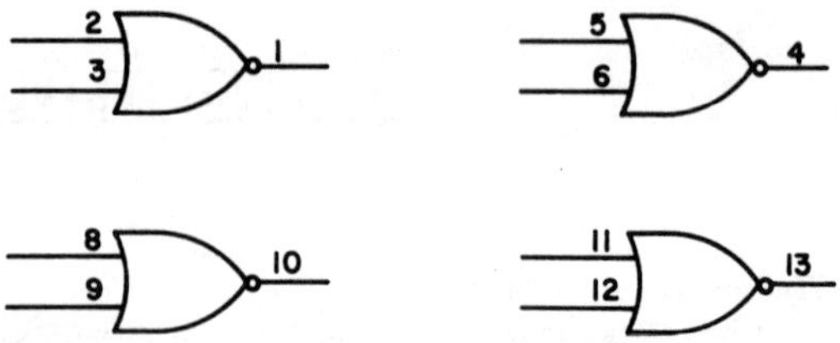

Fig. 2-46. Diagrams of the four independent 2-input NOR gates of the 7402 IC.

say about gates. In a subsequent chapter, you will study other important gate chips in the 7400 series.

We believe that one of the most important applications of single gates in digital electronics is to "gate" input data into counters, flip-flops, shift registers, memories, and the like. The experiments to be described should give you a good feeling for such applications.

EXPERIMENT NO. 1

Purpose

The purpose of this experiment is to demonstrate the operation of the 7408 2-input positive AND gate as a "gate" to the 7490 decade counter integrated circuit in Fig. 2-49.

Pin Configurations of Integrated Circuits (Figs. 2-47 and 2-48)

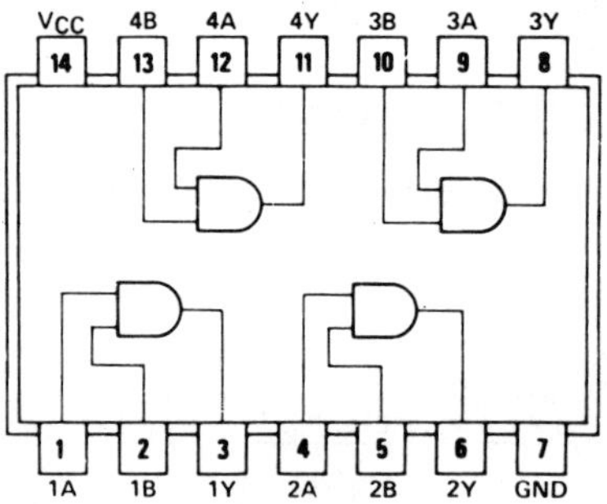

7408

Fig. 2-47.

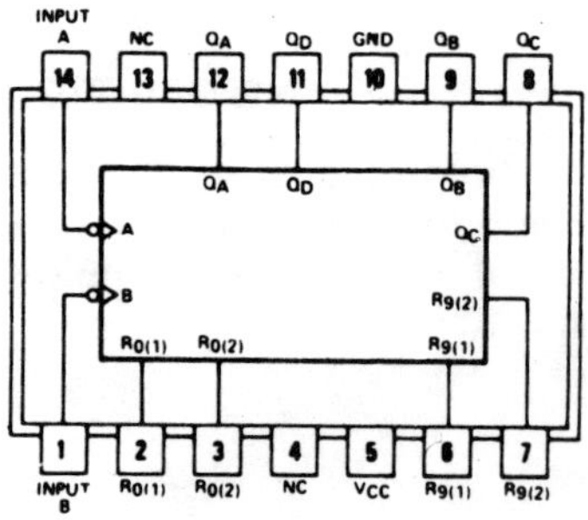

7490

Fig. 2-48.

Schematic Diagram of Circuit (Fig. 2-49)

In this schematic diagram, we have provided general representation for a display, which could either be a single LED lamp monitor, or else a seven-segment display wired as a 0/1 display, as shown in Fig. 2-50.

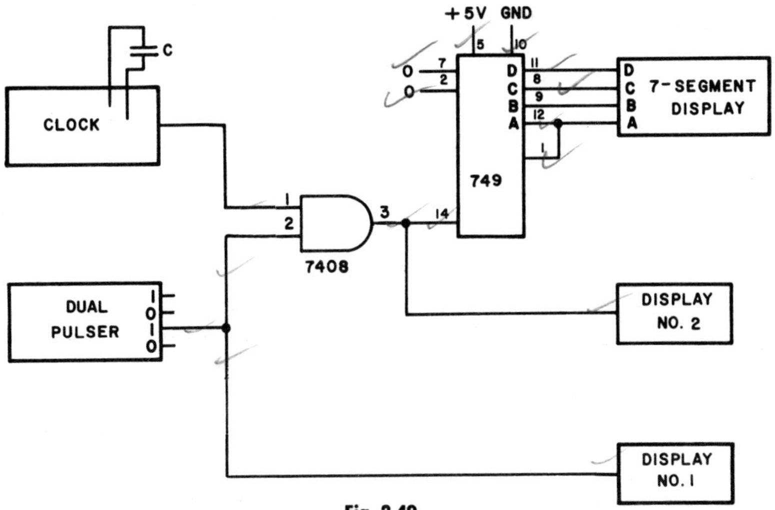

Fig. 2-49.

Fig. 2-50.

Step 1

Carefully inspect the schematic diagram of the circuit. To the circuit given in Experiment No. 9 in Chapter 1, we have added

- A pulser
- Two displays
- A single gate in the 7408 quad 2-input AND gate integrated circuit

Step 2

With no power applied to the breadboard, make the necessary wire connections shown in the schematic diagram for this experiment. The connections that will be required are as follows:

- Pin 14 on the 7408 chip is connected to +5 Volts.
- Pin 7 on the 7408 chip is connected to GROUND.
- Pin 3 on the 7408 chip is connected to pin 14 on the 7490 chip and also to lamp monitor B (Display No. 2).
- Pin 2 on the 7408 chip is connected to a "1" output on a pulser and also to lamp monitor A (Display No. 1).

- Pin 1 on the 7408 chip is connected to the output of a clock that provides approximately 1 Hz clock signals.

Step 3

Once the wiring of the circuit has been completely checked, apply power to the breadboard. Answer the following questions:

Does the seven-segment LED display count from 0 to 9 and back to zero again? Answer yes or no.

What is the logic state at pin 2 of the 7408 integrated-circuit chip, a "logic 0" or a "logic 1" state? Look at display #1 for your answer.

What behavior do you observe on display #2?

If you have wired this experiment correctly, you should indeed observe that the LED display counts from 0 to 9 and back to 0 again. If you do not observe such behavior, then your circuit is not functioning correctly and you must determine what is wrong. Have you connected pin 2 on the 7408 chip to the correct terminal on the pulser? If you have not, you can observe the counting of the "clock" simply by pressing the pulser button. You cannot answer the remaining two questions above until you observe the counting of the "clock."

You should observe a logic state of "logic 1" at pin 2 of the 7408 chip if the clock is counting. This particular pin on the 2-input positive AND gate is the "gating" input, and it must be at a "logic 1" state for the clock signal to pass through the gate to pin 14 of the 7490 chip.

You should observe that display #2 oscillates between a "logic 0" and a "logic 1" state. This oscillation is *in phase* with the clock signal that enters the AND gate at pin 1 on the 7408 chip. By "in phase," we mean that when pin 1 of the AND gate is at a "logic 0" state, pin 3 of the gate is also at a "logic 0" state, and when pin 1 of the gate is at a "logic 1" state, pin 3 is also at a "logic 1" state. In other words, the 7408 2-input AND gate transmits the clock signal directly and does not invert it.

Step 4

Now stop the 7490 decade counter from counting by disabling the 2-input AND gate. Before you do so, please answer the following question:

What gating signal, logic 0 or logic 1, is required to disable the gate at pin 2 of the 7408 chip?

The gate is enabled by a "logic 1" state. Therefore, the complement of logic 1, or logic 0, disables it. Push the pulser button and observe that the 7490 decade counter stops counting. You have successfully "disabled" or "inhibited" the clock input to pin 14 of the 7490 chip. This is a very common action in digital instruments such as digital counters and digital voltmeters.

Step 5

What is the logic state on display #1 and display #2 when the gate is disabled? Give your answers in the following chart.

	Observed Logic State
Display No. 1	logic Ø
Display No. 2	logic Ø

You should observe a "logic 0" state on both displays. Why do you observe a "logic 0" state on display #2, the output from the AND gate? an and needs 2 ones

Because a "logic 0" is the quiescent state output from the AND gate when it is disabled.

Step 6

If you have performed the above experiments correctly, you should now have the capability to enable and disable the AND gate quickly, certainly in less than one half a second. This means that you can start and stop the counter quickly. Therefore, you should now be able to perform more accurately Experiment No. 9 in Chapter 1, especially with small values of the timing capacitance. You might wish to repeat the experiment and employ the 7408 gate to "enable" and "disable" the clock input to pin 14 of the 7490 chip.

Save your circuit for the following experiment, in which you will replace the 7408 chip by a 7400 chip.

Questions

1. In your own words, define the noun, "gate."

a circuit uses a one signal to switch on or off another signal

2. Explain how you can "enable" or "disable" a clock input to a counter.

3. What is the logic state of the output from a disabled 2-input AND gate?

4. What possible outputs can you observe with an enabled 2-input AND gate?

5. What gating signal is required to disable a 2-input AND gate?

6. What gating signal is required to enable a 2-input AND gate?

7. Can an enabled 2-input AND gate faithfully transmit any input digital signal? Answer yes or no.

8. If the two inputs into a 2-input AND gate are both at a "logic 1" state, what is the logic state of the output from such a gate?

9. If the two inputs into a 2-input AND gate are both at a "logic 0" state, what is the logic state of the output from such a gate?

10. If one of the two inputs into a 2-input AND gate is at a "logic 0" state, what are the possible logic states of the output from such a gate?

EXPERIMENT NO. 2

Purpose

The purpose of this experiment is to demonstrate the operation of the 7400 2-input positive NAND gate as a "gate" to the 7490 decade counter integrated circuit. The experimental set-up is identical to Experiment No. 1 in this chapter; only the nature of the gate is changed.

Pin Configurations of Integrated-Circuit Chips (Figs. 2-51 and 2-52)

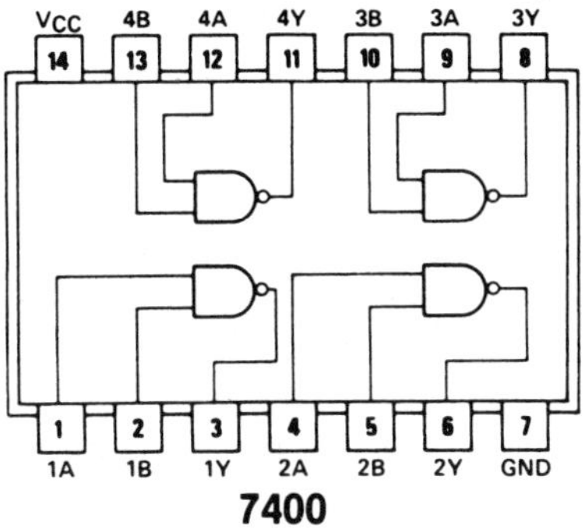

Fig. 2-51.

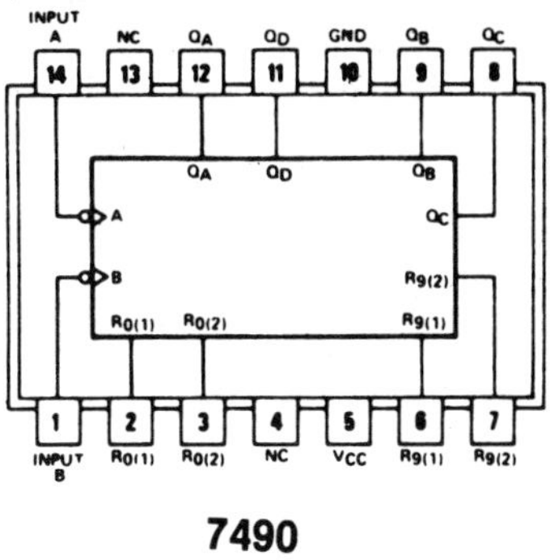

Fig. 2-52.

Schematic Diagram of Circuit (Fig. 2-53)

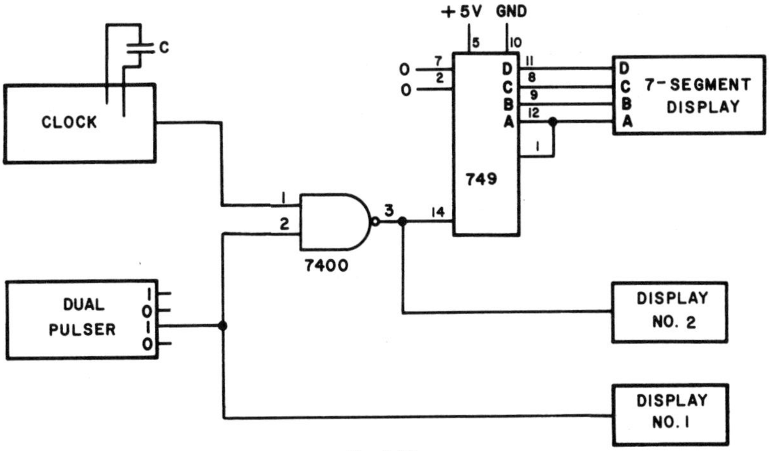

Fig. 2-53.

Step 1

Carefully study the schematic diagram of the circuit and observe that the only change is the type of gate that you use; otherwise, the diagram is identical to that given in Experiment No. 1 Note that even the pin connections are the same; the only difference is that a different integrated-circuit chip, in this case, the 7400 is used.

Step 2

Insert the 7400 IC chip on the breadboard, connect pin 14 of the chip to +5 Volts and pin 7 to GROUND, and finally, connect pins 1, 2, and 3 to the pulser, displays, clock outboard, and 7490 chip, as shown in the schematic diagram. You should not require any additional instructions other than these to wire the circuit correctly.

Step 3

Apply power to the breadboard and check to see that the seven-segment LED display is counting from 0 to 9 and back to 0. If it is not, you have not wired the 7400 quad 2-input positive NAND gate correctly. What logic state do you observe on display No. 1?

If you observe a "logic 1" state on display No. 1 and if the counter is counting, then the circuit is working perfectly.

Step 4

Push the pulser button. Does the counting on the seven-segment LED display stop? It should. What is the logic state that you observe with both display No. 1 and display No. 2 now that the display has stopped counting? Give your answers in the following chart.

	Observed Logic State
Display No. 1	
Display No. 2	

You should observe a "logic 0" on display No. 1 and a "logic 1" on display No. 2. Now answer the following important question:

Why do you observe a "logic 1" output state on display No. 2, the output from the NAND gate?

If you answered this type of question correctly in Experiment No. 1, you should be able to answer it correctly here. The output is "logic 1" because this is the quiescent state output from the NAND gate when it is disabled. *Save your circuit for the following experiment, in which you will replace the 7400 chip by a 7432.*

Questions

1. In the space provided, please compare the behavior of the 7408 2-input AND gate studied in Experiment No. 1 with the 7400 2-input NAND gate studied in this experiment. In what way do they behave similarly? In what way are they different?

2. What gating signal is required to enable a 2-input NAND gate?

3. What gating signal is required to disable a 2-input NAND gate?

4. What possible outputs can you observe with an enabled 2-input NAND gate?

5. What possible outputs can you observe with a disabled 2-input NAND gate?

6. Can an enabled 2-input NAND gate faithfully transmit any input digital signal? By "faithfully transmit," we mean "transmit without inverting." Answer yes or no.

7. If one of the two inputs into a 2-input NAND gate is at a "logic 0" state, what are the possible logic states of the output from such a gate?

8. If the two inputs into a 2-input NAND gate are both at a "logic 1" state, what is the logic state of the output from such a gate?

9. How do a 2-input AND gate and a 2-input NAND gate differ from each other?

10. Do a 2-input AND gate and a 2-input NAND gate require the same type of gating signal? If the answer is yes, what is the logic state of this gating signal that enables each gate?

EXPERIMENT NO. 3

Purpose

The purpose of this experiment is to demonstrate the operation of the 7432 2-input positive OR gate as a "gate" to the 7490 decade counter integrated circuit. The experimental set-up is identical to those in Experiments No. 1 and No. 2 in this chapter; only the nature of the gate is changed.

Pin Configurations of Integrated-Circuit Chips (Figs. 2-54 and 2-55)

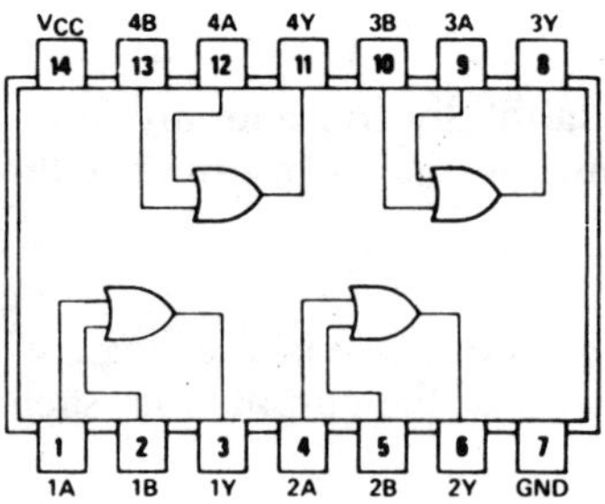

7432

Fig. 2-54.

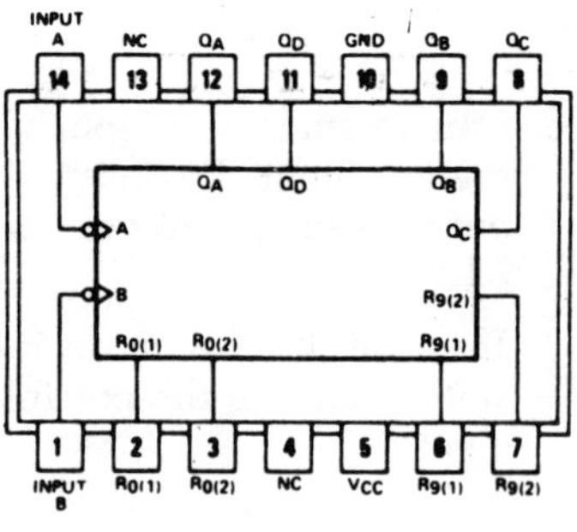

7490

Fig. 2-55.

Schematic Diagram of Circuit (Fig. 2-56)

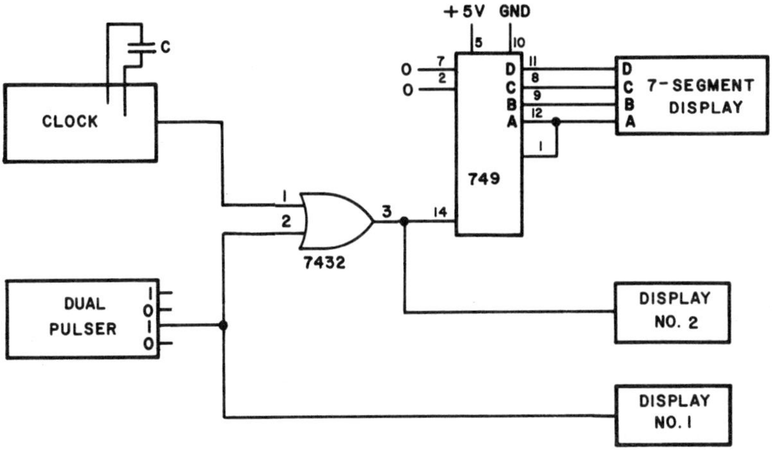

Fig. 2-56.

Step 1

Carefully study the schematic diagram of the circuit in this experiment and observe, as you did in the previous experiment, that the only change is in the nature of the gate. The pin connections are the same as they were in Experiments No. 1 and No. 2. The only difference is that here we are using a 7432 2-input positive OR gate.

Step 2

Insert the 7432 IC chip into the breadboard and connect pin 14 to +5 Volts and pin 7 to GROUND. Connect pins 1, 2, and 3 to the pulser, displays, clock outboard, and 7490 chip, as shown in the schematic diagram. Apply power to the breadboard and answer the following questions:

Do you observe counting on the seven-segment LED display? Answer yes or no.

What are the logic states of display No. 1 and display No. 2? Give your answers below.

	Observed Logic State
Display No. 1	
Display No. 2	

If you have wired this experiment correctly, you should not observe any counting on the LED display. If you do observe counting, please disconnect the power and check your wiring. You should observe a "logic 1" state on both displays. If you do, please answer the following questions:

Why do you observe a "logic 1" state on display No. 1?

Why do you observe a "logic 1" state on display No. 2?

You observe a "logic 1" on display No. 1 because of the way the pulser is wired in this experiment. You observe a "logic 1" on display No. 2 because this is the quiescent state for a disabled 2-input OR gate. You know that the gate is disabled because you do not observe any counting.

Step 3

In the space following, please explain why you do not observe counting in the above experiment. Remember, you now have a 2-input OR gate, which is different from either a 2-input AND gate or a 2-input NAND gate.

You do not observe counting because the 2-input OR gate is disabled. Why is it disabled? Because you are applying a "logic 1" signal at the gating input of the gate.

Step 4

Push the pulser button. You should observe that the seven-segment LED display counts from 0 to 9 and back to 0. What have you accomplished by pushing the pulser button?

The answer is that you have enabled the OR gate by applying a "logic 0" at the gating input.

Release the pulser button. Does the counting stop? It should. *Save your circuit for the following experiment, in which you will replace the 7432 chip by a 7402 chip.*

Questions

1. In the space following, please compare the behavior of the 7432 2-input OR gate studied in this experiment to the 2-input AND gate studied in Experiment No. 1. In what way are they similar? In what way are they different?

2. What gating signal is required to enable a 2-input OR gate?

3. What gating signal is required to disable a 2-input OR gate?

4. What possible outputs can you observe with an enabled 2-input OR gate?

5. What possible outputs can you observe with a disabled 2-input OR gate?

6. Can an enabled 2-input OR gate faithfully transmit any input digital signal? Answer yes or no.

7. If one of the two inputs to a 2-input OR gate is at a "logic 1" state, what are the possible logic states of the output from such a gate?

8. If the two inputs into a 2-input OR gate are both at a "logic 0" state, what is the logic state of the output from such a gate?

9. How do a 2-input OR gate and a 2-input NAND gate differ from each other?

10. Do a 2-input OR gate and a 2-input AND gate require the same type of gating signal? If the answer is yes, what is the logic state of this gating signal that enables each gate?

EXPERIMENT NO. 4

Purpose

The purpose of this experiment is to demonstrate the operation of the 7402 2-input positive NOR gate as a "gate" to the 7490 decade counter integrated circuit. The experiment set-up is similar to those in the three preceding experiments.

Pin Configurations of Integrated-Circuit Chips (Figs. 2-57 and 2-58)

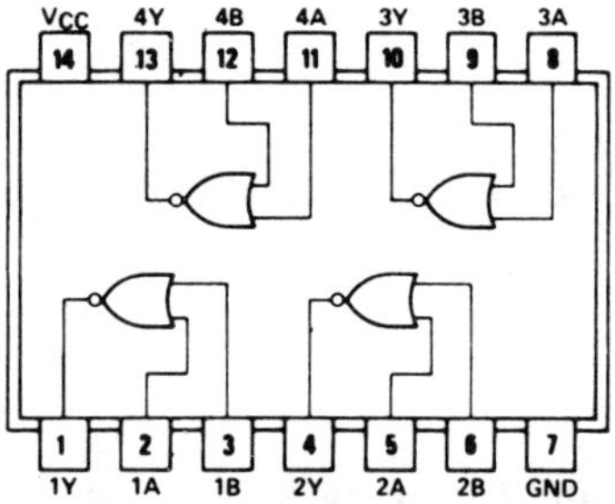

7402

Fig. 2-57.

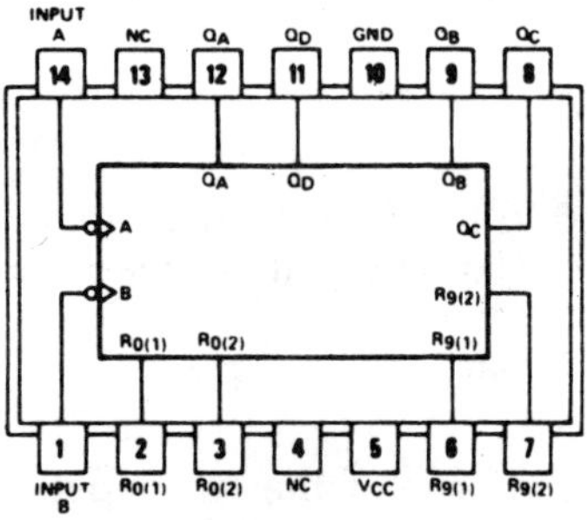

7490

Fig. 2-58.

Schematic Diagram of Circuit (Fig. 2-59)

Step 1

This is the last of a series of four experiments designed to compare the behavior of four different 2-input gates: the AND gate, the NAND gate, the OR gate, and now, the 2-input NOR gate. Keep in mind that

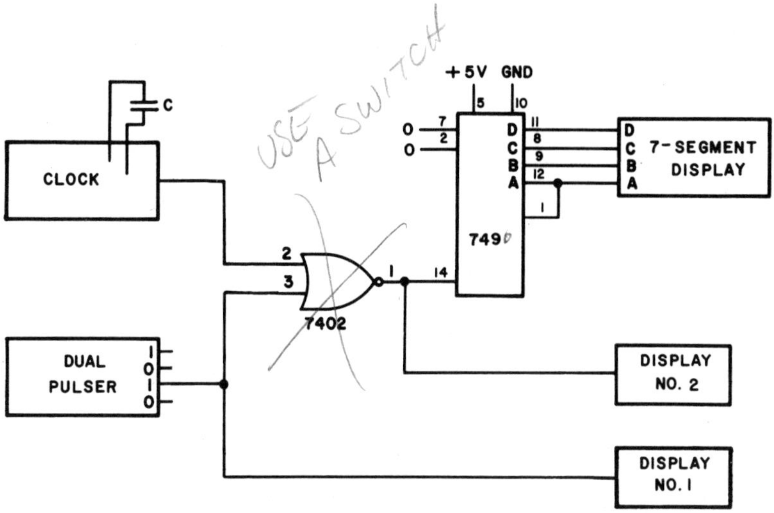

Fig. 2-59.

there are four such gates on the 7402 chip. The only change that you should observe on the schematic diagram is that we are now using a 7402 2-input NOR gate. However, *note that the pin connections to the gate are slightly different: pins 2 and 3 are now the inputs and pin 1 is the output.* If you overlook this point, your circuit will not operate correctly. Do not forget to change the display connections!

Step 2

Apply power to the breadboard. You should not observe counting on the seven-segment LED display. What are the logic states of displays No. 1 and No. 2? Give your answers in the following chart.

	Observed Logic State
Display No. 1	
Display No. 2	

You should observe a logic state of "logic 1" on display No. 1 and a "logic 0" state on display No. 2. Why do you observe a "logic 0" state on display No. 2?

Because "logic 0" is the quiescent state of the NOR gate when it is disabled. You know that the gate is disabled because you do not observe any counting.

Step 3

In the space provided in the chart, please compare the logic states on display No. 2 when each of the four different gates are disabled.

	Logic state observed on display No. 2 when gate is disabled
7408 2-input AND gate	
7400 2-input NAND gate	
7432 2-input OR gate	
7402 2-input NOR gate	

Refer to the three preceding experiments for experimental results that you can enter into the chart. Remember, we want the logic state of display No. 2 only when the respective gates are disabled.

Step 4

You should be interested in the display logic state when the respective gates are enabled, which means that you are able to observe counting on the seven-segment LED display. Naturally, you will have to push the pulser button to observe counting in this experiment with the NOR gate. If you haven't done so already, push the pulser button and observe the counting behavior.

In the space provided in the following chart, compare the logic states on display No. 1 when each of the four gates are enabled.

	Logic state observed on display No. 1 when gate is enabled
7408 2-input AND gate	
7400 2-input NAND gate	
7432 2-input OR gate	
7402 2-input NOR gate	

Step 5

Have you been able to observe whether or not each of the four different gates inverts the input clock signal? If your answer is yes, please explain how you were able to observe such behavior using your breadboarding system. The 2-input NOR gate is, of course, a gate that will invert the clock signal. You may have some trouble in answering this question.

If you are curious about the answer to this question, read the text in Chapter 4, which discusses the characteristics of the 7490 counter. A change of count on the 7490 occurs on a "negative edge," which is a logic 1 to logic 0 transition at pin 14 on the 7490. With this clue, you should now be able to answer this question. *Save your circuit for the following experiment, in which you will replace the 7490 counter by a 7493 counter.*

Questions

1. Which two 2-input gates require a "logic 0" gating signal?

2. Which two 2-input gates require a "logic 1" gating signal?

3. Which two 2-input gates give a "logic 0" output when they are disabled?

4. Which two 2-input gates give a "logic 1" output when they are disabled?

5. Which two 2-input gates invert the input clock signal at the output of the gates?

6. Which two 2-input gates transmit the input clock signal without inversion?

7. If one of the two inputs to a 2-input NOR gate is at a "logic 1" state, what are the possible logic states of the output from such a gate?

8. If the two inputs into a 2-input NOR gate are both at a "logic 0" state, what is the logic state of the output from such a gate?

9. Why, in your opinion, might it be useful to have four different gates? If you believe that it is not useful to have so many gates, give your reasoning.

10. How would you construct a 2-input NOR gate using a 2-input OR gate and an inverter? Just draw a schematic diagram of the circuit.

11. How would you construct a 2-input AND gate using a 2-input NAND gate and an inverter? Just draw a schematic diagram of the circuit.

12. How would you construct a 2-input OR gate using a 2-input NAND gate and two inverters? Just draw a schematic diagram of the circuit.

13. What is the symbol for an inverter?

14. Explain what an inverter does. Is it a gate?

15. How would you construct a 2-input AND gate using a 2-input NOR gate and two inverters? Just draw a schematic diagram.

16. How would you construct an inverter from a 2-input NAND gate? Draw schematic diagrams showing two distinctly different ways of doing so.

17. How would you construct an inverter from a 2-input NOR gate? Draw schematic diagrams showing two distinctly different ways of doing so.

EXPERIMENT NO. 5

Purpose

The purpose of this experiment is to demonstrate the operation of the 7402 2-input positive NOR gate as a "gate" to an entirely new counter, the *7493 4-bit binary counter*. The 7493 counter has the same pin configuration as the 7490 decade counter.

Pin Configurations of Integrated-Circuit Chips (Figs. 2-60 and 2-61)

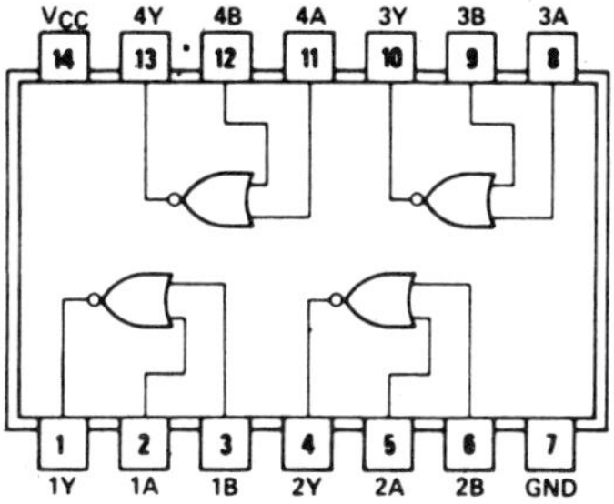

7402

Fig. 2-60.

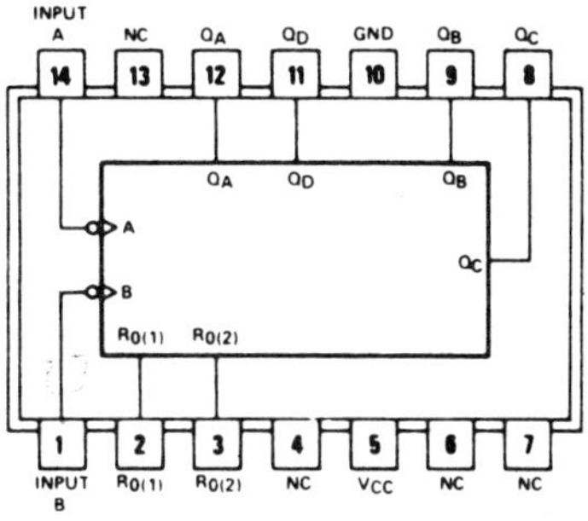

7493

Fig. 2-61.

Schematic Diagram of Circuit (Fig. 2-62)

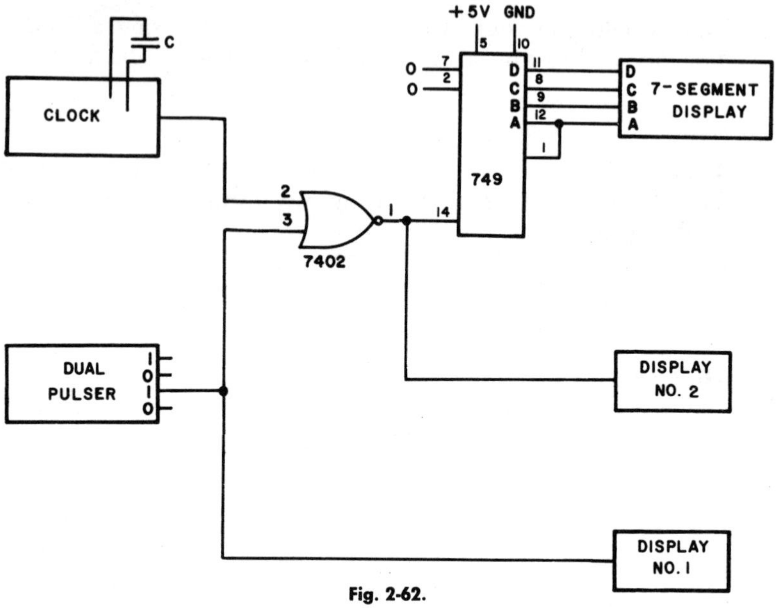

Fig. 2-62.

Step 1

Carefully examine the schematic diagram of the circuit in this experiment and compare it to the corresponding diagram for Experiment No. 4. You should observe a single change: the 7493 IC chip replaces the 7490 IC chip used in Experiment No. 4. Before you make the change, first connect the power to the Outboard and demonstrate the successful operation of the circuit of Experiment No. 4. You should observe counting on the seven-segment LED display once the pulser button is pushed. If you do, you are ready to proceed to Step 2.

Step 2

Disconnect the power from the breadboard and carefully remove the 7490 IC chip. This may be the first time that you have removed such a chip from the breadboard. Simply insert the tip of a small screwdriver underneath the bottom of the chip and gently lift the chip off of the board.

Gently insert the 7493 chip in exactly the same holes that accommodated the 7490 chip. Look for the notch on the 7493 IC chip and make sure that it is facing in the proper direction.

Step 3

Connect the power to the breadboard, push the pulser, and observe that now the seven-segment LED display counts between 0 and a blank display (if you are using the SLA-1 display and 7447 decoder). Observe the several unusual symbols that occur beyond decimal 9. Return to Step 7 of Experiment No. 5 in Chapter 1 and closely examine the table. Answer the following question:

Have you previously seen these unusual symbols on the seven-segment LED display?

Your answer should be yes, in Experiment No. 5 in Chapter 1. A *4-bit binary counter* counts from 0 to 1 and then back to 0 again, and has sixteen different states: 0, 1, 2, 3, 4, 5, 6, 7, 8, 9, 10, 11, 12, 13, 14, and 15. In contrast, a decade counter such as the 7490 has only ten different states: 0, 1, 2, 3, 4, 5, 6, 7, 8, and 9. It should now be clear why you never observed the blank display and the five unusual symbols when you employed the decade counter.

Step 4

In the space below, answer the following questions:

What is the seven-segment LED display symbol that corresponds to the decimal number ten? Write it in the following space.

The answer is .

What is the seven-segment LED display symbol that corresponds to the decimal number eleven?

The answer is .

What is the seven-segment LED display symbol that corresponds to the decimal number twelve?

The answer is .

What is the seven-segment LED display symbol that corresponds to the decimal number thirteen?

The answer is .

What is the seven-segment LED display symbol that corresponds to the decimal number fourteen?

The answer is .

And finally, what is the seven-segment LED display symbol that corresponds to the decimal number fifteen?

The answer is a blank display.

Step 5

Remember the above symbols and the decimal numbers to which they correspond. You will encounter them frequently in this book. In subsequent experiments, you will observe that many integrated-circuit chips in the 7400 series have four input pins that select different chip outputs, memory locations, and the like. The use of a 7493 binary counter will allow you to sequentially select each of the sixteen different outputs.

Questions

1. Why do you observe such unusual symbols in place of the decimal numerals 10, 11, 12, 13, 14, and 15?

2. What do we mean by the term, "pin-for-pin interchangeable?"

TEST

This test probes your understanding of the digital electronic concepts that you have encountered in this chapter. Please write your answers on a separate piece of paper.

1. Explain the difference between a switch and a gate.
2. In your own words, define the following terms:
   - to inhibit
   - to gate
   - to enable
   - to disable
   - to strobe

gate (noun)
gate, gating, and gated (adjectives)
AND gate
OR gate
NAND gate
NOR gate
positive logic
negative logic
clock

3. Write the symbols for the following digital electronic devices:
   2-input AND gate
   3-input NOR gate
   2-input NAND gate
   8-input NAND gate
   4-input NAND gate
   4-input AND gate
   2-input OR gate
   inverter
   buffer
   driver
   2-input AND gate with its output inverted
   2-input NAND gate with each of its two inputs inverted
   wire
4. Compare the gating signals to and the output from the following four different types of 2-input gates:
   2-input AND gate
   2-input NAND gate
   2-input OR gate
   2-input NOR gate

   You can assume that the "input data" into each of these gates is a train of clock pulses.

Your performance on this test will be acceptable of you answer all of the above questions correctly in a two-hour "closed book" examination. You will repeatedly encounter all of the above concepts in any book on digital electronics.

WHAT HAVE YOU ACCOMPLISHED IN THIS CHAPTER?

Review of Objectives

We have stated in the Introduction to this chapter that at the completion you should be able to do the following:

- Demonstrate, using appropriate integrated-circuit chips and auxiliary functions, how you can "enable" and "disable" a clock input to a 7490 decade integrated-circuit chip.

In Experiments Nos. 1 through 4, we have shown you how to do this using four different types of 2-input gates.

- Compare the gating signals to and the outputs from the following four different types of 2-input gates,

  2-input AND gate
  2-input NAND gate
  2-input OR gate
  2-input NOR gate

  when the "input data" to each of these gates is a train of clock pulses.

  We have asked you to do this on the test, in which you were expected to answer this question correctly. Many of the questions at the end of each experiment have probed your understanding of how the above gates compare.
- Pass a test in which you are asked to define in your own words, without the aid of this or any other text on digital electronics, the following terms:

to inhibit	OR gate
to gate	NAND gate
to enable	NOR gate
to disable	positive logic
to strobe	negative logic
gate (noun)	clock
AND gate	gate, gating, and gated (adjectives)

  You should be able to answer this question on the test correctly. The above terms are very common in digital electronics; you should know what they mean without the aid of a dictionary or book such as this one.
- Write the correct symbols for the following logic devices:

  2-input AND, NAND, OR, and NOR gates
  3-input AND, NAND, and NOR gates
  4-input AND and NAND gates
  inverter
  buffer
  driver

  We have also asked you to do this correctly in question 3 on the test.

3

Truth Tables

INTRODUCTION

Continuing our discussion of logic gates, in this chapter, you will learn how to construct and interpret a *truth table*.

OBJECTIVES

At the completion of this chapter, you will be able to do the following:

- Define, in your own words, the term, *truth table*.
- Write the correct truth table for the following logic gates:

  AND gate
  NAND gate
  OR gate
  NOR gate
  NOT gate
  Exclusive-OR gate
  AND-OR-INVERT gate

- Demonstrate the operation of some of the more common 7400-series integrated-circuit gate chips, including the 7402, 7410, 7411, 7404, 7420, 7421, 7427, 7430, 7432, 7451, and 7486 chips.
- Demonstrate, with the aid of a 7420 4-input NAND gate, that an unconnected pin on a 7400-series integrated-circuit chip is at a "logic 1" state.

DEFINITIONS

AND-OR-INVERT gate—A gating circuit that exhibits the characteristics of the AND gate and the NOR gate.

controllable complementer—A digital gating circuit that, upon being gated, can complement an input digital signal.

controllable inverter—A digital gating circuit that, upon being gated, can complement an input digital signal. See controllable complementer, which means the same thing.

data selector logic—In advanced logic design, a technique whereby a data selector/multiplexer is used to accurately simulate the behavior of a gating circuit.

data selector/multiplexer—A digital device that can select one of a number of inputs and pass the logic level of that input on to the output. Information for input-channel selection usually is presented to the device in binary weighted form and decoded internally. The device acts as a single-pole multiposition switch that passes digital information in one direction only.

Exclusive-OR gate—A binary circuit with two inputs and a single output, in which the output is logic 1 when the inputs are at different logic states, and the output is logic 0 if both inputs are at the same logic state.

folding technique—In advanced logic design, a technique whereby a data selector/multiplexer having only N data inputs is used to accurately simulate the behavior of a gating circuit having double the number, or 2N, of outputs.

gating circuit—A digital circuit in which a signal, called a "gating signal," is used to control the passage of a digital signal such as a clock pulse or train of clock pulses.

logic circuit—A circuit (usually electronic) that provides an input-output relationship corresponding to a Boolean-algebra logic function.

multiplexer—See data selector/multiplexer.

NOT gate—A binary ciricuit with a single output that is always the opposite of the single input. Also called an inverter circuit.[1] A binary circuit with one input and one output, in which the output is logic 1 when the input is logic 0, and the output is logic 0 when the input is logic 1.

truth table—A tabulation that shows the relationship of all output levels of a digital circuit to all possible combinations of input logic

Fig. 3-1. A hypothetical digital circuit with three inputs and four outputs.

levels in such a way as to characterize the circuit functions completely.

WHAT IS A TRUTH TABLE?

A *truth table* can be defined as *a tabulation that shows the relationship of all output logic levels of a digital circuit to all possible combinations of input logic levels in such a way as to characterize the circuit functions completely.*[1] In Fig. 3-1, we show a hypothetical digital circuit (shown as a typical "black box") that has three inputs—A, B, and C—and four outputs—Q1, Q2, Q3, and Q4. According to the definition, the object of a truth table is to completely characterize the relationships between these inputs and outputs. This is not as difficult a task as you may initially expect, since the inputs and outputs can only be either 0 or 1, corresponding to a "logic 0" or "logic 1" state, respectively. Since there are two possible states per input and three different inputs, there are a total of $2^3 = 8$ different sets of three inputs, which we have listed in the following truth table for a three input, four output digital circuit.

Inputs			Outputs			
C	B	A	Q1	Q2	Q3	Q4
0	0	0				
0	0	1				
0	1	0				
0	1	1				
1	0	0				
1	0	1				
1	1	0				
1	1	1				

On the left-hand side of the truth table, we show the eight different sets of three inputs. The question we can ask now is, How do we determine the logic states of the four outputs, Q1 through Q4, in the truth table? The answer to this question is relatively straightforward: It depends on the tpye of digital circuit. In other words, we cannot do anything further in the truth table until someone gives us the specific logic circuit that is being considered. Stated in another way, there is no such thing as a generalized truth table for a three-input digital circuit. In fact, in the above truth table, there are $2^{32} = 4{,}294{,}967{,}296$ different ways that the four outputs can be written for eight different sets of input conditions.

So we repeat: *We cannot complete a truth table for a given set of inputs until we are given the specific logic circuit that is being considered.*

WHY DO WE USE TRUTH TABLES?

This is a more difficult question to answer. Basically, we use truth tables because they are simple and convenient shorthand representations of digital circuits. It is much easier to write a truth table than to explain in words how a specific digital circuit operates. The more inputs and outputs that exist, the more obvious the advantage of a truth table over any verbal description. Basically, truth tables are to actual digital circuits as

electronic circuit diagrams	are to	actual electronic devices
blueprints of homes	are to	actual homes
two-dimensional molecular structures of chemical compounds	are to	actual chemical compounds

In other words, truth tables represent the actual behavior of their corresponding digital circuits. Remember the adage, "one picture is worth a thousand words." Well, one truth table is worth upwards of twenty words.

GATES: LOGIC CIRCUITS VS. GATING CIRCUITS

In the preceding chapter, when we talked about simple 2-input gates, we made a distinction between the use of gates to "gate" digital data and the use of gates in simple logic circuits. In the former case, there was generally a single "data input" to the gate and one or more "gating inputs" that controlled whether or not the "data input" would ever appear at the output of the gate. A train of clock pulses was used as "data input" in experiments that probed this type of gate function.

In this chapter, we shall be concerned with the use of gates in *logic circuits,* as opposed to *gating circuits.* A *logic circuit* can be defined as an electronic circuit that provides an input-output relationship corresponding to a Boolean-algebra logic function.[1] We shall not digress at this point into the subject of Boolean algebra. Suffice to say that we are simply interested in discerning relationships between inputs and outputs that exist for simple logic gates such as AND, NAND, OR, NOR, and NOT as well as slightly more complicated logic gates called Exclusive-OR and AND-OR-INVERT gates.

TRUTH TABLES FOR SIMPLE LOGIC GATES

In this section, we shall provide the correct truth tables for all the common logic gates. We shal restrict our attention basically to 2- and 3-input gates, although we shall write the truth table for a 4-input NAND gate since the specific chip is widely used in digital electronics.

The AND Gate

Recall that the symbol for a simple 2-input AND gate is shown in Fig. 3-2. In this symbol, A and B are the two inputs and Q is the

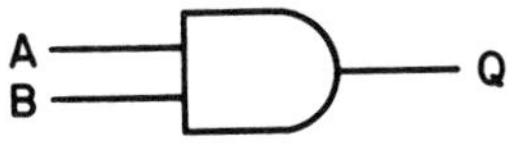

Fig. 3-2. Symbol for a 2-input AND gate.

single output. A, B, and Q can only take one of two possible values: 0 or 1. There are thus $2^2 = 4$ different input conditions and, consequently, four possible output responses, Q, to these input conditions. The following truth table is for a 2-input AND gate.

B	A	Q
0	0	0
0	1	0
1	0	0
1	1	1

The Oxford English Dictionary defines the word *unique* as "*That is or forms the only one of its kind; having no like or equal; standing alone in comparison with others; of which there is only one; one and no other; sole, or solitary.*" In the sense of the word given by the dictionary, there clearly is a unique output state for the 2-input AND gate: that output state corresponding to the A = 1 and B = 1 input conditions. With the simple logic gates, including AND, NAND, OR, and NOR, this is the important question that we must always ask, i.e., What is the unique output state? For a 3-input positive AND

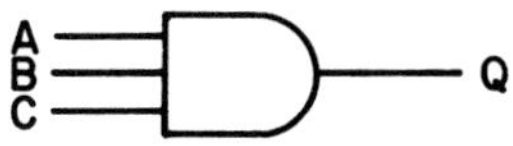

Fig. 3-3. Symbol for a 3-input AND gate.

gate represented by the symbol in Fig. 3-3, we still have A = B = C = 1 and Q = 1, a unique output state, as shown by the following truth table.

C	B	A	Q
0	0	0	0
0	0	1	0
0	1	0	0
0	1	1	0
1	0	0	0
1	0	1	0
1	1	0	0
1	1	1	1

Thus, we can define an AND gate as *a binary circuit with two or more inputs and a single output, in which the output is logic 1 only when all inputs are logic 1, and the output is logic 0 if any one of the inputs is logic 0.*[1]

The NAND Gate

The symbol for a simple 2-input NAND gate, otherwise known as a "negated" 2-input AND gate, is shown in Fig. 3-4. The little circle to the right of the AND gate symbol represents the "negation" or

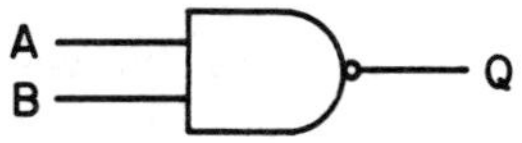

Fig. 3-4. Symbol for a 2-input NAND gate.

NOT function, and means that we have inverted, or complemented, the output of a simple AND gate to produce the correct output of a NAND gate. The following truth table for the 2-input NAND gate resembles the corresponding table for the AND gate.

B	A	Q
0	0	1
0	1	1
1	0	1
1	1	0

Here again, we have a unique output state, Q = 0, that corresponds to the A = 1 and B = 1 simultaneous input conditions. If we would compare this truth table with the one previously given for a 2-input AND gate, we would observe that the outputs of one table are the complements of the outputs from the other table. Note that both gates have unique output states for identical input conditions.

Fig. 3-5. Symbol for a 3-input NAND gate.

For the 3-input positive NAND gate, which is represented by the symbol in Fig. 3-5, we have the following truth table.

C	B	A	Q
0	0	0	1
0	0	1	1
0	1	0	1
0	1	1	1
1	0	0	1
1	0	1	1
1	1	0	1
1	1	1	0

For the 4-input positive NAND gate, we have a still more complicated truth table. The unique output state occurs when all four inputs are at logic 1.

D	C	B	A	Q
0	0	0	0	1
0	0	0	1	1
0	0	1	0	1
0	0	1	1	1
0	1	0	0	1
0	1	0	1	1
0	1	1	0	1
0	1	1	1	1
1	0	0	0	1
1	0	0	1	1
1	0	1	0	1
1	0	1	1	1
1	1	0	0	1
1	1	0	1	1
1	1	1	0	1
1	1	1	1	0

Boring, isn't it? It is clear that a NAND gate can be defined as *a binary circuit with two or more inputs and a single output, in which the output is logic 0 when all inputs are logic 1, and the output is logic 1 if any one of the inputs is logic 0.*

The OR Gate

A gate that is not widely used in digital electronic circuits is the 2-input OR gate, whose symbol is shown in Fig. 3-6. The truth table

Fig. 3-6. Symbol for a 2-input OR gate.

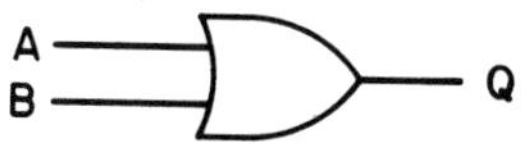

for the 2-input OR gate is somewhat different than that for either of the previous gates. The truth table would be as follows:

B	A	Q
0	0	0
0	1	1
1	0	1
1	1	1

Again, there is a unique output state, Q = 0, but this time the unique state occurs only when both A and B are at a logic 0 simultaneously. 3-input OR gates are not in common use, so we shall not discuss them here. An OR *gate* can be defined as *a binary circuit with two or more inputs and a single output, in which the output is logic 0 only when all inputs are logic 0, and the output is logic 1 if any of the inputs is logic 1.*

The NOR Gate

A gate that is closely related to the 2-input OR gate is the 2-input NOR gate, otherwise known as the "negated" 2-input OR gate (Fig. 3-7). Note the little circle to the right of the OR gate symbol. As

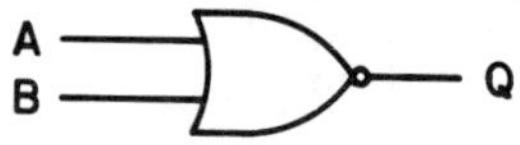

Fig. 3-7. Symbol for a 2-input NOR gate.

in the case of the NAND gate, this circle represents "negation" of the normal OR gate output states. The 2-input NOR gate truth table is simply

B	A	Q
0	0	1
0	1	0
1	0	0
1	1	0

The unique state, Q = 1, occurs only when A and B are both at a logic 0 state. The most popular NOR gate is the 2-input NOR gate; we shall not discuss multiple input NOR gates here. A NOR gate can be defined as *a binary circuit with two or more inputs and a single output, in which the output is logic 1 only when all inputs are logic 0, and the output is logic 0 if any one of the inputs is logic 1.*

The INVERTER or NOT Gate

The final simple gate is the inverter or NOT gate; the symbol is shown in Fig. 3-8. The truth table for an inverter is the simplest of all.

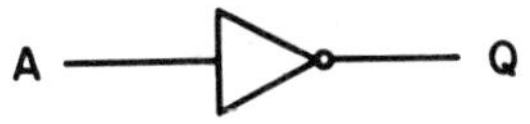

Fig. 3-8. Symbol for an inverter or NOT gate.

A	Q
0	1
1	0

There is no such thing as a multiple-input inverter. An *inverter,* or NOT *gate,* can be defined as *a binary circuit with one input and one output, in which the output is logic 1 when the input is logic 0, and the output is logic 0 when the input is logic 1.*

TRUTH TABLES FOR MORE COMPLEX LOGIC CIRCUITS

Using inverters plus any of the above n-input logic gates, we can build exceedingly complex logic circuits having many inputs and outputs. In the early days of digital electronics, when MSI chips were rare and even gate chips were expensive, it was quite common to construct such circuits. Today there is considerably less incentive to do so. MSI and LSI chips, including microprocessor chips, are rapidly replacing "rat's nest" logic circuits with multiple inputs and outputs. Despite the information that is given in many current textbooks on digital electronics, we recommend that you mainly focus attention on the simple aspects of the most important logic gates and not try to get involved in constructing complicated logic circuits. In keeping with this admonition, the experiments that accompany this unit will be relatively simple and straightforward.

It is, however, useful to discuss two additional gates that are available in integrated-circuit chips: the Exclusive-OR gate and the AND-OR-INVERT gate.

The Exclusive-OR Gate

The Exclusive-OR gate has not been previously discussed since it is not a simple gate. The symbol for the commonly used 2-input Exclusive-OR gate is shown in Fig. 3-9. It has the following truth table:

B	A	Q
0	0	0
0	1	1
1	0	1
1	1	0

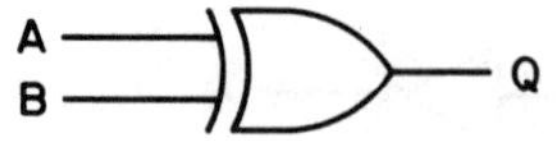

Fig. 3-9. Symbol for an Exclusive-OR gate.

We observe a very important difference between this gate and the four 2-input gates that we previously discussed: there is no unique output state!

Exclusive-OR gates have a variety of special uses. They can be used to construct *half adders* and *full adders*, dc-to-ac converters, data scramblers, and controllable complementers. This last application can be best seen if we write the symbol in the data input/gating signal form used in Chapter 2 (Fig. 3-10). The use of a

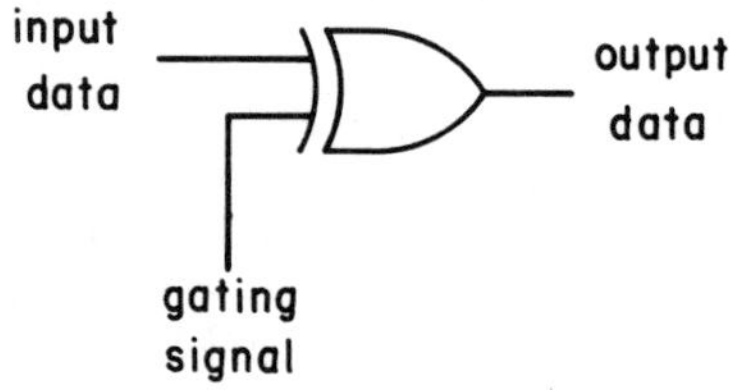

Fig. 3-10. Diagram of an Exclusive-OR gate used to gate a digital signal.

"logic 1" gating signal will cause inverted input data to appear at the output of the gate, as shown in Fig. 3-11, whereas the use of a "logic 0" gating signal will cause input data to be transmitted through the gate and appear at the output without inversion as shown in Fig. 3-12.

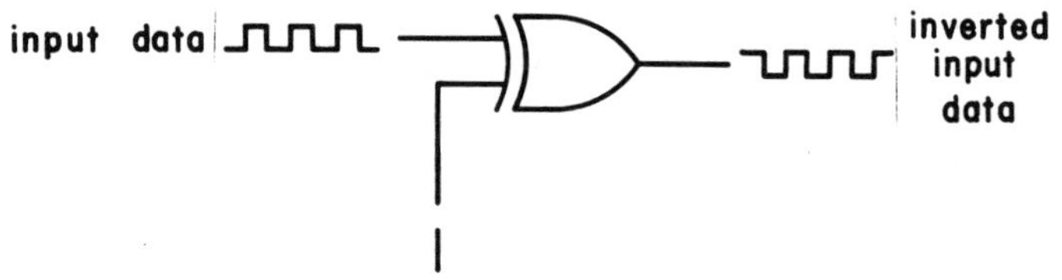

Fig. 3-11. Inversion of digital data by an Exclusive-OR gate.

We thus have a "controllable complementer" or a "controllable inverter," with which we can invert an input digital signal simply by maintaining a "logic 1" state at the gating input.

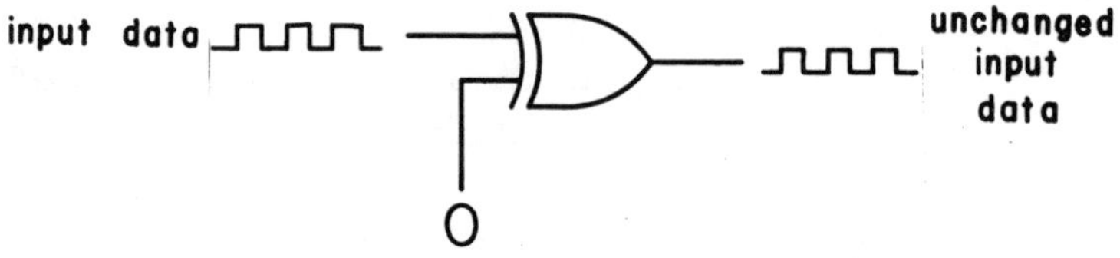

Fig. 3-12. Transmission of digital data by an Exclusive-OR gate.

The AND-OR-INVERT Gate

The simplest AND-OR-INVERT gate that is available in integrated circuit form is given by the symbol in Fig. 3-13.

Note that we have two 2-input AND gates whose outputs are fed into a 2-input NOR gate. We thus have the AND function, the OR function and the INVERT function (the OR gate is "negated" or inverted). The following truth table for the AND-OR-INVERT gate can be written from a knowledge of the truth tables for the component gates:

D	C	B	A	Q
0	0	0	0	1
0	0	0	1	1
0	0	1	0	1
0	0	1	1	0
0	1	0	0	1
0	1	0	1	1
0	1	1	0	1
0	1	1	1	0
1	0	0	0	1
1	0	0	1	1
1	0	1	0	1
1	0	1	1	0
1	1	0	0	0
1	1	0	1	0
1	1	1	0	0
1	1	1	1	0

As can be seen, the truth table for the AND-OR-INVERT gate is complex.

PIN CONFIGURATIONS FOR GATE INTEGRATED CIRCUITS

In Chapter 2, we gave the pin configurations for four different 7400-series gates: the 7408 quad 2-input positive AND gate, the 7400

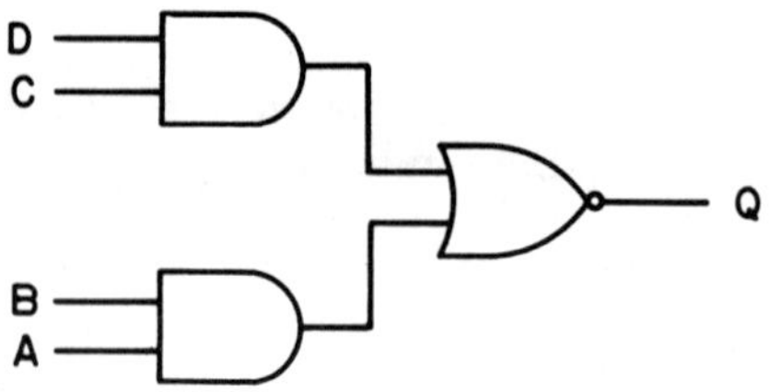

Fig. 3-13. An AND-OR-INVERT gate available commercially as the 7451 IC chip.

quad 2-input positive NAND gate, the 7432 quad 2-input positive OR gate, and the 7402 quad 2-input positive NOR gate.

In this chapter, we shall perform experiments using the following gate chips:

- The 7404 hex inverter
- The 7410 triple 3-input positive NAND gate
- The 7411 triple 3-input positive AND gate
- The 7420 dual 4-input positive NAND gate
- The 7421 dual 4-input positive AND gate
- The 7427 triple 3-input positive NOR gate
- The 7430 8-input positive NAND gate
- The 7451 dual 2-wide 2-input AND-OR-INVERT gate
- The 7846 quad 2-input Exclusive-OR gate

It is appropriate to comment on them briefly in this section.

The 7404 Hex Inverter Integrated Circuit (Fig. 3-14)

The pin configuration for the 7404 hex inverter is shown in Fig. 3-14. There are six *independent* inverters that can be schematically represented as shown in Fig. 3-15.

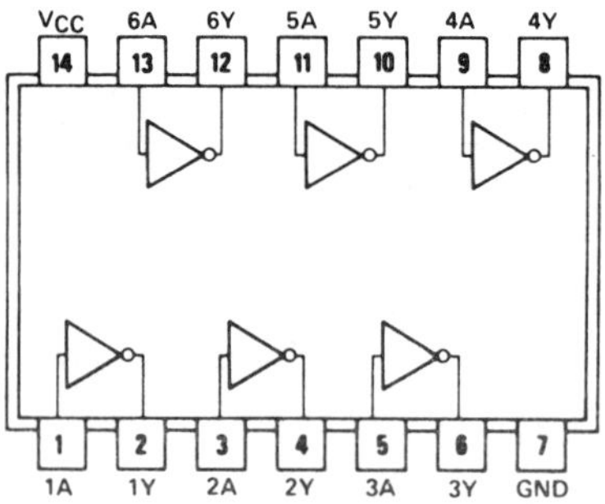

Fig. 3-14.

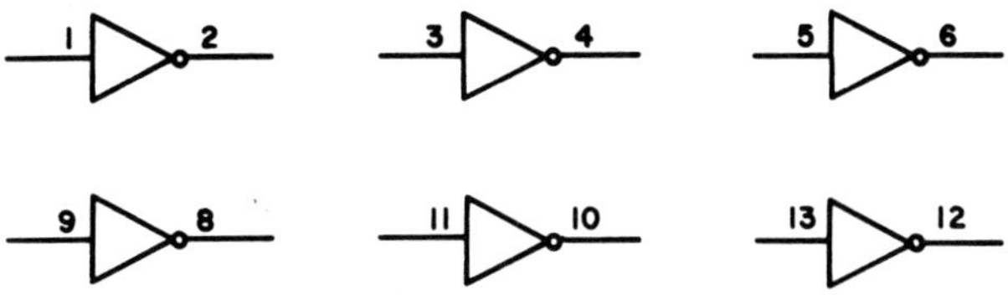

Fig. 3-15. Diagrams of six independent inverters on the 7404 IC.

The 7410 Triple 3-Input NAND Gate Integrated Circuit (Fig. 3-16)

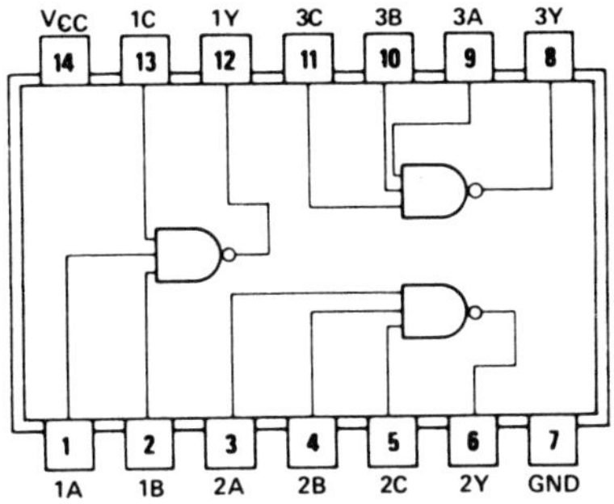

Fig. 3-16.

The pin configuration (Fig. 3-16) is for the 7410 triple 3-input positive NAND gate, where it is observed that there are three *independent* 3-input positive NAND gates, as shown in Fig. 3-17.

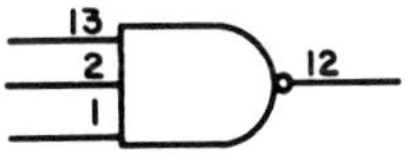

Fig. 3-17. Diagrams of the three independent 3-input NAND gates on the 7410 IC.

The 7420 Dual 4-Input Positive NAND Gate Integrated Circuit (Fig. 3-18)

The pin configuration (Fig. 3-18) is for the 7420 dual 4-input positive NAND gate. There are two *independent* 4-input positive NAND gates, which can be represented as shown in Fig. 3-19.

The 7430 8-Input Positive NAND Gate Integrated Circuit (Fig. 3-20)

The final NAND gate (Fig. 3-20) is the 7340 8-input NAND gate, which is schematically represented in Fig. 3-21.

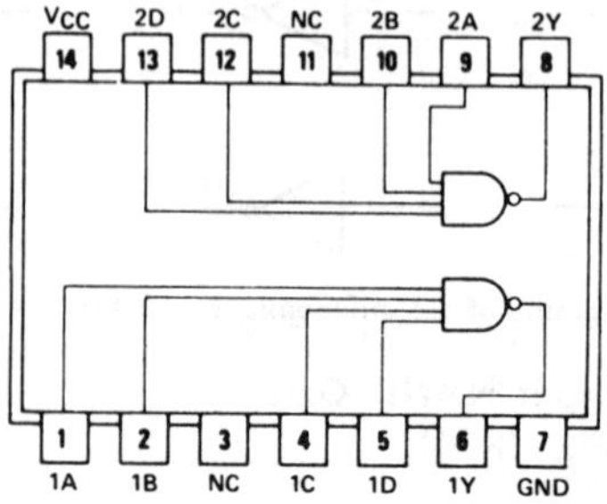

Fig. 3-18.

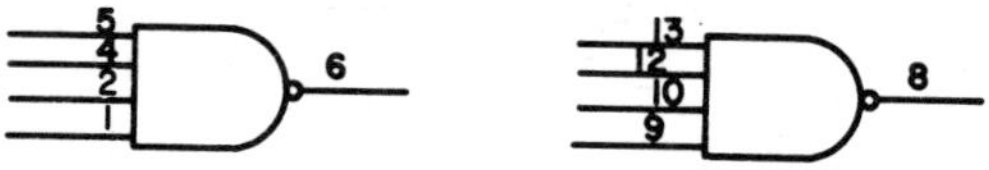

Fig. 3-19. Diagrams of the two independent 4-input NAND gates on the 7420 IC.

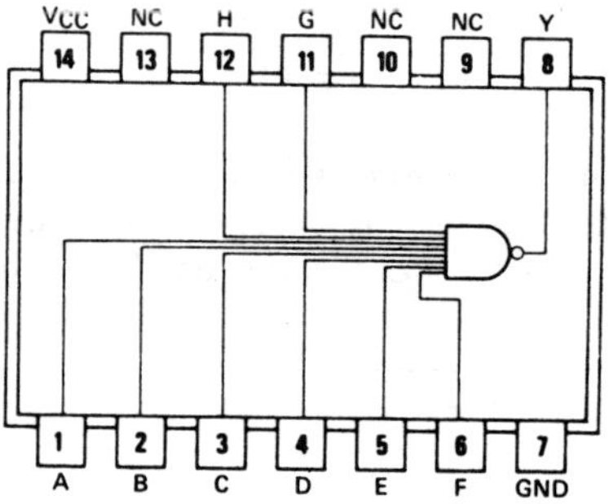

Fig. 3-20.

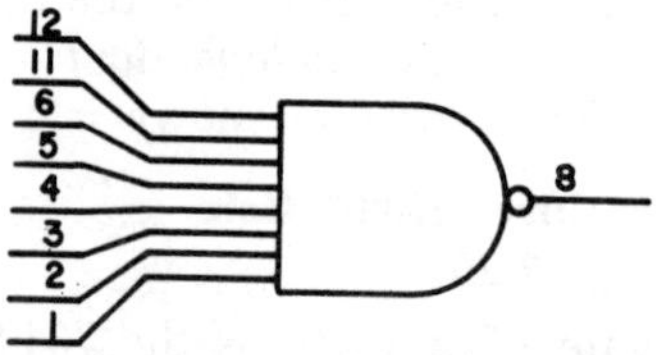

Fig. 3-21. Diagram of the single 8-input NAND gate on the 7430 IC.

The 7411 Triple 3-Input Positive AND Gate Integrated Circuit (Fig. 3-22)

The 7411 triple 3-input positive AND gate has a pin configuration that is essentially identical to the 7410 IC shown in Fig. 3-22,

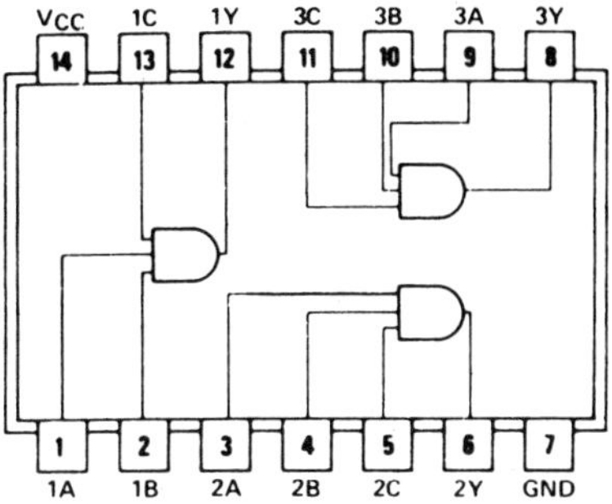

Fig. 3-22.

There are three *independent* 3-input positive AND gates, as shown on the chip in Fig. 3-23.

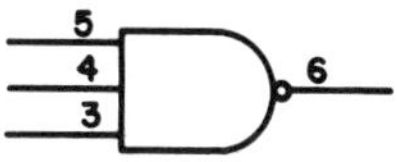

Fig. 3-23. Diagrams of the three independent 3-input AND gates on the 7411 IC.

The 7421 Dual 4-Input Positive AND Gate Integrated Circuit (Fig. 3-24)

In a similar manner, the following 7421 (74H21) dual 4-input positive AND gate is identical in pin configuration to the 7420 IC shown in Fig. 3-24. The H in 74H21 indicates that this is a high-speed high-power chip, a result of the special TTL technology used to fabricate the chip.

There are two *independent* 4-input positive AND gates on the chip diagram in Fig. 3-25.

The 7427 Triple 3-Input Positive NOR Gate Integrated Circuit (Fig. 3-26)

The 7427 triple 3-input positive NOR gate has a pin configuration that is identical to both the 7410 and 7411 integrated circuits, and is shown in Fig. 3-26. There are three *independent* 3-input positive NOR gates on this chip as shown in Fig. 3-27.

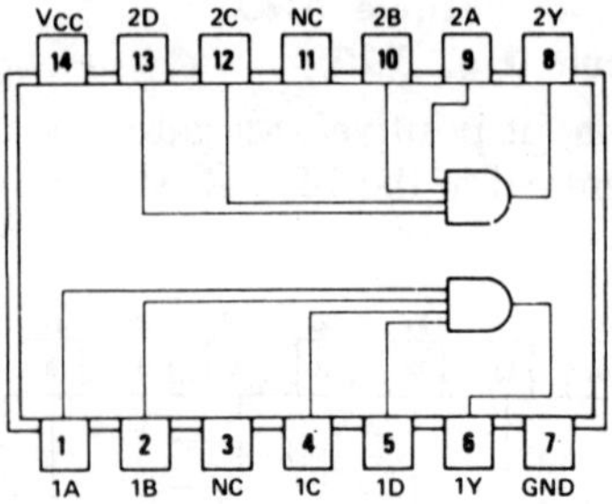

Fig. 3-24.

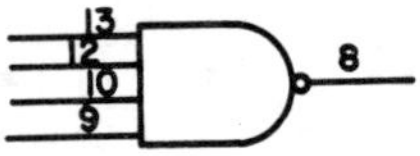

Fig. 3-25. Diagrams of the two independent 4-input AND gates on the 7421 IC.

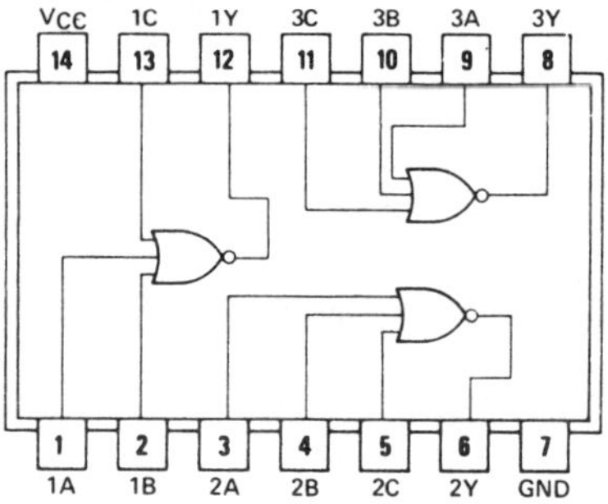

Fig. 3-26.

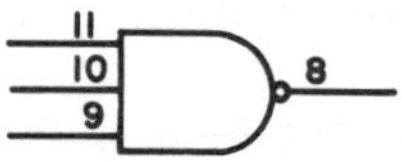

Fig. 3-27. Diagrams of the three independent 3-input NOR gates on the 7427 IC.

The 7451 Dual 2-Wide 2-Input AND-OR-INVERT Gate Integrated Circuit (Fig. 3-28)

The 7451 dual 2-wide 2-input AND-OR-INVERT gate IC in Fig. 3-28 has a somewhat more unusual pin configuration. There are two *independent* AND-OR-INVERT gates on the chip, as shown in Fig. 3-29.

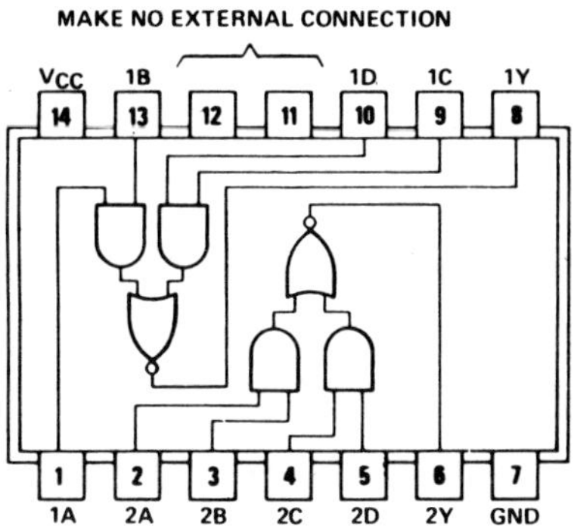

Fig. 3-28.

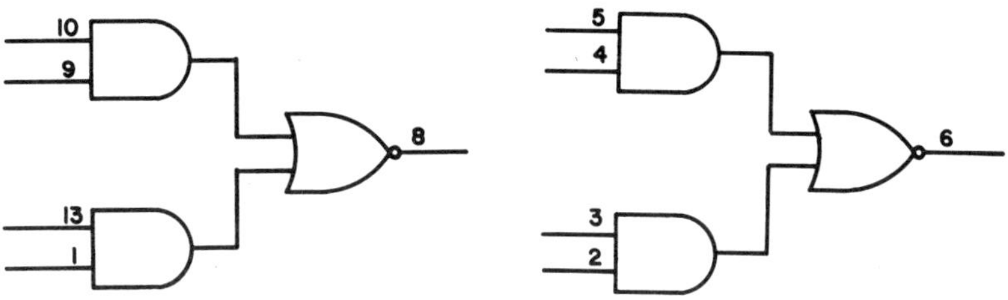

Fig. 3-29. Diagrams of the two independent 2-wide 2-input AND-OR-INVERT gates on the 7451 IC.

The 7486 Quad 2-Input Exclusive-OR Gate Integrated Circuit (Fig. 3-30)

Finally, the pin configuration in Fig. 3-30 is for the 7486 quad 2-input Exclusive-OR gate. There are four *independent* Exclusive-OR gates on this chip, as represented by the diagrams in Fig. 3-31.

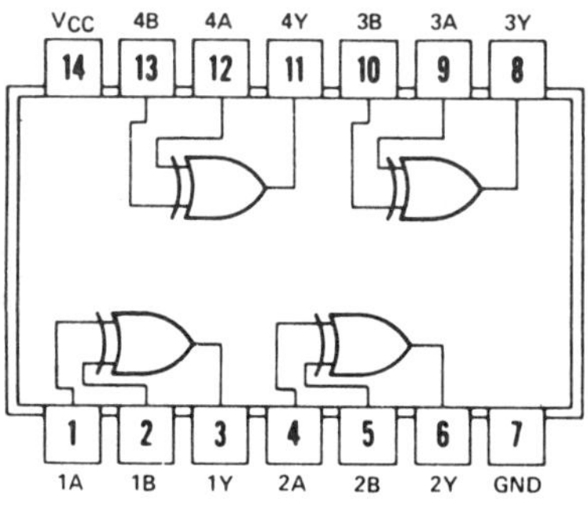

Fig. 3-30.

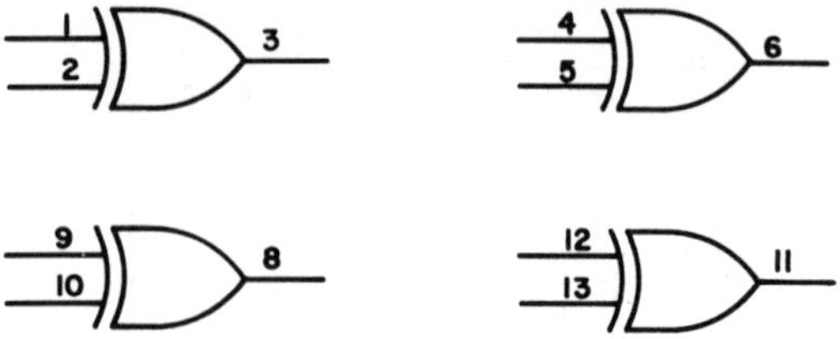

Fig. 3-31. Diagrams of the four independent Exclusive-OR gates on the 7486 IC.

INTRODUCTION TO THE EXPERIMENTS

You are now ready to conduct experiments using the various integrated-circuit chips that have been described in the preceding sections in this chapter. With the exception of the 7486 2-input Exclusive-OR gate, your main interest in this chapter should be to study how the chips function as *logic gates* in situations where you cannot ascribe the terms "input data" or "gating input" to specific input pins.

We have provided more experiments than you should need to understand how to apply logic gates to your digital electronics problems. Consequently, we recommend that you do not perform every experiment in this chapter. You should, at the very least, do the following experiments:

- Experiment No. 1: The 7411 3-input positive AND gate.
- Experiment No. 2: The 7410 3-input positive NAND gate.
- Experiment No. 4: The 7420 4-input positive NAND gate with unconnected input pins.
- Experiment No. 6: The 7432 OR gate, the 7402 NOR gate, and the 7486 Exclusive-OR gate.
- Experiment No. 8: The 7451 2-wide 2-input AND-OR-INVERT gate.

and should read through the remaining experiments. Experiments Nos. 1 and 2 help to develop your ability to wire circuits directly from schematic diagrams. Experiments Nos. 6 and 8 introduce you to two new gates: the 7486 Exclusive-OR gate and the 7451 AND-OR-INVERT gate. Finally, Experiment No. 4 demonstrates what happens when you leave input pins unconnected on a gate chip.

EXPERIMENT NO. 1

Purpose

The purpose of this experiment is to determine the truth table for each of three 3-input positive AND gates on the 7411 integrated circuit.

Pin Configuration of Integrated Circuit (Fig. 3-32)

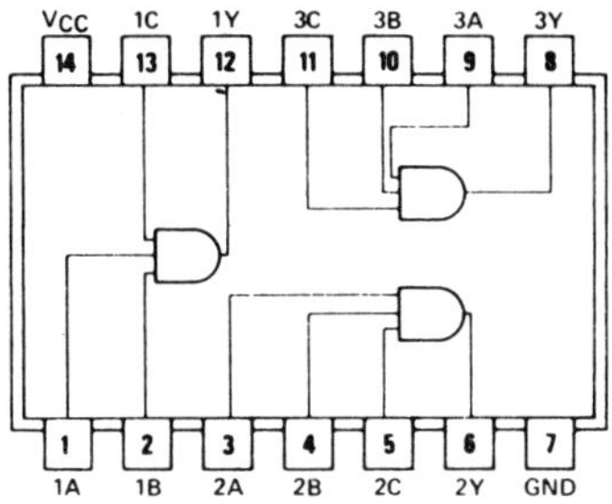

74H11

Fig. 3-32.

Schematic Diagram of Circuit (Fig. 3-33)

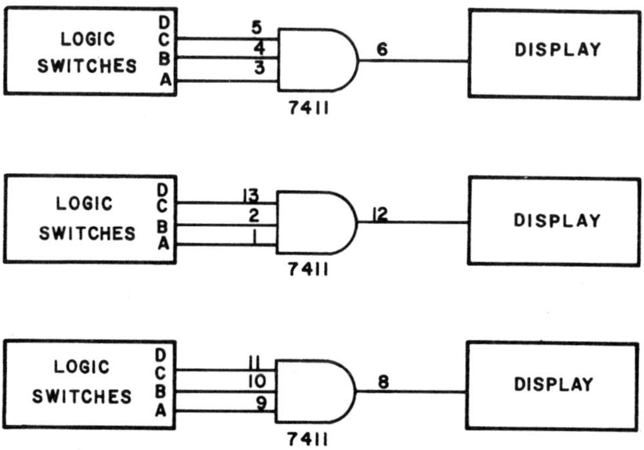

Fig. 3-33.

Step 1

Study the three schematic diagrams of the circuits that will be implemented in this experiment. There are three independent 3-input positive AND gates on the 7411 chip, and you will demonstrate that all three work in an identical manner.

Step 2

Make certain that the power to the breadboard is *off*. You should develop the habit of always disconnecting the power whenever you make connections to integrated circuits.

Make the required wire connections shown on the first schematic diagram for this experiment. The connections include the following:

- Pin 14 on the 7411 chip is connected to +5 Volts. You can use a very short length of wire for this connection, preferably red in color.
- Pin 7 on the 7411 chip is connected to GROUND. You can use a very short length of wire for this connection as well. Try to use a black wire.
- Pin 3 on the 7411 chip is connected to logic switch A.
- Pin 4 on the 7411 chip is connected to logic switch B.
- Pin 5 on the 7411 chip is connected to logic switch C.
- Pin 6 on the 7411 chip is connected to a suitable display, such as a single LED monitor.

Step 3

Check the wiring of the circuit. If it is correct, set logic switches A, B, and C to a "logic 0" state and apply power to the breadboard. Answer the following question:

Does the reading on the display correspond to a "logic 0" or a "logic 1" state?

The correct answer to this question is, a "logic 0" state. If you are using a lamp monitor for a display, then it should be unlit. If you are using a zero/one LED display, then it should give a "0" readout value.

Step 4

Shown below is a partially completed truth table for a 3-input positive AND gate.

C	B	A	Q
0	0	0	0
0	0	1	
0	1	0	
0	1	1	
1	0	0	
1	0	1	
1	1	0	
1	1	1	

Note that we have filled in the output column, Q, for A = B = C = 0. The value that we have written is a "logic 0," which corresponds to the result that we obtained in Step 3 above. Now set A = 1 and B = C = 0. In other words, logic switch A is now set at a "logic 1" state, whereas logic switches B and C each remain at a "logic 0" state. Write below the logic state that you observe on the display.

Step 5

If you have performed Step 4 correctly, you should still observe a "logic 0" on the display. If you did, write a "0" immediately below the 0 in column Q opposite the A = 1 and B = C = 0 row. This is your second entry to the truth table.

Step 6

Repeat the above experiments for the remaining six combinations of logic switch settings A, B, and C. In the appropriate row in the truth table in Step 4, write the logic state of the display that you observe in each case.

Step 7

You have now completed an experimental measurement for a 3-input positive AND gate, and have filled out the appropriate truth table. Compare the preceding truth table to that given in the truth table for a 3-input AND gate discussed earlier in this chapter. The two truth tables should be identical. If they are not, repeat the above experiment. Make certain that your wire connections to the 7411 IC are correct. You have taken an important step in this experiment, since you have demonstrated to yourself the ability to determine a truth table for any logic gate or logic system!

Step 8

Repeat Steps 2 through 7 for the 3-input positive AND gate located at pins 1, 2, 13, and 12 on the 7411 integrated circuit. See the schematic diagram for details.

Do you observe an identical truth table as you observed for the 3-input AND gate located at pins 3, 4, 5, and 6? You should.

Step 9

Repeat Steps 2 through 7 for the 3-input positive AND gate located at pins 9, 10, 11, and 8 on the 7411 IC. See the schematic diagram for details. Do you observe an identical truth table as with the preceding two gates? Your answer should be yes.

If others have used the 7411 integrated-circuit chip before you, it would be appropriate for you to be alert to the fact that one of the three gates on the chip might be burned out. Always be aware of this possibility whenever you work with integrated-circuit chips. We have observed that some persons become quite frustrated by their inability to correctly wire and operate a digital circuit. They check their wiring and schematic diagram and everything looks fine. Yet their circuit doesn't work. In most cases, the answer to this dilemma is a burned out gate or chip.

Questions

1. What do we mean by a "unique output state" in a truth table? Explain this term in your own words in the following space.

   Only one type of input will give a specified output

2. Is there a unique output state in the truth table for a 7411 3-input positive AND gate? If the answer is yes, give the values for A, B, and C that correspond to this unique output state. Yes

   A=1 C=1
   B=1

3. If A = B = C = 1 for a 3-input positive AND gate, what would be the value of the output, Q, from the gate? Write the logic state in the space below.

   1

EXPERIMENT NO. 2

Purpose

The purpose of this experiment is to determine the truth table for a 3-input positive NAND gate on the 7410 integrated circuit.

Pin Configuration of Integrated Circuit (Fig. 3-34)

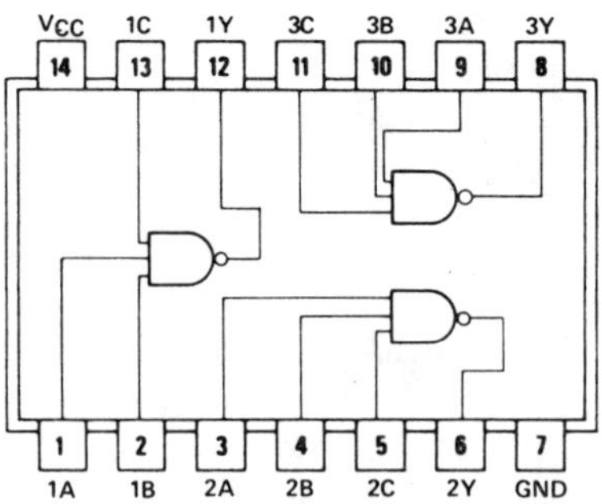

7410

Fig. 3-34.

Schematic Diagram of Circuit (Fig. 3-35)

Step 1

This experiment is quite similar to Experiment No. 1 in this chapter. The only difference is that you now are using the 7410 3-input positive NAND gate rather than the 7411 3-input positive AND gate. With the power to the breadboard off, exchange the 7410 IC for the 7411 IC on the breadboard. The circuit should be wired so that pin 3 is connected to logic switch A, pin 4 is connected to logic switch B, pin 5 is connected to logic switch C, and pin 6 is connected to the display. The +5 Volts and GROUND connections to pins 14 and 7, respectively, remain as before.

Step 2

Set logic switches A, B, and C each to a "logic 0" state. Apply power to the breadboard and observe the reading on the display. Does the display give a "logic 0" or a "logic 1" output?

If you have performed this experiment correctly, you should observe that the display gives a "logic 1" reading.

Step 3

A partially completed truth table for a 3-input positive NAND gate follows.

C	B	A	Q
0	0	0	1
0	0	1	
0	1	0	
0	1	1	
1	0	0	
1	0	1	
1	1	0	
1	1	1	

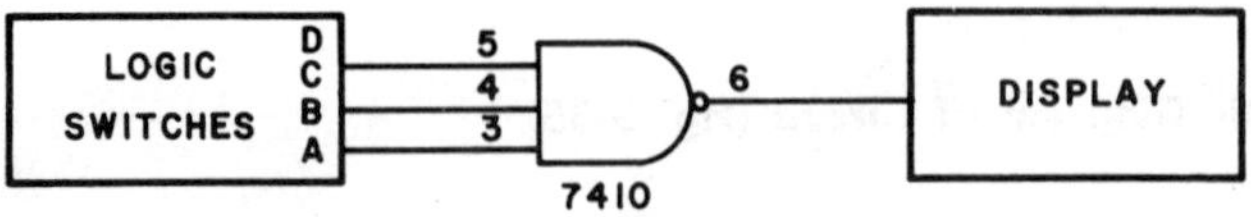

Fig. 3-35.

In the row corresponding to A = B = C = 0, note that we now have a value of Q = 1. This result is in distinct contrast to the result observed in Experiment No. 1 in this chapter, where Q = 0 for similar input conditions on a 3-input positive AND gate. Now set A = 1 and B = C = 0. What display reading do you observe, a "logic 0" state or a "logic 1" state? Write this state in the truth table opposite the appropriate row of inputs.

If you have performed this second step correctly, you should now observe a "logic 1" state.

Step 4

Repeat Step 3 for the remaining six logic switch combinations and thus complete the truth table. Answer the following questions:

(a) Is there a unique output state in the truth table in Step 3? yes

(b) What is the value of Q for this unique output state?

Q = 0

(c) What are the values of A, B, and C that create this unique output state?

A = 1
B = 1
C = 1

Your answers to these questions should be (a) yes, there is a unique state, (b) the value of Q for this unique state is logic 0, and (c) the values of A, B, and C to create this unique state are A = B = C = 1.

Questions

1. Compare the truth tables for a 3-input positive AND gate and a 3-input positive NAND gate. Though the outputs from these two gates may seem completely different, you should note a fundamental "relationship" between the two sets of outputs. In your own words, describe this "relationship." (HINT: It is very easy to convert an AND gate into a NAND gate, and vice versa.)

the two truth tables are completely opposite

2. How many 3-input positive NAND gates are there on a 7410 IC?

3. How would you convert a 3-input positive NAND gate into a 3-input positive AND gate? Do not give a complicated answer here.

4. If A= B = C = 1 for a 3-input positive NAND gate, what would be the logic state for the output, Q, from such a gate?

5. Could you guess what the output, Q, would be if you left all three inputs on the 7410 integrated-circuit chip unconnected? Your answer here will represent your "hypothesis" (what you expect to happen) for an experiment in which you would actually determine what would happen if the inputs to the 7410 chip were left unconnected. You will shortly perform this experiment with a 4-input positive NAND gate.

6. A NAND gate is also known as a "negated" AND gate. What do we mean by the term, "negated?"

7. What type of gate do we obtain if we "negate" the output from a NAND gate? Write your answer in the following space.

EXPERIMENT NO. 3

Purpose

The purpose of this experiment is to determine the truth table for a 7420 4-input positive NAND gate.

Pin Configuration of Integrated Circuit (Fig. 3-36)

Schematic Diagram of Circuit (Fig. 3-37)

Step 1

By this point, you should have reached the first level of digital electronics maturity: the ability to wire simple digital electronic circuits directly from the schematic diagram of the circuit. Accord-

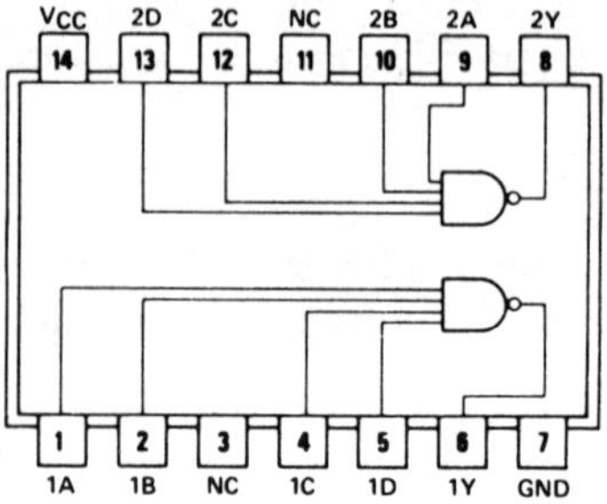

7420

Fig. 3-36.

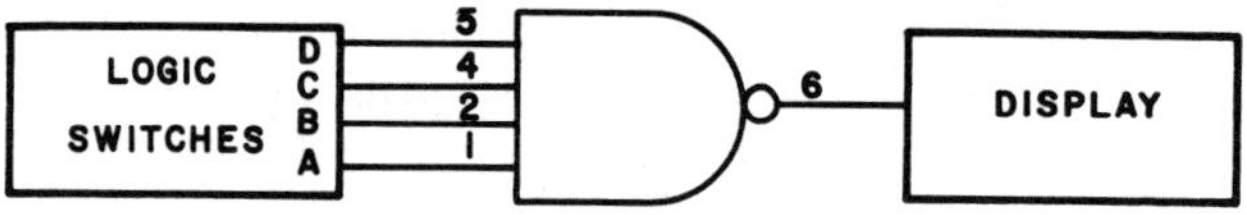

7420

Fig. 3-37.

ingly, we shall determine your prowess in this regard by requesting that you wire up the above 4-input positive NAND gate to four logic switches and a suitable display.

Using the above schematic diagram, wire the circuit indicated. Don't forget the power connections at pins 7 and 14 of the 7420 IC. Leave the power disconnected until you have wired and checked the circuit.

Step 2

Apply power to the breadboard and set all four logic switches, A, B, C, and D, each to a "logic 0" state. The display should be at a "logic 1" state.

Step 3

Vary the logic state settings of the four logic switches and fill in the following truth table based on the outputs that you observe on the display.

Step 4

Answer the following questions:

(a) Is there a unique output state in the above truth table?

(b) If the answer is yes, what is the value of Q for this unique output state?

$Q =$

D	C	B	A	Q
0	0	0	0	1
0	0	0	1	
0	0	1	0	
0	0	1	1	
0	1	0	0	
0	1	0	1	
0	1	1	0	
0	1	1	1	
1	0	0	0	
1	0	0	1	
1	0	1	0	
1	0	1	1	
1	1	0	0	
1	1	0	1	
1	1	1	0	
1	1	1	1	0

(c) What is the value of A, B, C, and D for this unique output state?

A =
B =
C =
D =

Do not disconnect the circuit; it will be used in Experiment No. 4.

Questions

1. How many 4-input positive NAND gates are there on the 7420 IC?

2. If A = B = C = 0 and D = 1 for a 4-input positive NAND gate, what would be the logic state of the output, Q, from such a gate?

3. What digital device would you require to convert a 4-input NAND gate into a 4-input AND gate?

4. Assume that A = B = C = 1 and D were left unconnected on the 7420 4-input NAND gate. What would be the logic state of the output, Q, from such a gate? Keep in mind that you are considering the behavior of the NAND gate when only three of the four inputs are actually connected to a specific logic level on the breadboard.

Question No. 4 raises the issue, as did Question No. 5 in Experiment No. 2, concerning what happens when one (or more) of the inputs to a logic gate are left unconnected. You will discover the answer in Experiment No. 4.

EXPERIMENT NO. 4

Purpose

The purpose of this experiment is to determine the logic state of the output, Q, from a 7420 4-input positive NAND gate when one, two, three, or all four of the inputs are not connected to a specific logic state.

Pin Configuration of Integrated Circuit (Fig. 3-38)

Schematic Diagram of Circuit (Fig. 3-39)

Step 1

The circuit for this experiment is essentially the same as that used in the preceding experiment. Set all four logic switches to a "logic 1" state and observe that the display gives a "logic 0" output state.

Step 2

Disconnect the wire to logic switch D; either remove it from the breadboard or let it dangle in mid-air. What logic state on the display do you observe now?

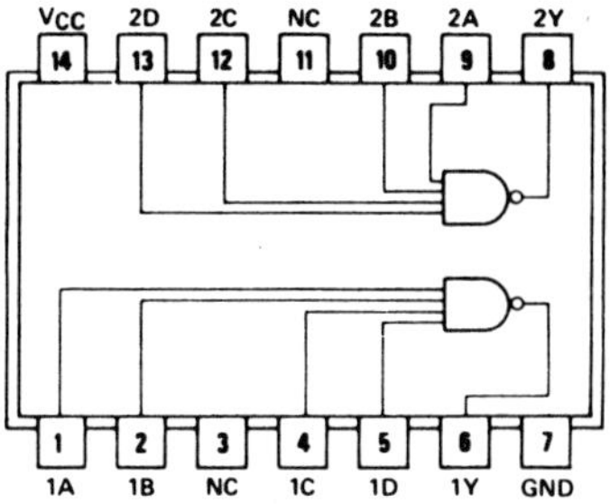

7420

Fig. 3-38.

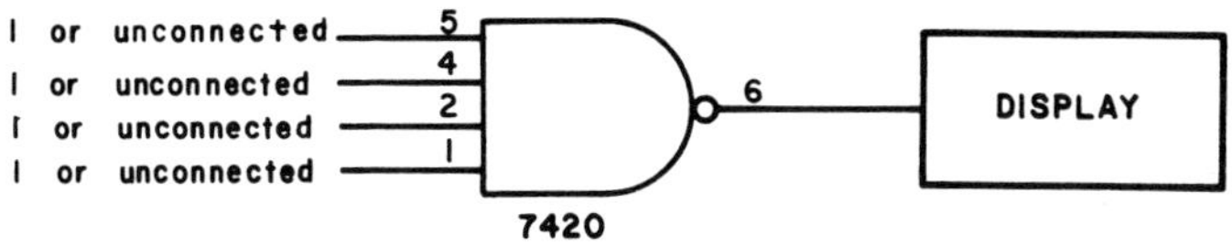

Fig. 3-39.

If you did this step correctly, you should observe a "logic 0" state on the display.

Step 3

Now also disconnect the wire to logic switch C and remove it from the breadboard. What logic state appears on the display after you have removed this wire?

If you did this third step correctly, you should observe that the display still gives a "logic 0" reading.

Step 4

Disconnect and remove the wire to logic switch B. What logic state appears on the display now?

Finally, disconnect and remove the wire to logic switch A. If you have done this step correctly, there should be no wires connected to the four input pins of the 4-input positive NAND gate,. What is the logic state reading on the display for this final case?

Step 5

If you did all of the above steps correctly, you should have observed that the display always read a "logic 0." Rather than let you puzzle about this behavior, we would like to emphasize the important application of this experiment: *The logic state for an unconnected input pin on the 7420 4-input positive* NAND *gate is always a "logic 1."* This is the only conclusion that is consistent with the experimental results observed above!

You might now inquire, what about other chips in the 7400 series of integrated-circuit chips? You will observe that THE LOGIC STATE FOR ANY UNCONNECTED INPUT PIN IS ALWAYS A "LOGIC 1" FOR ALL OF THE INTEGRATED-CIRCUIT CHIPS IN THE 7400 SERIES. To repeat, *All unconnected input pins on chips of the 7400 series are at a "logic 1" state!* We must not underestimate the importance of this observation. You will encounter such behavior repeatedly in this book. Keep in mind that we have only made a statement about the input pins in the 7400 series of ICs. The output pins can, and frequently do, exist in a "logic 0" state even if they are unconnected.

Remember:

> *All unconnected input pins on chips of the 7400 series are at a "logic 1" state.*

You should note that, from an engineering design point of view, "unconnected" inputs on printed-circuit boards are usually wired to +5 Volts (i.e., a logic 1 state) through a 1000-ohm "pull-up resistor" (a term that will be explained in a later chapter).

Questions

1. The input pins on a 3-input positive AND gate present on the 7411 IC are unconnected. What is the logic state of the output, Q, from such a gate?

2. The input pins on a 7410 3-input positive NAND gate are unconnected. What is the logic state of the output, Q?

3. The input pins on a 7400 2-input positive NAND gate are unconnected. What is the output logic state from the gate?

4. The input pins on a 7340 8-input positive NAND gate are all left unconnected. What is the logic state of the output, Q, from such a gate?

5. The input pins on a 7402 2-input positive NOR gate are unconnected. What is the output logic state from the gate?

6. The input pin on a 7404 inverter is left unconnected. What is the logic state of the output from such an inverter?

7. The input pins on a 7432 2-input positive OR gate are unconnected. What is the ouput logic state from the gate?

8. The input pins on a 7486 2-input Exclusive-OR gate are left unconnected. What logic state would you observe for the output from such a gate?

EXPERIMENT NO. 5

Purpose

The purpose of this experiment is to determine the truth table for the following 7400-series integrated circuits:

- The 7404 hex inverter
- The 7421 dual 4-input positive AND gate
- The 7427 triple 3-input positive NOR gate
- The 7430 8-input positive NAND gate

Pin Configurations of Integrated Circuits (Figs. 3-40 through 3-43)

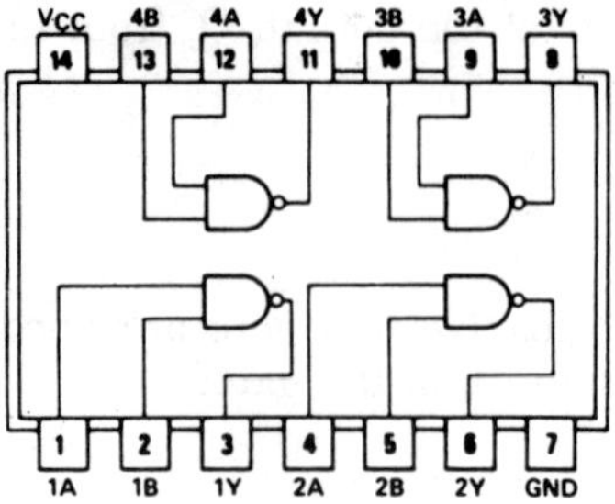

7400

Fig. 3-40.

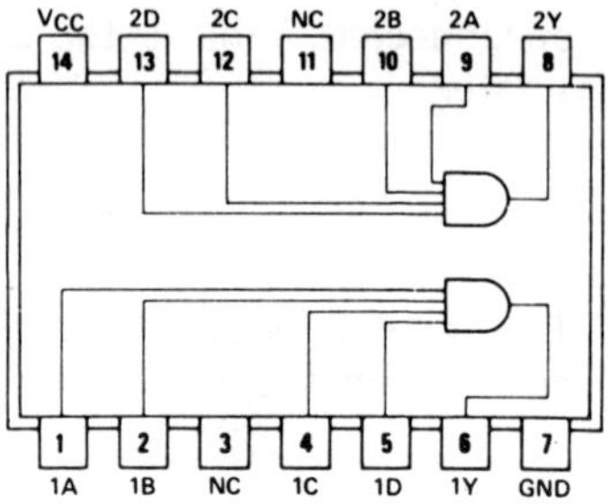

74H21

Fig. 3-41.

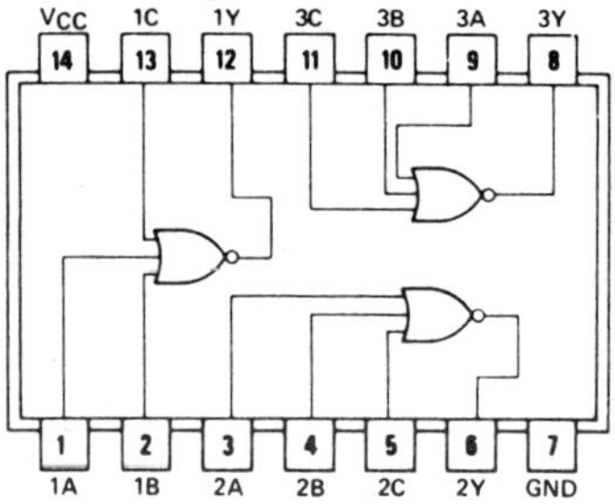

7427

Fig. 3-42.

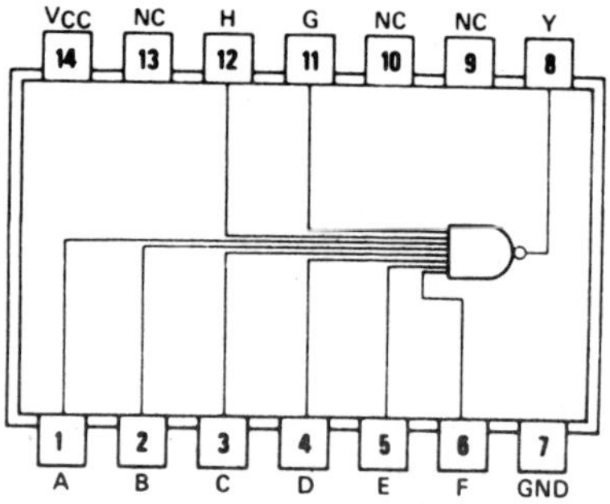

7430

Fig. 3-43.

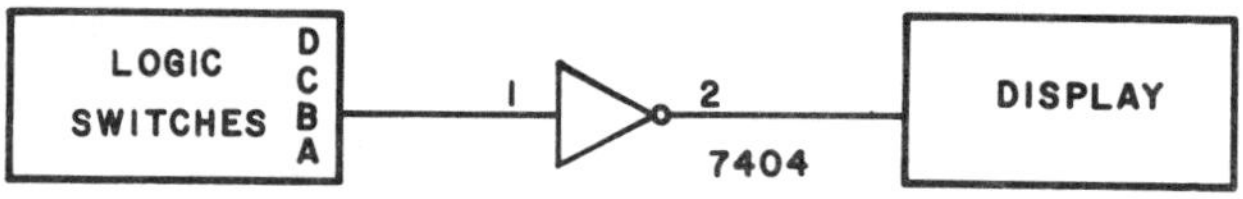

Schematic Diagrams of Circuits (Figs. 3-44 through 3-47)

Fig. 3-44.

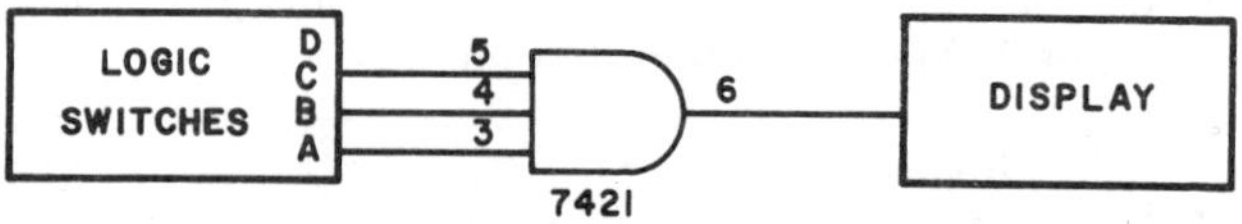

Fig. 3-45.

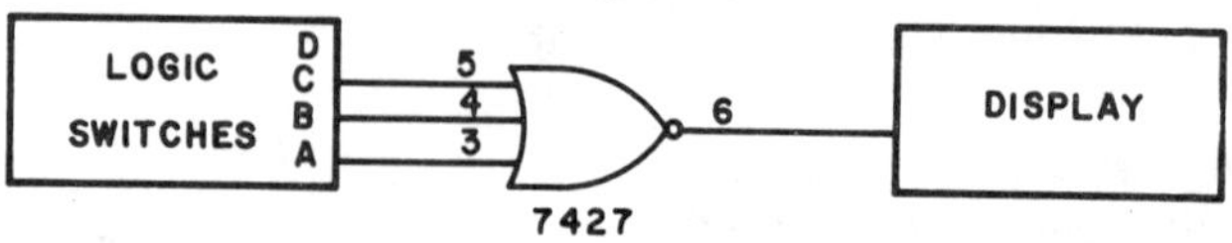

Fig. 3-46.

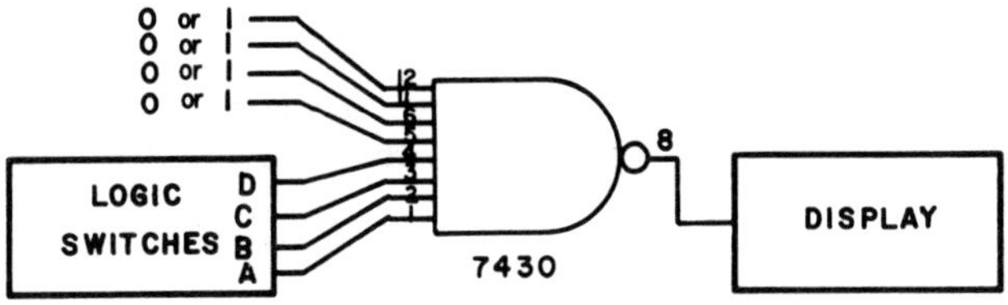

Fig. 3-47.

Step 1

Wire the circuit for the 7404 hex inverter (use only one of the six possible inverters on the IC), as shown in the above schematic diagram, and fill in the following truth table:

A	Q
0	
1	

Compare your answer to that given in the truth table for an inverter earlier in this chapter.

Step 2

Wire the circuit shown for the 7421 4-input positive AND gate. Using the results from your experiment, fill in the following truth table:

C	B	A	Q
0	0	0	0
0	0	1	
0	1	0	
0	1	1	
1	0	0	
1	0	1	
1	1	0	
1	1	1	

Compare your answer to that given in the truth table for a 3-input NAND gate earlier in this chapter.

Step 3

Wire the circuit given for the 3-input positive NOR gate. Using the results from your experiment, fill in the following truth table:

C	B	A	Q
0	0	0	
0	0	1	
0	1	0	
0	1	1	
1	0	0	
1	0	1	
1	1	0	
1	1	1	

Compare your answer with the truth table for a 2-input NOR gate earlier in this chapter; it should provide the necessary clues for you to determine the truth table for a 3-input NOR gate.

Step 4

Wire the 7430 8-input positive NAND gate in the manner shown in the schematic diagram. If you leave input pins 5, 6, 11, and 12 unconnected, they will be at a "logic 1" state. Rather than determine the entire truth table, fill out the abbreviated truth table given below:

H	G	F	E	D	C	B	A	Q
0	0	0	0	0	0	0	0	
0	0	0	0	0	0	0	1	
0	0	0	0	1	1	1	1	
1	1	1	1	0	0	0	0	
1	1	1	1	1	1	0	0	
1	1	1	1	1	1	1	0	
1	1	1	1	1	1	1	1	

To satisfy your curiosity about this 8-input gate, you might try other combinations of inputs that are not listed on the above abbreviated truth table. You should observe an output of Q = 1 only when all eight inputs are at a "logic 1" state.

Step 5

Finally, fill in the following table based on your results for this and preceding experiments:

Type of Gate	Logic State of Unique Output State	Logic State of All Inputs for Unique Output State
2-input positive AND	1	1
3-input positive AND		
4-input positive AND		
n-input positive AND		
2-input positive NAND		
3-input positive NAND		
4-input positive NAND		
8-input positive NAND		
n-input positive NAND		
2-input positive OR		
3-input positive OR		
n-input positive OR		
2-input positive NOR		
3-input positive NOR		
n-input positive NOR		

Questions

1. How can you convert an OR gate into an AND gate? For your answer, draw a circuit diagram involving a 2-input OR gate and 7404 inverters.

2. How can you convert a NOR gate into a NAND gate? Draw a circuit diagram involving a 2-input NOR gate and 7404 inverters.

3. How can you convert a NAND gate into a NOR gate? Draw a circuit diagram based upon a 2-input NAND gate and 7404 inverters.

4. In the space to the right, draw the appropriate circuit diagram for the conversion of the following 2-input gates:
   (a) An AND gate into a NOR gate

   (b) An OR gate into a NAND gate

   (c) A NAND gate into an AND gate

   (d) A NAND gate into an OR gate

   (e) A NOR gate into an AND gate

It is possible to convert any gate into any other type of gate with the use of inverters. In Questions 2, 3, and 4, keep trying until you succeed.

5. You should memorize the truth tables for the four basic 2-input positive gates: AND, NAND, OR, and NOR. Test your knowledge by filling in the following truth tables.

7408			7400			7432			7402		
B	A	Q	B	A	Q	B	A	Q	B	A	Q
0	0		0	0		0	0		0	0	
0	1		0	1		0	1		0	1	
1	0		1	0		1	0		1	0	
1	1		1	1		1	1		1	1	

EXPERIMENT NO. 6

Purpose

The purpose of this experiment is to determine the truth table for three 7400-series OR-type gates:

- The 7432 OR gate
- The 7402 "negated" OR or NOR gate
- The 7486 Exclusive OR gate

Pin Configurations of Integrated Circuits (Figs. 3-48 through 3-50)

Schematic Diagrams of Circuits (Figs. 3-51 through 3-53)

Step 1

You have already encountered both the 7402 quad 2-input positive NOR gate and the 7432 quad 2-input positive OR gate in the preceding chapter, so it should be easy for you to wire and determine a truth table for each of these integrated-circuit chips.

Follow the schematic diagram and wire a 7432 2-input positive OR gate. Fill in the following truth table for the gate.

B	A	Q
0	0	
0	1	
1	0	
1	1	

Compare your results with those given in the truth table for the 2-input positive OR gate given earlier in this chapter. Wire a 7402 2-input positive NOR gate and fill in the following truth table:

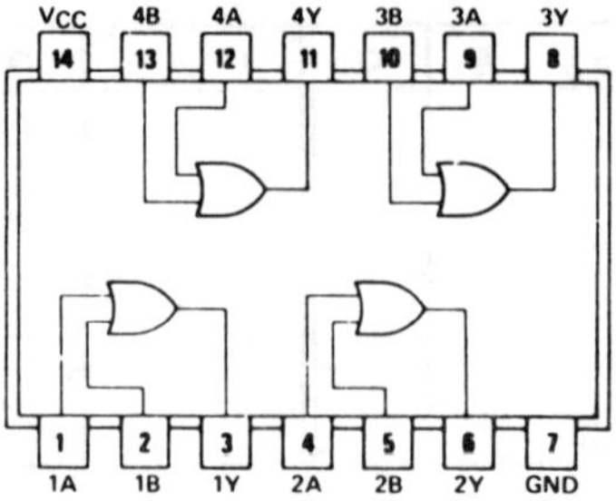

7432

Fig. 3-48.

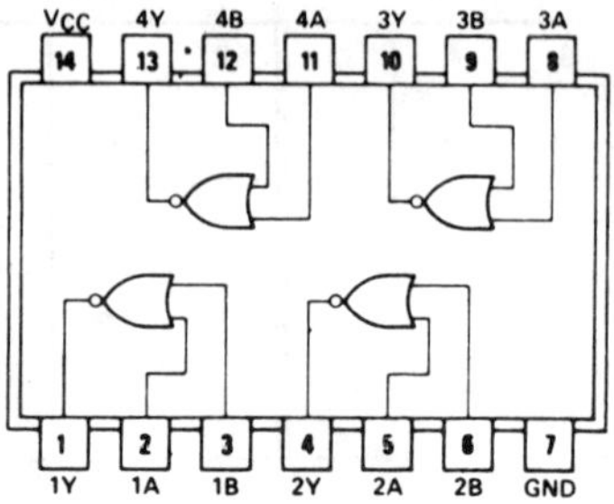

7402

Fig. 3-49.

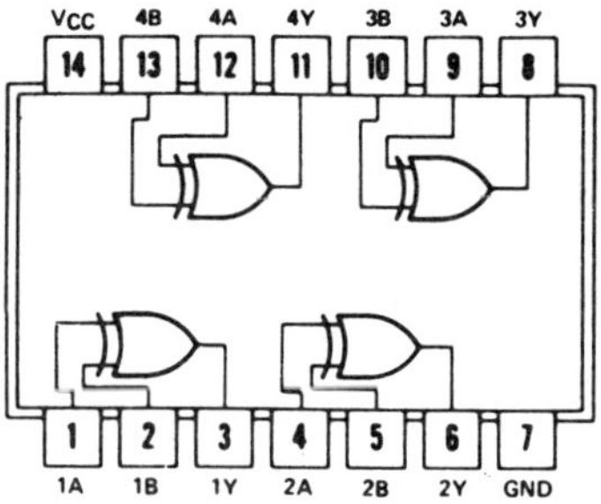

7486

Fig. 3-50.

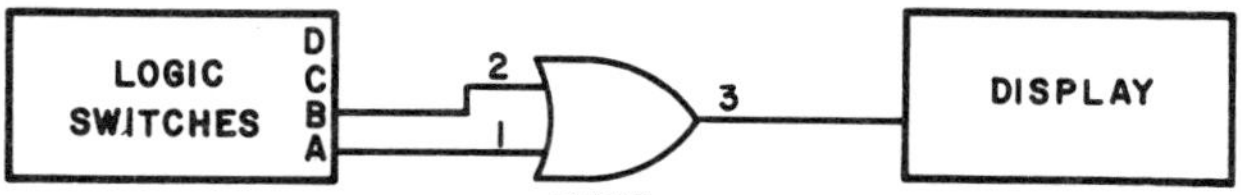

7432

Fig. 3-51.

7402

Fig. 3-52.

7486

Fig. 3-53.

B	A	Q
0	0	
0	1	
1	0	
1	1	

Compare your results to those given earlier in the truth table for a 2-input positive NOR gate.

Step 2

Perhaps the most interesting gate of the three is the 7486 2-input Exclusive-OR gate. Wire the circuit shown in the schematic diagram and fill in the following truth table:

B	A	Q
0	0	
0	1	
1	0	
1	1	

Compare your observations to the results found earlier in the truth table for a 2-input Exclusive-OR gate.

The 7486 Exclusive-OR gate can be viewed as a "controllable inverter" (see definitions section). Such a gate has the capability of transmitting a data input signal to the output either with or without inversion. You will demonstrate such a capability in the following experiment.

Questions

1. Give the pin numbers for the outputs from the four independent 2-input positive OR gates on the 7432 IC.

2. Give the pin numbers for the outputs from the four independent 2-input positive NOR gates on the 7402 IC.

3. Give the pin numbers for the outputs from the four independent 2-input Exclusive-OR gates on the 7486 IC.

4. What is the chief difference between a 2-input OR gate and a 2-input Exclusive-OR gate?

EXPERIMENT NO. 7

Purpose

The purpose of this experiment is to demonstrate the use of a 7486 2-input Exclusive-OR gate as a "controllable inverter."

Pin Configuration of Integrated Circuit (Fig. 3-54)

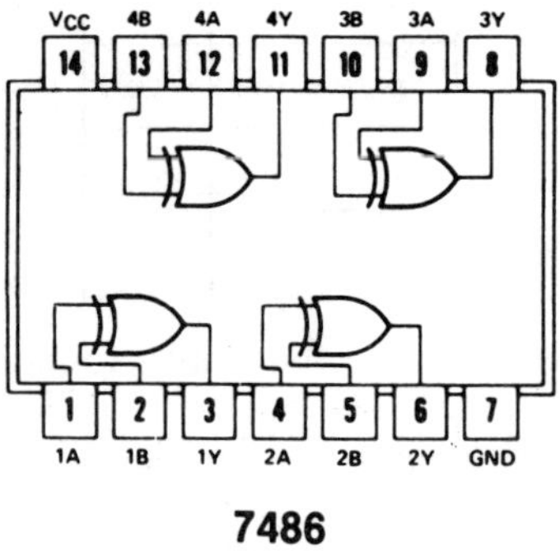

Fig. 3-54.

Schematic Diagram of Circuits (Figs. 3-55 and 3-56)

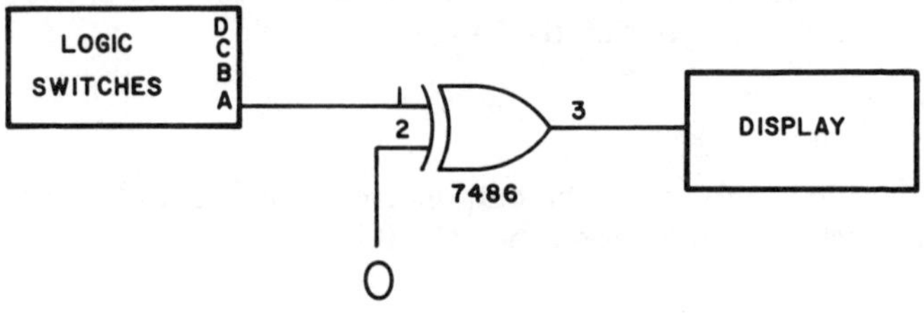

Fig. 3-55.

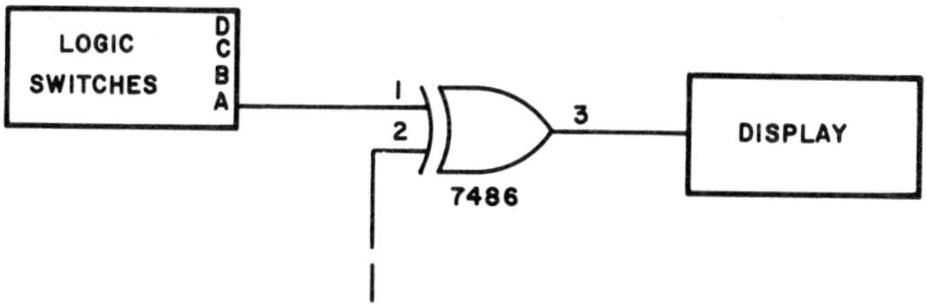

Fig. 3-56.

Step 1

In the above schematic diagrams, the logic switch A is used to provide a "data input" signal to the 7486 2-input Exclusive-OR gate.

Wire the indicated circuit. Remember, leave the power off when wiring! Don't forget the connections to pin 14 (+5 Volts) and pin 7 (GROUND).

Step 2

Apply power to the breadboard. Connect pin 2 on the 7486 IC to GROUND. Vary logic switch A between the logic 0 and logic 1 states and determine whether or not the input data signal is inverted at the output pin 3. With a gating signal at a logic 0 state, does the 2-input Exclusive-OR gate invert the data input?

Step 3

Now connect pin 2 to a logic 1 state, i.e., to +5 Volts. Vary logic switch A between the logic 0 and logic 1 states and again determine whether or not the input data signal is inverted at output pin 3. With a gating signal at a logic 1 state, does the 2-input Exclusive-OR gate invert the data input?

If you did this experiment correctly, you should have observed that a "logic 1" gating input causes the 7486 to act as a controllable inverter.

Questions

1. What do we mean by a "controllable inverter"? Give your own definition based on this experiment.

2. What do we mean by a "controllable complementer"? Is this term identical to "controllable inverter"? Attempt to answer both questions before looking at the definitions section.

3. Can you suggest how to construct a 2-input Exclusive-OR gate using three 2-input positive NAND gates and two inverters? Show the schematic diagram below. This is not an easy question.

4. In your own words, define the term, "truth table."

EXPERIMENT NO. 8

Purpose

The purpose of this experiment is to determine the truth table for a 7451 2-input 2-wide AND-OR-INVERT gate.

Pin Configuration of Integrated Circuit (Fig. 3-57)

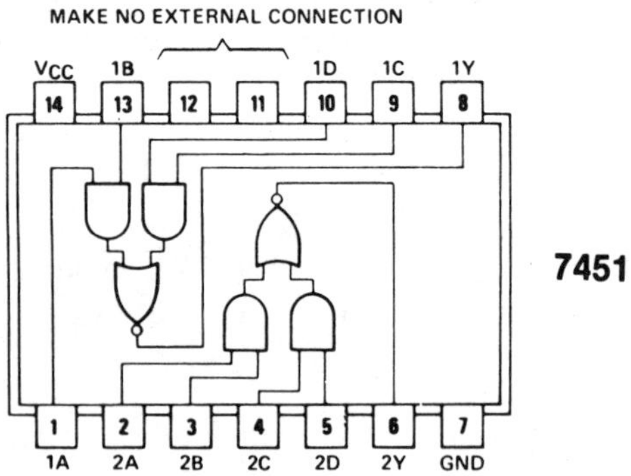

Fig. 3-57.

Schematic Diagram of Circuit (Fig. 3-58)

Step 1

Wire the digital circuit shown in the schematic diagram. Pin 7 is the GROUND connection and pin 14 is the +5 Volts connection on the 7451 integrated circuit.

Set the four logic switches, A, B, C, and D, all to a logic 0 state and apply power to the breadboard.

Step 2

By varying the settings of the four logic switches, perform the necessary experiments that will allow you to fill in the truth table shown below for the 7451 2-input 2-wide AND-OR-INVERT gate.

D	C	B	A	Q
0	0	0	0	1
0	0	0	1	
0	0	1	0	
0	0	1	1	
0	1	0	0	
0	1	0	1	
0	1	1	0	
0	1	1	1	
1	0	0	0	
1	0	0	1	
1	0	1	0	
1	0	1	1	
1	1	0	0	
1	1	0	1	
1	1	1	0	
1	1	1	1	

Answer the following question:

What is the minimum requirement for one to observe a "logic 0" output from this type of gate?

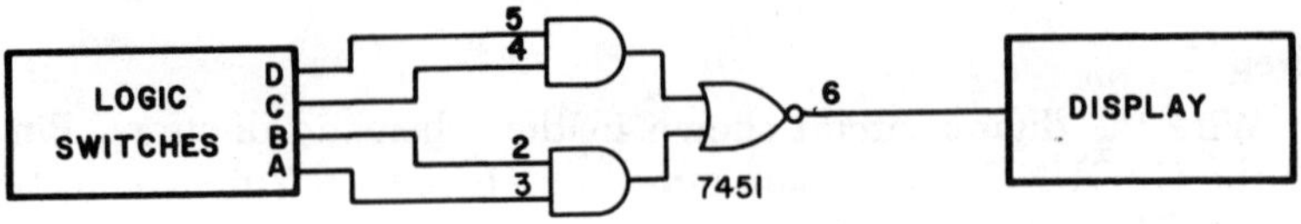

Fig. 3-58.

The answer to this question is that one of the inputs to the NOR gate must be a logic 1 in order to observe a logic 0 output. Compare the truth table that you measured with that given earlier in the truth table for a 2-input 2-wide AND-OR-INVERT gate.

Questions

1. Consider the following 2-input 2-wide AND-OR-INVERT gate. Fill in the following truth table when inputs C and D are left unconnected.

B	A	Q
0	0	
0	1	
1	0	
1	1	

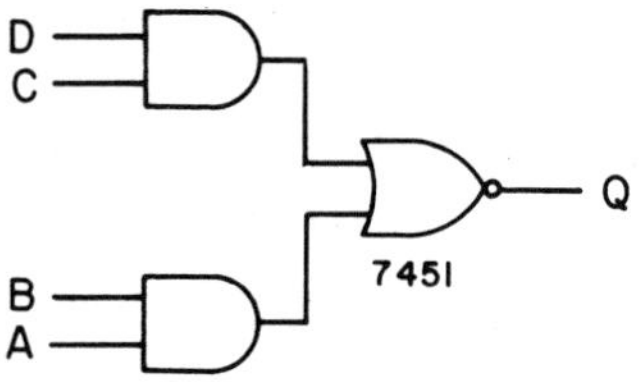

Fig. 3-59.

2. Refer to the 2-input 2-wide AND-OR-INVERT gate shown above. Fill in the following truth table when C = 0 and D = 1.

B	A	Q
0	0	
0	1	
1	0	
1	1	

3. Refer once more to the 2-input 2-wide AND-OR-INVERT gate shown above. Fill in the following truth table when B = 1, C = 0, and D = 1.

A	Q
0	
1	

4. For a final time, refer to the 2-input 2-wide AND-OR-INVERT gate shown above. Fill in the following truth table when D = 1 .

C	B	A	Q
0	0	0	
0	0	1	
0	1	0	
0	1	1	
1	0	0	
1	0	1	
1	1	0	
1	1	1	

5. How many independent 2-input 2-wide AND-OR-INVERT gates exist on the 7451 IC?
6. Can you suggest some uses for the AND-OR-INVERT gate?
7. The diagram for a 2-input 2-wide AND-OR-INVERT gate follows.

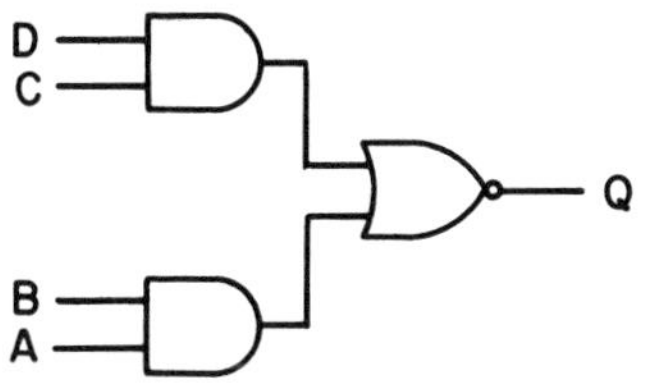

Fig. 3-60.

The "2-wide" refers to the two AND gates that act as inputs to the NOR gate; the "2-input" term refers to the fact that each of the two AND gates has two inputs. Based on this discussion, draw a schematic diagram for (a) a 3-input 2-wide AND-OR-INVERT gate, and (b) a 2-input 4-wide AND-OR-INVERT gate. Use the space below for your answer.

(a) (b)

8. Consider the following 2-input 2-wide AND-OR-INVERT gate, in which one of the inputs to each of the two AND gates is at a logic 1 state. When we have only two inputs, it is customary to use the letters A and B, rather than A and C. Fill in the following truth table for this gate:

B	A	Q
0	0	
0	1	
1	0	
1	1	

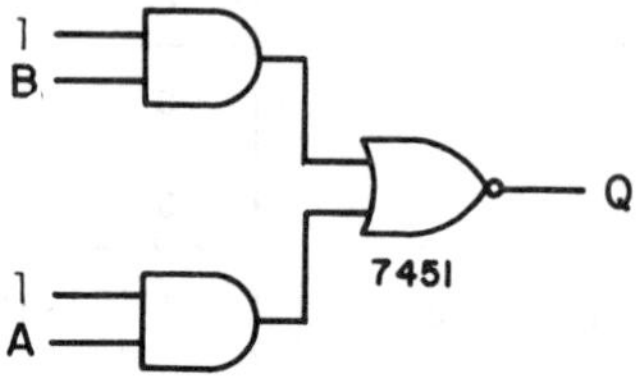

Fig. 3-61.

WHAT HAVE YOU ACCOMPLISHED IN THIS CHAPTER?

Review of Objectives

We stated in the Introduction to this chapter that at the end you should be able to do the following:

- Define, in your own words, the term, *truth table*.
  You have been asked to provide such a definition in Question No. 5 in Experiment No. 7. A definition of the term, truth table, has been given in the section entitled, "What is a Truth Table?"
- Write the correct truth table for the following logic gates:
  AND gate
  NAND gate
  OR gate
  NOR gate
  NOT gate
  Exclusive-OR gate
  AND-OR-INVERT gate
  The appropriate truth tables were given in the general introduction sections and you were requested to derive them in the experiments in the chapter. Questions at the end of each experiment further probed your understanding of truth tables. The largest truth table that you were asked to complete was the truth table for a 4-input gate.
- Demonstrate the operation of some of the more common 7400-series integrated-circuit gate chips, including the 7402, 7404, 7410, 7411, 7420, 7421, 7427, 7430, 7432, 7451, and 7486 chips. the operation of all of these simple integrated-circuit chips. In the eight experiments in this chapter, you have demonstrated the operation of all of these simple integrated-circuit chips.

Hopefully, you have built up confidence in your ability to wire a digital circuit from a schematic diagram alone.

- Demonstrate, with the aid of a 7420 4-input NAND gate, that an unconnected pin on a 7400-series integrated-circuit chip is at a "logic 1" state.

  You have demonstrated this result in the very important Experiment No. 4 in this chapter.

4

Multidecade Counters

INTRODUCTION

You will probably admit that the filling of truth tables is not one of the most exciting activities that you can perform using integrated circuits. In this section, you will take a brief respite from the "science" of the subject and instead will construct several counting and timing circuits that employ one or more seven-segment LED displays and 7490 decade counters. Such circuits simulate the action of many important digital electronic instruments: counters, timers, frequency meters, digital voltmeters, capacitance meters, and the like. Some of the experiments are not easy; we strongly recommend Experiments 7 through 9. Many of the circuits that we shall describe in this chapter are among the most complicated in this book.

OBJECTIVES

At the completion of this chapter, you will be able to do the following:

- Construct and operate a four-decade counter consisting of four cascaded 7490 counters.
- Employ a four-decade counter to measure both frequency and capacitance over a 50,000-fold range.
- Count contact bounces when you attempt to ground a wire.
- Count contact bounces in a single-pole double-throw (spdt) switch.

- Construct a circuit that eliminates contact bounces in an spdt switch.
- Employ a 555 Timer as a source of clock pulses at frequencies up to 70 kHz.
- Employ a 74121 monostable multivibrator integrated circuit as a source of single gating pulses.
- Pass a test in which you are asked to distinguish between different types of clock pulses and clock pulse transitions.
- Explain, in your own words, what is meant by the term, digital waveform, and give an example.

DEFINITIONS

bounce—The uncontrolled making and breaking of a contact when the switch or relay contacts are closed.[1] An important problem in digital circuits, where bounces can act as clock pulses.

cascade—An arrangement of two or more similar circuits in which the output of one circuit provides the input of the next.[1]

to cascade, cascading—The connecting of two or more circuits in series so that the output from one provides the input to the next.[1]

clock input—That flip-flop terminal whose condition or change of conditions controls the admission of data through the synchronous inputs and thereby controls the output of the flip-flop. The clock signal permits data signals to enter the flip-flop to change state accordingly.[1]

to contact—To join two conductors or conducting objects in order to provide a complete path for current flow.[1]

contact—The juncture point to provide the complete path. One of the current-carrying parts of a relay, switch, or connector that are engaged or disengaged to open or close the associated electrical circuits.[1]

contact bounce—The uncontrolled making and breaking of contact when the switch or relay contacts are closed.[1]

counter—A circuit that counts input pulses. One specific type produces one output pulse each time it receives some predetermined number of input pulses. The same term may also be applied to several such circuits connected in cascade to provide digital counting. Also called a divider.[1]

to debounce—To eliminate contact bounce in a switch or relay.

debounced—Refers to a switch or relay that no longer exhibits contact bounce.

digital waveform—A graphical representation of a digital signal, showing the variations in logic state as a function of time.

edge—The transition from logic 0 to logic 1, or from logic 1 to logic 0, in a clock pulse.

edge-triggered flip-flop—A type of flip-flop in which some minimum clock-signal rate of change is one necessary condition for an output change to occur.[1]

electrolytic capacitor—A capacitor consisting of two conducting electrodes, with the anode having a metal oxide film formed on it. The film acts as the dielectric or insulating medium. The capacitor is operable in the presence of an electrolyte, usually an acid or salt.[1]

farad—The capacitance of a capacitor in which a charge of 1 coulomb produces a change of 1 volt in the potential difference between its terminals.[1] The farad is the unit of capacitance in the SI system.

flip-flop—Also called bistable multivibrator, Eccles-Jordan circuit, or trigger circuit. A two-stage multivibrator circuit having two stable states. A trigger signal changes the circuit from one state to the other, and the next trigger signal changes it back to the first state. For counting and scaling purposes, a flip-flop can be used to deliver one output pulse for each two input pulses.[1]

frequency—The number of recurrences of a periodic phenomenon in a unit of time. Electrical frequency is specified as so many cycles per second, or hertz.[1]

hertz—Abbreviated Hz. A unit of frequency equal to one cycle per second.[1]

leading edge—The transition of a pulse that occurs first.[1]

microswitch—A small switch.

monostable multivibrator—Also called one-shot multivibrator, single-shot multivibrator, or start-stop multivibrator. A circuit having only one stable state, from which it can be triggered to change the state, but only for a predetermined interval, after which it returns to the original state.[1]

Mylar® capacitor—A capacitor in which Mylar® film, either alone or in combination with paper, is the dielectric.[1] Trademark of DuPont Corporation.

negative clock pulse—A complete logic cycle in which there is a transition from logic 1 to logic 0 and back to logic 1.

negative edge, negative edge transition—The transition from logic 1 to logic 0 in a clock pulse.

ohm—Symbolized by the letter omega (Ω). The unit of resistance. It is defined as the resistance, at 0°C, of a uniform column of mercury 106.300 cm long and weighing 14.4521 grams. One ohm is the value of resistance through which a potential difference of one volt will maintain a current of one ampere.[1]

positive clock pulse—A complete logic cycle in which there is a transitional change from logic 0 to logic 1 and then changing back to logic 0.

positive edge, positive edge transition—The transition from logic 0 to logic 1 in a clock pulse.

pull-up resistor—A resistor connected to the positive supply voltage of a transistor circuit, as from the collector supply to the output collector.[1] Also used occasionally with mechanical switches to ensure the voltage of one or more switch positions.

pulse width, pulse duration—Also called pulse length. The time interval between the points at which the instantaneous value on the leading and trailing edges bears a specified relationship to the peak pulse amplitude.[1]

single-pole, double-throw switch—Abbreviated spdt switch. A three-terminal switch or relay contact for connecting one terminal to either of two other terminals.[1]

time constant—The time required for an exponential quantity to change by an amount equal to 0.632 times the total change that will occur. Specifically, in a capacitor-resistor (RC) circuit, the number of seconds required for the capacitor to reach 63.2% of its full charge after a voltage is applied. The time constant of a capacitor having a capacitance C in farads in series with a resistance R in ohms is equal to R × C, the result having units of seconds.[1]

timer—A special clock mechanism used to perform switching operation at predetermined time intervals.[1]

555 Timer—An integrated-circuit chip containing eight pins that is used to perform switching operations at predetermined time intervals. It is used for precision timing, pulse generation, sequential timing, time-delay generation, pulse-width modulation, pulse-position modulation, etc.

timing capacitance, timing resistance—Passive components used to adjust the frequency or pulse duration of digital signals.

trailing edge—The transition of a pulse that occurs last, such as the high-to-low transition of a positive clock pulse.[1]

transition—The instance of changing from one state to a second state.

trigger—To cause, by means of one circuit, action to start in another circuit, which then functions for a certain length of time under its own control. A pulse that starts an action.

undebounced—Refers to a switch that still exhibits contact bounce.

THE 7490 DECADE COUNTER INTEGRATED CIRCUIT

Perhaps the most important single operation in digital electronic instrumentation is that of *counting*, i.e., the measurement of how many individual digital clock pulses occur during a specified period of time. Physical quantities such as resistance, capacitance, voltage, current, frequency, and the like are almost invariably transformed into a series of clock pulses whose number over a specified period of time is in some way proportional to the physical quantity being measured.

In this chapter you will perform experiments using a counting circuit that is capable of counting clock pulses from 0000 to 9999, i.e., a counting circuit that can count 10,000 clock pulses. This circuit consists of four *cascaded* 7490 decade counters, where the term, *cascade*, means that the output of one 7490 decade counter is connected to the input of the next. In addition, each 7490 decade counter contains an associated seven-segment LED display. You thus have a four-digit display that can directly read from 0000 to 9999.

Shown in Figs. 4-1 and 4-2 are the pin configuration and diagram of the 7490 decade counter integrated circuit. You may recall that you have already used this integrated-circuit chip in Experiments

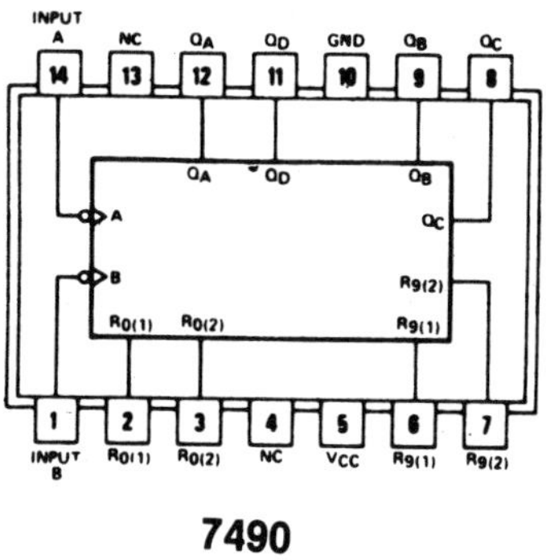

Fig. 4-1. Pin configuration of the 7490 decade counter IC.

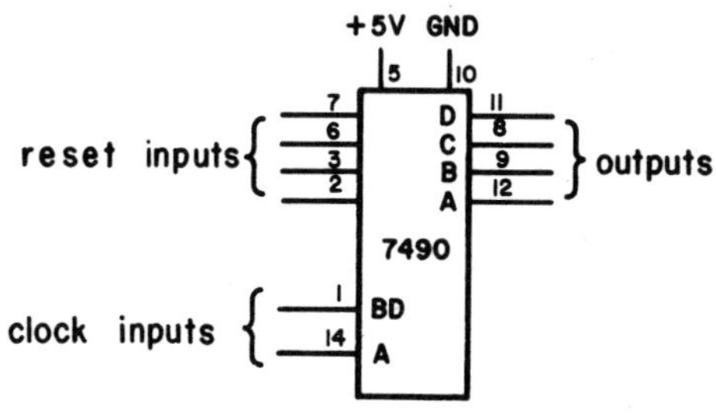

Fig. 4-2. Diagram showing various input/output pins on the 7490 decade counter IC.

Nos. 7 and 8 in Chapter 1. The circuit that you wired is shown in Fig. 4-3. Note that we have indicated that pins 3 and 6, which were left unconnected in Experiments Nos. 7 and 8, are at a logic 1 state.

A clock pulse has been previously defined as a complete logic cycle, i.e., a transition from logic 0 to logic 1 and back to logic 0 (*positive clock pulse*) or a transition from logic 1 to logic 0 and back to logic 1 (*negative clock pulse*). The symbol for a positive clock pulse is shown in Fig. 4-4, whereas the symbol for a negative clock pulse is shown in Fig. 4-5.

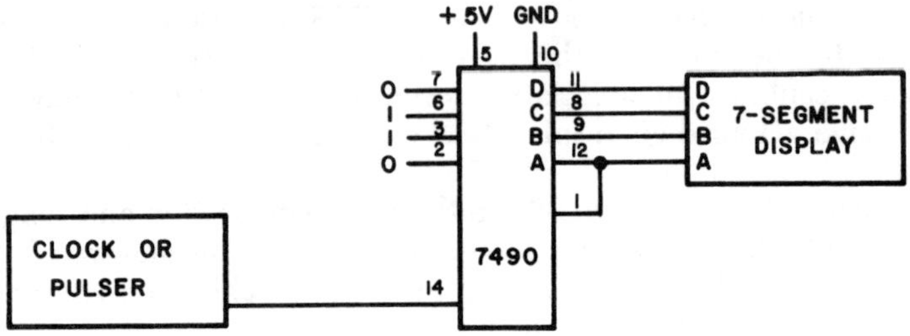

Fig. 4-3. A simple counting circuit used in several previous experiments.

When we apply a single positive clock pulse using the output "0" from a pulser that is connected to pin 14 of the 7490 decade counter chip, we observe that the decimal numeral on the seven-segment LED display changes *only when we release the pulser*. The significant point here is that the decimal numeral changes only on an *edge of* a clock pulse rather than in the middle of the pulse. The 7490 decade counter is said to be *edge-triggered*, which means that

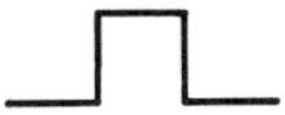

Fig. 4-4. Symbol for a positive clock pulse.

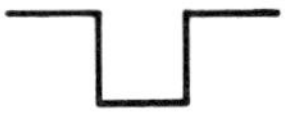

Fig. 4-5. Symbol for a negative clock pulse.

the *flip-flops* within the counter change state only on the *negative edge* of a clock pulse. This is an extremely important characteristic of this counter, for it permits you to cascade as many 7490 counters as you desire. Before we discuss the cascading of counters, we would first like to discuss the concepts of *negative* and *positive edges* of a clock pulse.

POSITIVE AND NEGATIVE EDGES OF A CLOCK PULSE

Many important digital electronic devices contain flip-flops that change state upon the application of a clock pulse. It is usually quite important to distinguish at which edge of a clock pulse that the state transition actually occurs. Shown in Figs. 4-6 and 4-7 are a positive and a negative clock pulse, respectively. In these clock pulses, there are two types of transitions,

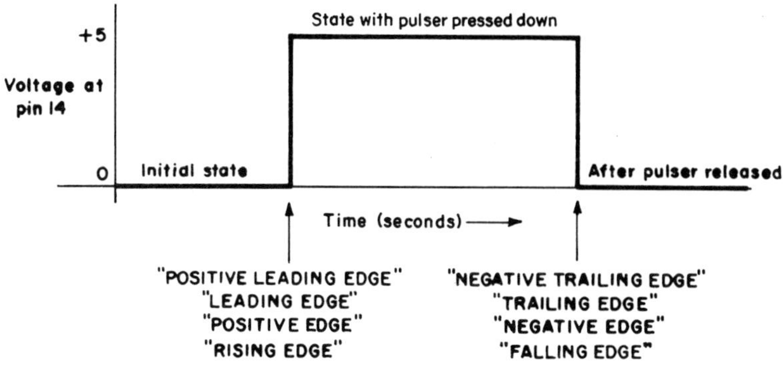

Fig. 4-6. Diagram of a positive clock pulse (initial state is a logic 0).

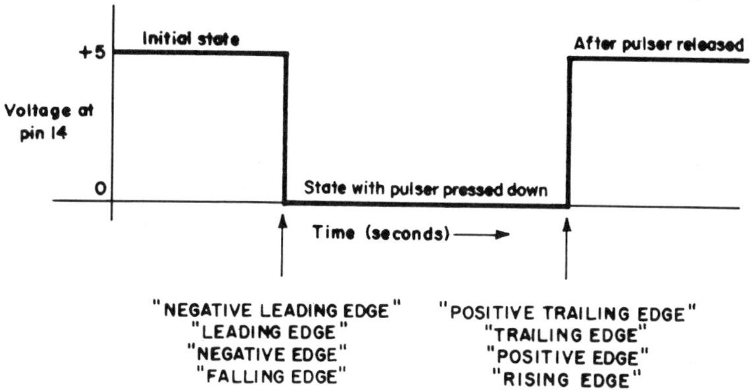

Fig. 4-7. Diagram of a negative clock pulse (initial state is a logic 1).

1. A *positive edge transition,* in which the transition is from a logic 0 to a logic 1 state, and
2. A *negative edge transition,* in which the transition is from a logic 1 to a logic 0 state.

It is observed in Figs. 4-6 and 4-7 that the adjectives, "leading," "rising," "positive," "negative," "trailing," and "falling" are all applied to edge transitions in a clock pulse. The terms, "leading" and "trailing" are quite misleading, since a "leading" edge on a positive clock pulse is not the same as a "leading" edge on a negative clock pulse.

Throughout this text, we shall try to avoid confusion by restricting our description of clock pulse edges to only six terms, *negative edge, positive edge, negative leading edge, negative trailing edge, positive leading edge, and positive trailing edge.* Several definitions are in order:

edge-triggered flip-flop—A type of flip-flop in which some minimum clock-signal rate of change is one necessary condition for an output change to occur.[1]

trigger—A pulse that starts an action. May also be the edge of a pulse.

positive edge—The transition from logic 0 to logic 1 in a clock pulse. In a positive clock pulse, it is a positive leading edge, whereas in a negative clock pulse, it is a positive trailing edge.

negative edge—The transition from logic 1 to logic 0 in a clock pulse. In a positive clock pulse, it is a negative trailing edge, whereas in a negative clock pulse, it is a negative leading edge.

In the 7400-series of integrated circuits, many edge-triggered transitions occur on the negative edge of clock pulses. Such is the case with the 7490 decade counter shown on the preceding page. There are a few 7400-series chips that change on the positive edge of clock pulses; such chips are highly valued for this characteristic.

Returning to Figs. 4-6 and 4-7, please note that the ordinate is voltage (or logic state), whereas the abscissa is time. Such a plot is called a *digital waveform* and is in common use in digital electronics. We encourage you to understand the type of information that such a plot provides.

DIGITAL WAVEFORMS FOR A 7490 DECADE COUNTER

A *waveform* can be defined as the graphic representation of an electromagnetic wave, showing the variations in amplitude with time. In digital electronics, the concept of a waveform specifically applies to the variation of logic state as a function of time at a specific point in a digital circuit. In a complex digital circuit, it is clear that we may need many different waveforms, corresponding to different points in the digital circuit, in order to understand and characterize the behavior of the circuit completely as a function of time.

We have shown five digital waveforms in Fig. 4-8 for a 7490 decade counter that counts from 0 to 9 and then resets itself back to 0, where it again resumes its count to 9. The specific points on the 7490 chip where we have measured these waveforms are, proceeding from top to bottom:

- The A CLOCK INPUT at pin 14 on the 7490 chip.
- The A output at pin 12 on the 7490 chip.
- The B output at pin 9 on the 7490 chip.

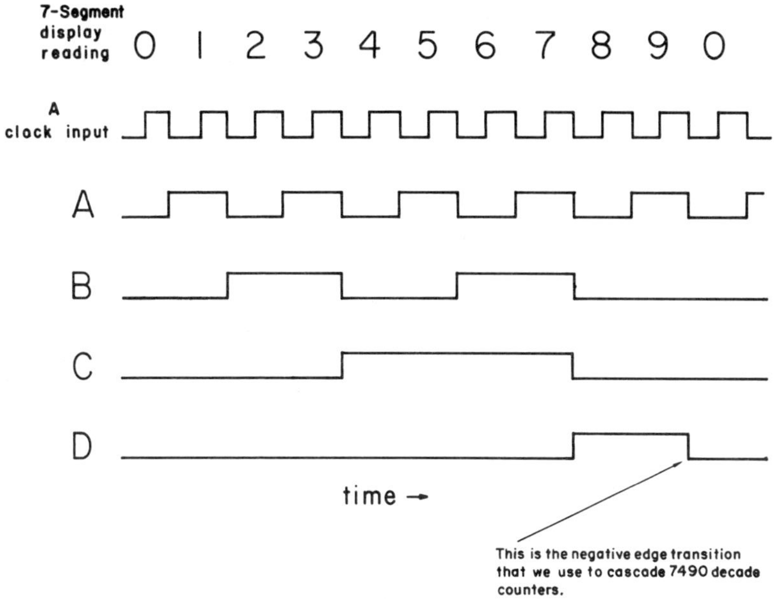

Fig. 4-8. Digital waveforms for the A CLOCK INPUT and the four outputs A, B, C, and D on a 7490 decade counter.

- The C output at pin 8 on the 7490 chip.
- The D output at pin 11 on the 7490 chip.

Let us take the situation when the seven-segment LED display gives a readout of decimal 3. In Fig. 4-8, it can be seen that the four outputs are at the following logic states: A = 1, B = 1, C = 0, and D = 0. These logic states correspond to binary-coded-decimal 0011, which, of course, is decimal three. Now note what happens at the negative edge of the clock pulse at the A CLOCK INPUT when the display reads decimal 3: both the A and B outputs make a rapid transition to a "logic 0" state while the C output at the same time makes a transition to a "logic 1" state. When the clock pulse terminates, we now have a binary-coded-decimal reading of 0100, or decimal four. Thus, we have just analyzed the behavior of the 7490 decade counter when it counts from decimal three to decimal four. All other transitions of the 7490 counter should be understandable with the aid of these five digital waveforms.

CASCADING 7490 DECADE COUNTERS

To cascade means to arrange two or more similar digital circuits or integrated-circuit chips in a manner such that the output of one circuit or chip provides the input of the next. In Fig. 4-9, we have shown two cascaded 7490 decade counters. With such a circuit, you are now able to count from decimal 0 to decimal 99 and back to decimal 0 again. We have also indicated the key connection that is needed to cascade the two 7490 counters: the connection from pin 11, the D output, on 7490 counter No. 1 to the A CLOCK INPUT pin 14 on 7490 counter No. 2. In other words, the D output from

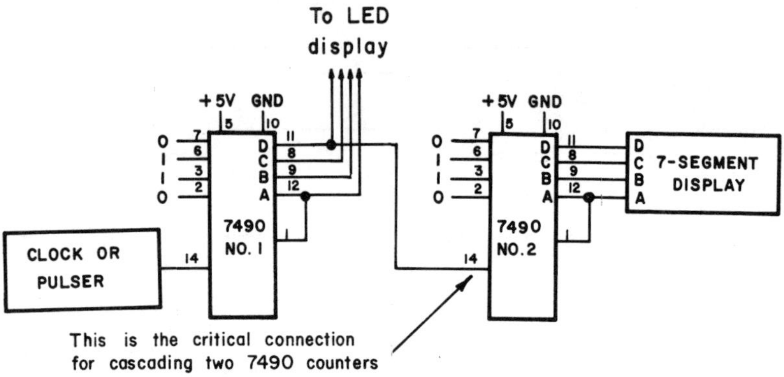

Fig. 4-9. Two cascaded 7490 decade counters, each connected to its own seven-segment LED display.

counter No. 1 is acting as the clock input to counter No. 2. The particular negative edge transition that is required for such a cascading operation is shown in Fig. 4-8; it is the negative edge transition at the D output.

In Fig. 4-10, we show four cascaded 7490 decade counters each of which is connected to its own seven-segment LED display. We have arranged the schematic diagram so that the 1000's position readout is on the far left and the 1's position readout is on the far right. In this chapter, you shall be asked to wire up this circuit. More than any other circuit that you have wired, the four cascaded decade counters bring the excitement and vitality of digital electronics to your immediate attention. It is particularly interesting to watch a transition from 9999 to 0000; the four LED displays change simultaneously.

Study counter No. 1 in Fig. 4-10. Note that at the "A CLOCK INPUT" (pin 14), we can have any source of clock pulses. As shown in Fig. 4-11, we usually employ a gate, such as the 7400 2-input NAND gate, between the source of clock pulses and pin 14; this allows us to gate the clock pulses into the counter with the aid of a gating input.

The use of a gate permits you to make two types of experimental measurements:

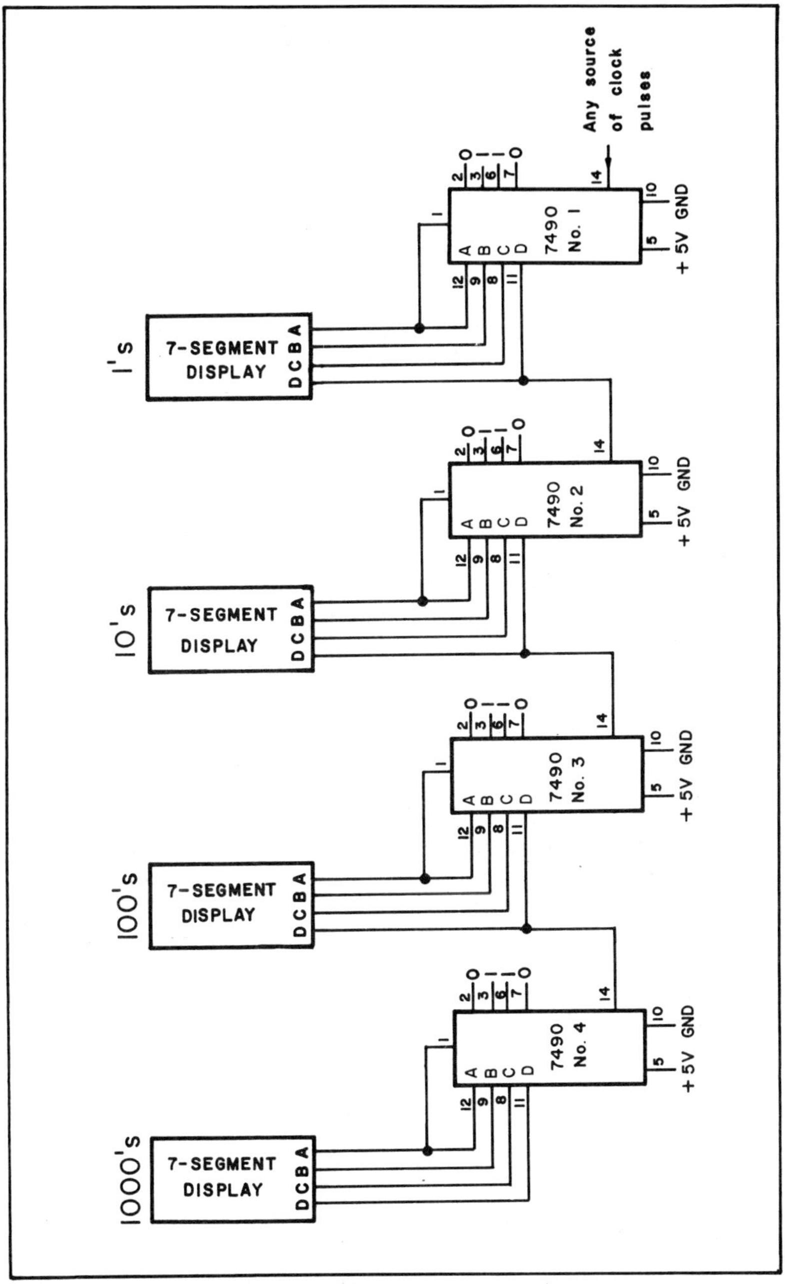

Fig. 4-10. Four cascaded 7490 decade counters, each connected to its own seven-segment LED display.

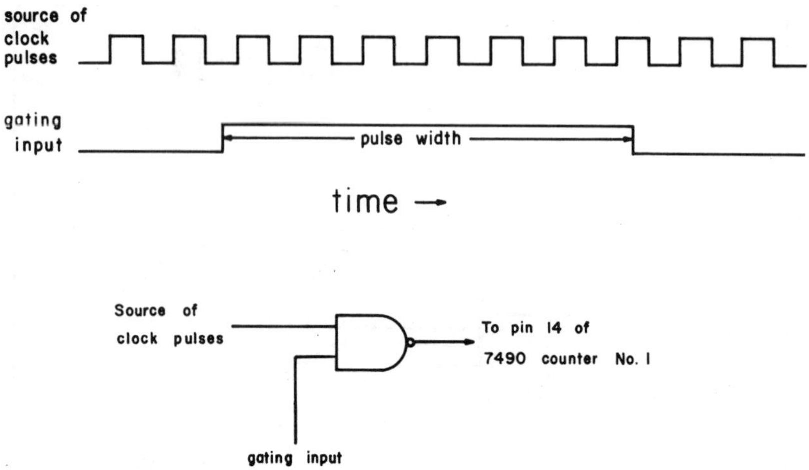

Fig. 4-11. Gating circuit can be used with multidecade counting circuits such as those in Figs. 4-9 and 4-10.

- Given a source of clock pulses of known frequency, measure the *pulse width,* i.e., the time duration of a pulse, of the gating input.
- Given a gating input of known pulse width, measure the number of clock pulses that occur during a single gating pulse.

These two types of measurements are best understood with the aid of the digital waveforms shown in Fig. 4-11. Assuming that the counter is negative-edge triggered, you can count six negative-edge transitions before the gating input returns to a "logic 0" state. If the source of clock pulses had a higher frequency, you would observe more counts for the same pulse width. On the other hand, if the pulse width were shorter, you would observe fewer counts. With the proper adjustment of the pulse width and the frequency of the source of clock pulses, a wide variety of measurements can be made. The source of clock pulses can be:

- A clock operating at frequencies ranging as low as 0.01 Hz or as high as several hundred thousand hertz.
- A pulser.
- A nondebounced mechanical switch.
- A transducer or voltage-to-frequency converter that converts any physical quantity into a series of clock pulses the frequency of which is in some way related to the magnitude of the physical quantity. Using such a technique, we can measure frequency, voltage, current, resistance, capacitance, events, velocity, etc.

The source of gating input pulses can be:

- A clock operating at a much lower frequency.
- A monostable multivibrator.
- A debounced pulser.
- Other systems that generate a single clock pulse.

INTRODUCTION TO THE EXPERIMENTS

In this chapter, you will conduct a variety of experiments that are more complicated than many of the experiments that you will encounter in remaining chapters. However, experiments contained in this chapter are among the most interesting in this book. They are worth your attention and the time it takes to wire them correctly. Since four 7490 counters are cascaded, the connections are somewhat repetitive. This should simplify the task of wiring the circuits.

We can classify the experiments in terms of the source of clock pulses and the source of the gating clock signal. Thus, we have the following:

Experiment No.	*Source of Clock Pulses*	*Source of Gating Signal*
1	Clock module	Pulser
2	Clock module	None
3	Clock module	Pulser
4	Clock module	None, or pulser
5	Clock module	Pulser
6	Connection to ground, undebounced spdt switch, or debounced spdt switch	None
7	Clock module	74121 monostable
8	Clock module	555 monostable
9	555 astable	74121 monostable

In the experiments that follow, you will encounter experimental data that we obtained while we were writing or modifying the experiments. We used two different clock Outboard modules, one with a no-capacitor frequency of 72.3 kHz, and the other with a no-capacitor frequency of 62.31 kHz.

We did encounter problems when we used the 74121 monostable integrated circuit. They were associated with a low power supply voltage. When a proper power voltage was used, the 74121 behaved perfectly.

EXPERIMENT NO. 1

Purpose

The purpose of this experiment is to demonstrate the operation of a gated counter using a single 7490 decade counter chip.

Pin Configurations of Integrated-Circuit Chips (Figs 4-12 and 4-13)

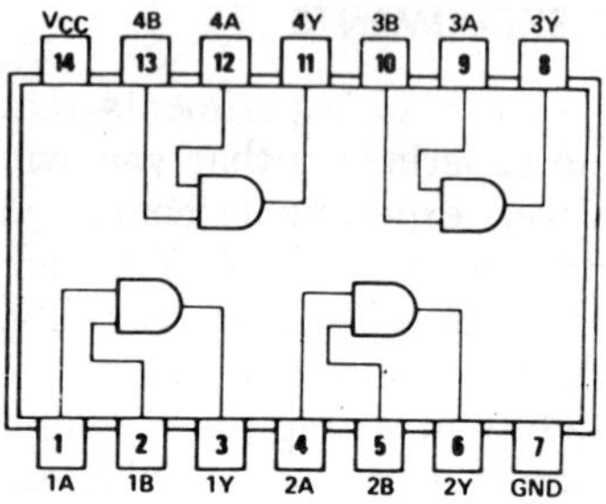

7408

Fig. 4-12.

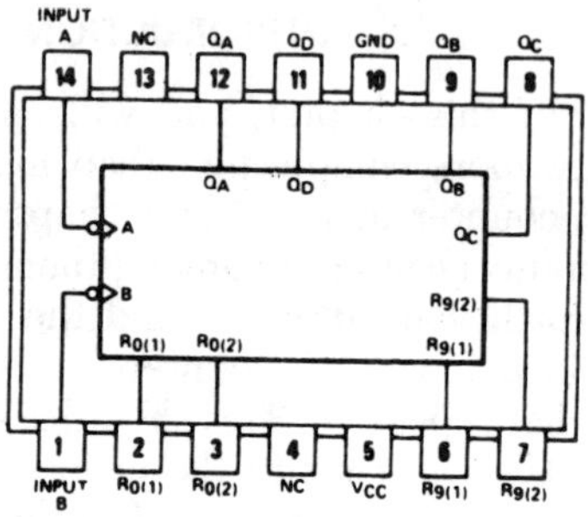

7490

Fig. 4-13.

Schematic Diagram of Circuit (Fig. 4-14)

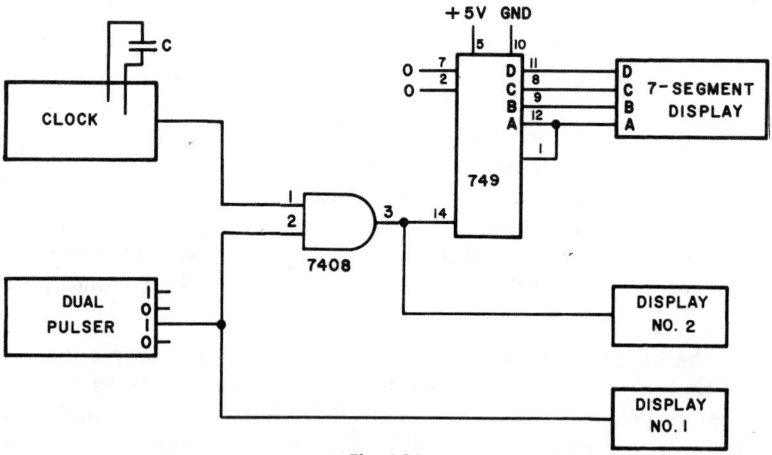

Fig. 4-14.

Step 1

In the schematic diagram, a display can be either a single lamp monitor or a zero/one display constructed from a seven-segment display Outboard module. Wire the circuit shown. As wired, is the AND gate enabled or disabled?

It is enabled. The gating input is from the pulser. Since the logic 1 output of the pulser is used, pin 2 on the 7408 chip is at logic 1 unless you press the pulser in.

Step 2

Apply power to the circuit. Use a slow clock frequency, in the vicinity of 1 Hz. What do you observe on the seven-segment display and display No. 2?

We observed a count sequence on the seven-segment display and a blinking lamp monitor for display No. 2.

Step 3

Now press the pulser. What happens?

The decade counter stops counting, display No. 2 remains at a "logic 0" state, and display No. 1 remains at logic 0. We have disabled the gate.

Step 4

Perform the same experiment for the 7400 NAND gate and the 7432 OR gate, and fill in the chart below.

Gate	Logic State of Gating Input	Gate enabled or disabled?	Counting or no counting?	Logic State of Display No. 1	Logic State of Display No. 2
7408	0				
	1	enabled	counting	1	clock pulses
7400	0				
	1				
7432	0				
	1				

The pin configurations for the 7400 and 7432 integrated-circuit chips are shown in Figs. 4-15 and 4-16, respectively. You should

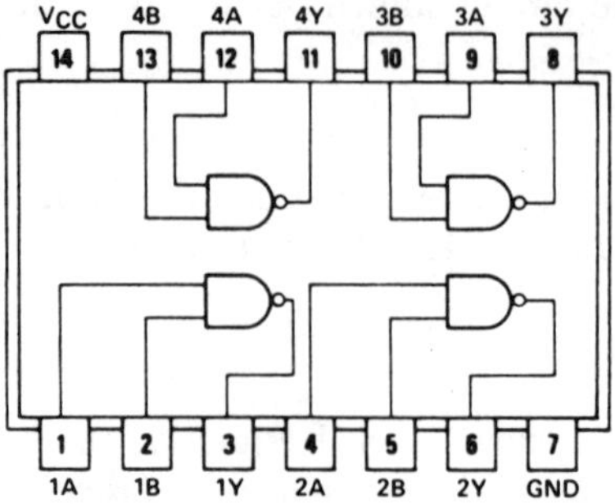

7400

Fig. 4-15.

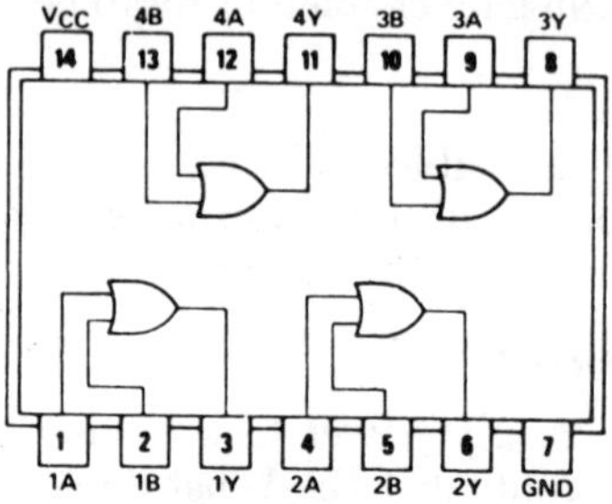

7432

Fig. 4-16.

observe that they are pin-for-pin compatible with the 7408 chip. To change from one gate to another, all you need do is remove the power and substitute the 7400 or 7432 for the 7408 chip. When we performed this experiment, we observed the following results:

Gate	Logic State of Gating Input	Gate enabled or disabled?	Counting or no counting?	Logic State of Display No. 1	Logic State of Display No. 2
7408	0	disabled	no counting	0	0
	1	enabled	counting	1	clock pulses
7400	0	disabled	no counting	0	1
	1	enabled	counting	1	clock pulses
7432	0	enabled	counting	0	clock pulses
	1	disabled	no counting	1	1
7402	0				
	1				

Fill in your predicted behavior for the 7402 NOR gate. Based on the pin configuration of the chip shown in Fig. 4-17, what wiring changes must you make to the circuit shown in the schematic diagram?

The 7402 chip is not pin-for-pin compatible with the 7400, 7408, and 7432 chips. The two inputs are now pins 2 and 3 and the output is now pin 1.

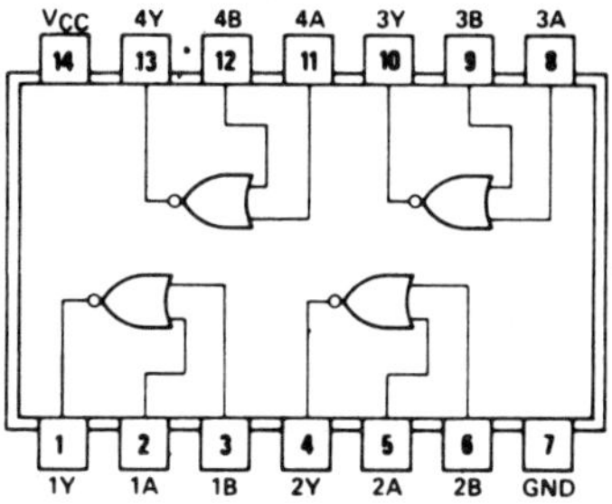

7402

Fig. 4-17.

Step 5

The 7493 binary counter chip, shown in Fig. 4-18, is pin-for-pin compatible with the 7490 decade counter chip. Substitute the 7493 for the 7490 and repeat Steps 2 and 3 for your choice of gate. Does the binary counter behave in substantially the same way as the decade counter?

Yes. We observed sixteen unique counting states rather than just ten. With the seven-segment display, we observed five unusual symbols and a blank display in addition to the numerals 0 through 9.

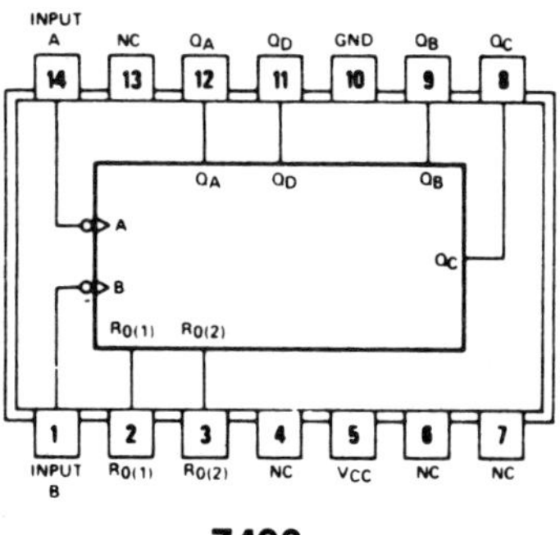

7493

Fig. 4-18.

EXPERIMENT NO. 2

Purpose

The purpose of this experiment is to construct a two-decade, three-decade, or four-decade counter and use it to measure a wider range of clock frequencies.

Pin Configuration of Integrated-Circuit Chip (Fig. 4-19)

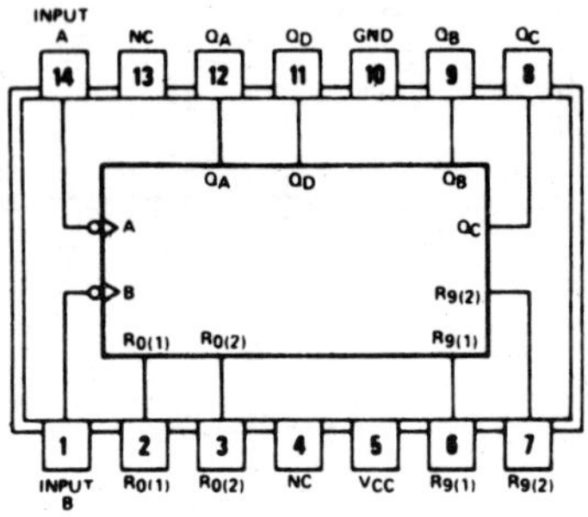

7490

Fig. 4-19.

Schematic Diagram of Circuit (Fig. 4-20)

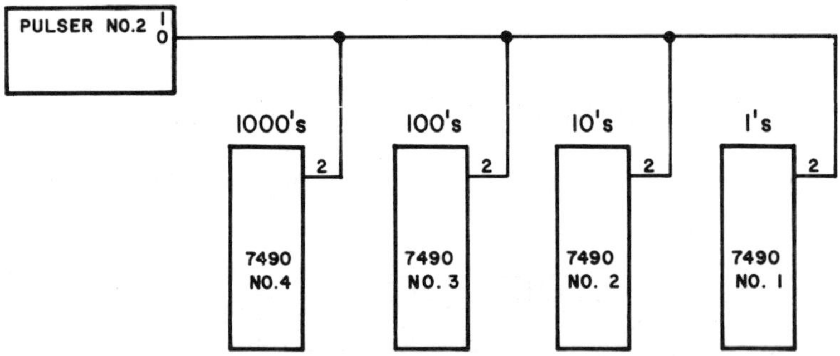

Fig. 4-20.

The diagram for a four-decade counter is given in Fig. 4-20. You may wish to simultaneously clear all four decade counters. To do so, connect pin 2 on each counter to a pulser, as shown in Fig. 4-21.

Step 1

Wire the circuit shown. You should be able to fit it on a single breadboarding socket. Since the purpose of this experiment is to

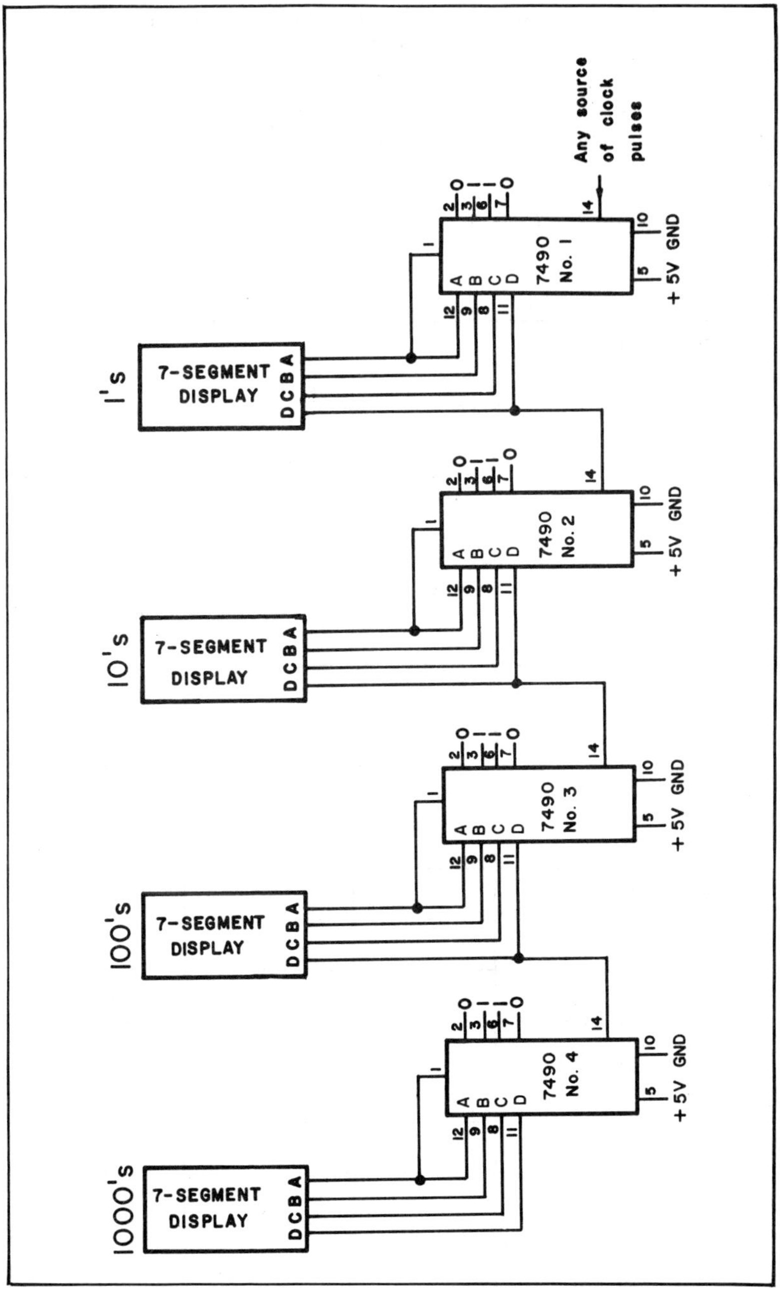
Any source of clock pulses
7490 No. 1
7490 No. 2
7490 No. 3
7490 No. 4
7-SEGMENT DISPLAY
DCBA
1's
10's
100's
1000's
+ 5V GND

Fig. 4-21.

demonstrate how a multidecade counter operates, a two-decade counter would be perfectly acceptable for your measurements in this experiment. A three-decade counter would be better. If you are working in teams, each team member may wish to wire a two-decade counter on a separate solderless breadboard. You can then consolidate the two breadboards into a single four-decade counter circuit. Power the breadboards from only a single +5-volt power supply. *Save this circuit for Experiment No. 3.*

Timing Capacitance, μF	Frequency, Counts/min	Counts/sec	Frequency^{-1}, Seconds/count	Frequency × Capacitance 10^{-8} Farads/sec
none				
0.00001				
0.00002				
0.00005				
0.0001				
0.0002				
0.0005				
0.001				
0.002				
0.005				
0.01				
0.02				
0.05				
0.1				
0.22				
0.47				
1.0				
2.2				

Step 2

This experiment is similar to Experiment No. 9 in Chapter 1. You will measure the frequency associated with different timing capacitors. Connect the clock input to pin 14 of the first 7490 counter (the 1's counter). With the aid of a stop watch or wristwatch, measure the number of counts in a given time interval, such as one minute. To start the counting, press and then release the pulser that clears the counters. Fill in the following chart with your results.

If the clock frequency is too high, skip the measurement. With two decades, you should be able to measure the frequency associated with a timing capacitance of 0.01 μF. With three decades, you should be able to go to 0.001 μF. With four decades, you should be able to measure the frequency associated with a timing capacitance of 0.0001 μF or even lower.

In the following chart, we used a six-decade counter to obtain these results:

Timing Capacitance, μF	Frequency, Counts/min	Counts/sec	Frequency^{-1}, Seconds/count	Frequency × Capacitance 10^{-8} Farads/sec
none		72,300.	0.0000138	
0.00001		33,300.	0.0000300	33.3
0.00002		23,300.	0.000043	46.5
0.00005		11,900.	0.000084	59.5
0.0001		6,540.	0.000153	65.4
0.0002		3,510.	0.000285	70.2
0.0005		1,540.	0.00065	76.9
0.001		766.	0.00131	76.6
0.005		141.	0.00708	70.7
0.01		73.5	0.0136	73.5
0.02		39.6	0.0253	79.2
0.05		17.7	0.0565	88.5
0.1	400.	6.67	0.15	66.6
0.33	74.		0.81	40.7
1.0	19.5		3.08	32.5
2.2	8.38		7.16	30.7
3.3	6.36		9.43	35.0
0.33	75.		0.80	41.6

The product of frequency and capacitance should be a constant above a capacitance of 0.0005 μF. We do not understand why the electrolytic capacitors in the above table gave consistently low frequency-capacitance products. Rather than hide the data points that did not fit on the expected curve, we decided to include them in the above chart as an example of possible errors in capacitance values. We believe that the frequency data up to a capacitance value of 0.1 μF are reliable. Tolerances for electrolytics are sometimes ±50%.

We can calculate the interval capacitance of the 555 Timer present on the clock Outboard by dividing the frequency with no capacitor, 72,300 Hz, into the frequency-capacitance product. Thus,

$$\text{Internal capacitance} = \frac{76.0 \times 10^{-8}\ \text{farads/sec}}{72{,}000\ \text{sec}^{-1}}$$
$$= 10.5\ \text{pF}$$

This internal capacitance should be added to the external capacitance to obtain the total capacitance, from which the timer frequency can be calculated. Thus, for a 10-pF external capacitor, the total capacitance is 20.5 pF and the predicted frequency is:

$$\text{Predicted frequency} = \frac{76.0 \times 10^{-8}\ \text{farads/sec}}{20.5 \times 10^{-12}\ \text{farads}} = 37{,}100\ \text{Hz}$$

Save this circuit and continue to Experiment No. 6.

Questions

1. Can you suggest several different sources of clock pulses that you could wire to the four cascaded 7490 decade counters? List them.

2. What if you wanted to count from 00000 to 99999_{10}. What changes in the circuit would be required? In the space below, show the additional wire connections that would be needed (Do not repeat the entire circuit of Fig. 4-1; just show the additional connections required).

3. We call the quantity, RC, a time constant since it has units of seconds when the resistance R is in ohms and the capacitance C is in farads. What do we mean by the term, "time constant"? This is a question that requires some knowledge of simple electrical circuit theory.

4. According to the manufacturer of the 555 Timer integrated-circuit chip, the clock frequency is expressed by the following equation:

$$\text{Clock frequency} = \frac{1.443}{(R_A + 2R_B)C} \text{ counts/second}$$

where,

R_A = 100K in the circuit diagram for this experiment,
R_B = 1 Meg in the circuit diagram for this experiment,
C = timing capacitance, in farads.

Calculate the predicted clock frequency for C = 0.33 μF. The measured value for typical clock Outboard modules is in the vicinity of 0.7 Hz.

EXPERIMENT NO. 3

Purpose

The purpose of this experiment is to use a multidecade *gated counter* to measure the frequency of the clock Outboard as a function of timing capacitance. This experiment is essentially identical to the previous experiment.

Pin Configurations of Integrated Circuit Chips (Figs. 4-22 and 4-23)

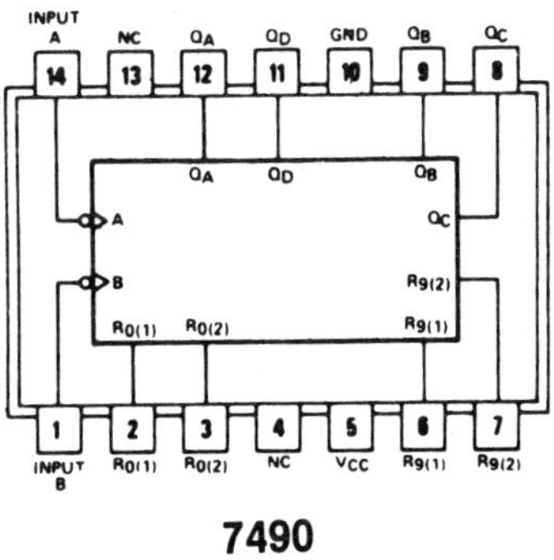

Fig. 4-22.

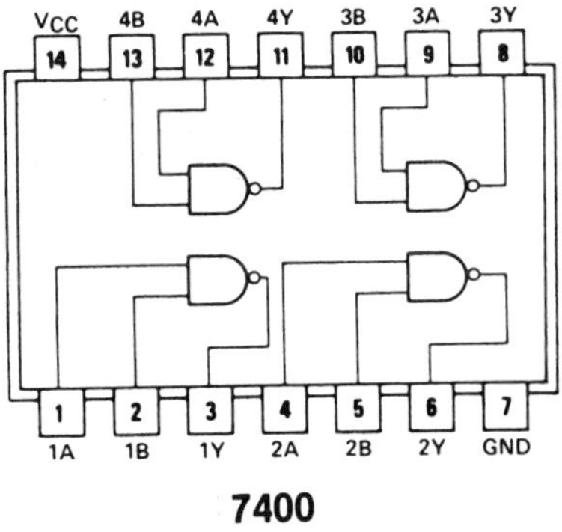

Fig. 4-23.

Schematic Diagram of Circuit (Fig. 4-24)

The diagram for the four-decade counter is shown in Fig. 4-24. You may wish to simultaneously clear all of the 7490 decade counters, as shown in Experiment No. 2. The basic gating circuit is shown.

Step 1

Wire the gating and multidecade counter circuits shown. The more decades you have, the wider the range of frequencies you can measure.

Timing Capacitance, μF	Frequency, Counts/min	Frequency, Counts/sec	Frequency^{-1}, Seconds/count	Frequency × Capacitance 10^{-8} Farads/sec
If you have only three decades in your counter, the following frequencies may be too high to measure:				
none				
0.00001				
0.00002				
0.00005				
0.0001				
0.0002				
The following should be easier to measure:				
0.0005				
0.001				
0.002				
0.005				
0.01				
0.02				
0.05				
0.10				
0.22				
0.47				
1.0				
2.2				

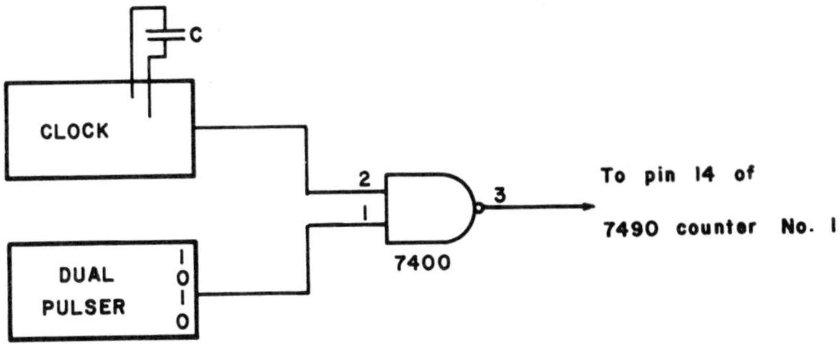

Fig. 4-24.

Step 2

Vary the timing capacitance on the clock Outboard, and fill in the following chart with your results.

Based on your measurements, calculate the internal capacitance of the clock that you are using. You cannot make this calculation if you are unable to measure the clock frequency with no timing capacitor. Save this circuit and continue to Experiment No. 6.

EXPERIMENT NO. 4

Purpose

The purpose of this experiment is to construct a three-decade counter using the Hewlett-Packard 5082-7300 numeric indicator or the LR-28 Three Digit Latch/Display Outboard module.

Schematic Diagrams of Circuits (Figs. 4-25 through 4-27)

Step 1

Wire the circuit shown and repeat the experiment given in Experiment No. 2. Fill in the following chart with your results. Perform measurements on as many different capacitors as you can.

You may wish to implement the counter clearing circuit given in Experiment No. 1. *If you have not already wired the circuit in Experiment No. 2, save this circuit and use it in Experiment No. 6.*

EXPERIMENT NO. 5

Purpose

The purpose of this experiment is to demonstrate how a pair of 74153 data selectors/multiplexers can be used to multiplex the ABCD outputs from four cascaded 7490 decade counters into a single seven-segment LED display.

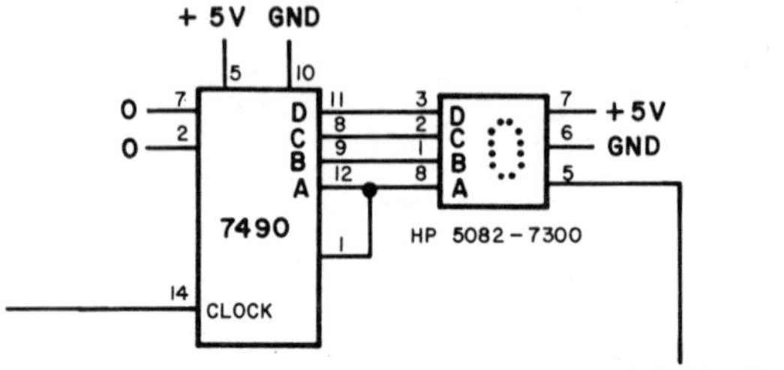

Fig. 4-25.

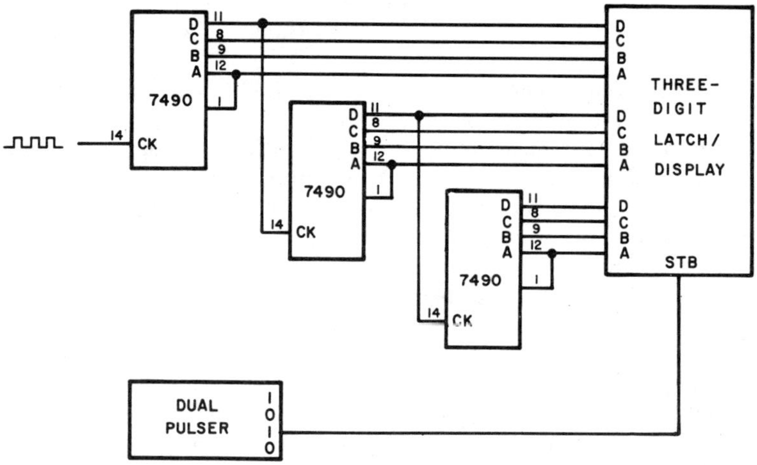

Fig. 4-26.

Notes and Hints

- This experiment requires quite a bit of wiring.
- Two breadboards will be required for this experiment.
- This experiment is similar to Experiment No. 2.

Pin Configurations of Integrated-Circuit Chips (Figs. 4-28 and 4-29)

Schematic Diagram of Circuit (Fig. 4-30)

To reduce the complexity of the circuit diagram, three different diagrams are given:

- A circuit diagram of the four cascaded 7490 counters (Fig. 4-30).
- A circuit diagram of the connections from the four 7490 counters to the pair of 74153 data selector/multiplexer chips (Fig. 4-31).

HEWLETT hp PACKARD

COMPONENTS

NUMERIC and HEXADECIMAL INDICATORS

5082-7300 SERIES

FEATURES

- Numeric 5082-7300/-7302
  - 0-9, Test State, Minus Sign, Blank States
  - Decimal Point
    - 7300 Right Hand D.P.
    - 7302 Left Hand D.P.
- Hexadecimal 5082-7340
  - 0-9, A-F, Base 16 Operation
  - Blanking Control, Conserves Power
  - No Decimal Point
- DTL – TTL Compatible
- Includes Decoder/Driver with Memory
  - 8421 Positive Logic Input
- 4 X 7 Dot Matrix Array
  - Shaped Character, Excellent Readibility
- Standard .600 inch X .400 inch Dual-in-Line Package including Contrast Filter
- Categorized for Luminous Intensity
  - Assures Uniformity of Light Output from Unit to Unit within a Single Category

DESCRIPTION

The HP 5082-7300 series solid state numeric and hexadecimal indicators with on-board decoder/driver and memory provide a reliable, low-cost method for displaying digital information.

The 5082-7300 numeric indicator decodes positive 8421 BCD logic inputs into characters 0-9, a "–" sign, a test pattern, and four blanks in the invalid BCD states. The unit employs a right-hand decimal point. Typical applications include point-of-sale terminals, instrumentation, and computer systems.

The 5082-7302 is the same as the 5082-7300, except that the decimal point is located on the left-hand side of the digit.

The 5082-7340 hexadecimal indicator decodes positive 8421 logic inputs into 16 states, 0-9 and A-F. In place of the decimal point an input is provided for blanking the display (all LED's off), without losing the contents of the memory. Applications include terminals and computer systems using the base-16 character set.

The 5082-7304 is a "±1" overrange character, including decimal point, used in instrumentation applications.

PACKAGE DIMENSIONS

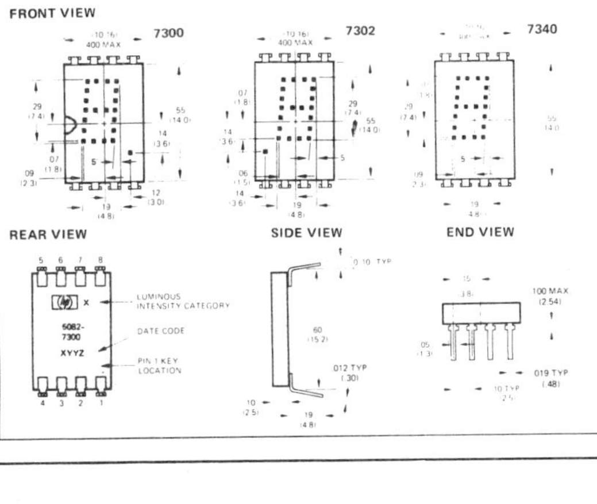

PIN	FUNCTION 5082-7300 and 7302 Numeric	5082-7340 Hexadecimal
1	Input 2	Input 2
2	Input 4	Input 4
3	Input 8	Input 8
4	Decimal point	Blanking control
5	Latch enable	Latch enable
6	Ground	Ground
7	V_{CC}	V_{CC}
8	Input 1	Input 1

NOTES: 1. Dimensions in inches and (millimeters).
2. Unless otherwise specified, the tolerance on all dimensions is ±.015 inches.

Fig. 4-27.

Timing Capacitance, μF	Frequency, Counts/min	Counts/sec	Frequency^{-1} Seconds/count	Frequency × Capacitance, 10^{-8} Farads/sec
0.0001				
0.0002				
0.0005				
0.001				
0.002				
0.005				
0.01				
0.02				
0.05				
0.1				
0.22				
0.47				
1.0				
2.2				
4.7				

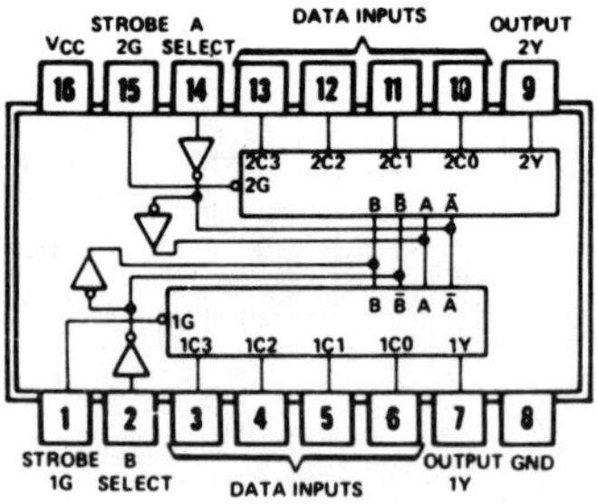

74153

Fig. 4-28.

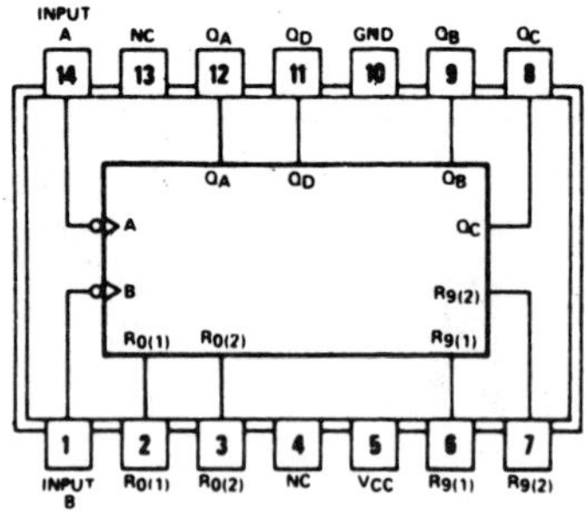

7490

Fig. 4-29.

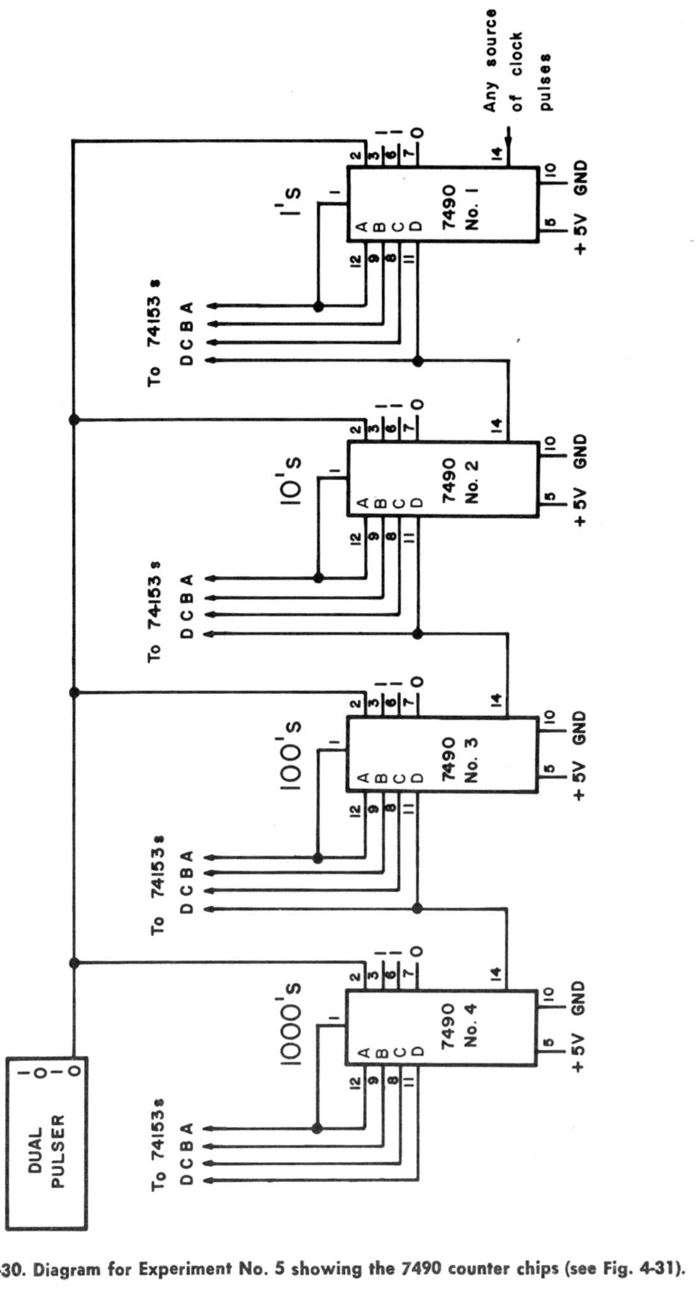

Fig. 4-30. Diagram for Experiment No. 5 showing the 7490 counter chips (see Fig. 4-31).

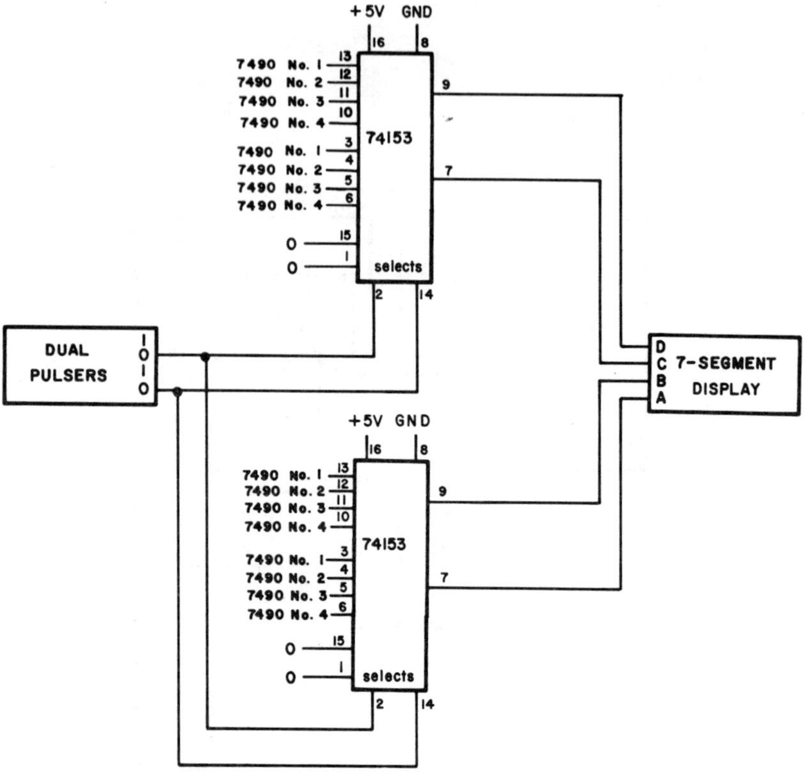

Fig. 4-31. Diagram for Experiment No. 5 showing two 74153 multiplexer chips (see Fig. 4-30).

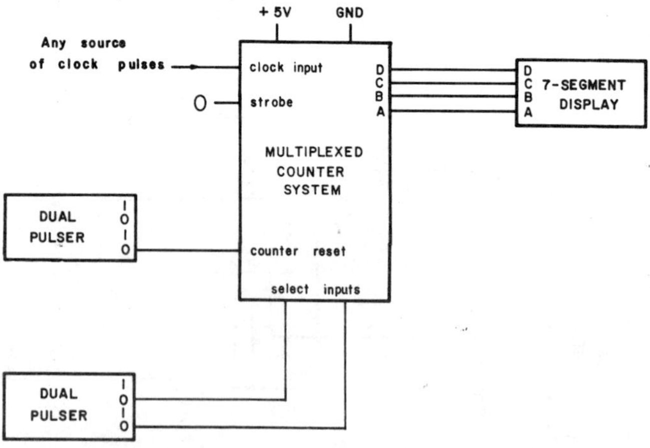

Fig. 4-32. A counter/multiplexer system based on the circuits in Figs. 4-30 and 4-31.

- An *Input/Output circuit diagram* of Figs. 4-30 and 4-31 combined. In this diagram, which is shown in Fig. 4-32, we show only the input and output connections, and not the many connections between the four counters and two multiplexers. This type of diagram is assuming increasing significance today since it points the way to how a more complex integrated-circuit chip can be manufactured. Note that in this diagram, we only require eleven connections; such a circuit, assuming that it can be manufactured by the sophisticated semiconductor techniques now available, would fit on a single 14-pin DIP chip! The system itself is called the *counter/multiplexer system.*

Step 1

With the power to the two breadboards off, make the wire connections shown in Figs. 4-30 and 4-31. You will be unable to prevent a "jungle" of wires from developing. Use your long-nosed pliers to help make some of the wire connections. Feel free to place the pulsers and seven-segment LED display anywhere on the two breadboarding sockets.

Step 2

You now require a "source of clock pulses." Any source will do, but we recommend the gated clock circuit given in Experiment No. 3. With the power off, add this simple circuit to your breadboard system.

Step 3

Adjust the frequency of the module Clock Outboard with the aid of a 0.03-microfarad capacitor, which should set the clock frequency to approximately six counts/second. Turn the power on and push the reset pulser in. (The reset pulser applies a "logic 1" state to pin 2 of the four cascaded 7490 chips, thus resetting the four counters to 0000.) Release the reset pulser and observe that the seven-segment LED display begins to count at a rate of approximately six counts per second.

Step 4

Now apply a "logic 1" state at the two input pins 14 to the pair of 74153 data selector/multiplexer chips. You should now observe a much slower counting rate, approximately six counts every ten seconds. If you have performed this experiment correctly, you should observe the output from 7490 counter No. 2 as long as you continue to press only one of the dual pulsers, the pulser to pins 14 of the 74153 chips.

Step 5

It is possible to select any one of the four 7490 decade counters by applying appropriate logic states to the select inputs at pins 2 and 14 of the 74153 chips. A simple truth table follows:

	Select Inputs	
	pin 2	*pin 14*
7490 counter No. 1, 1's count	0	0
7490 counter No. 2, 10's count	0	1
7490 counter No. 3, 100's count	1	0
7490 counter No. 4, 1000's count	1	1

Thus, if you leave the dual pulsers alone, you automatically select 7490 counter No. 1 since the pulser outputs apply "logic 0" states to pins 2 and 14. On the other hand, if you press both pulsers, you select 7490 counter No. 4, the 1000's count. In this way, you eliminate the necessity of purchasing four separate seven-segment LED readouts; only one will do provided that you have a pair of 74153 chips handy.

Select 7490 counter No. 3 and determine its counting rate. If you have performed this experiment correctly, you should observe approximately six counts every 100 seconds, or approximately four counts a minute.

Step 6

Select 7490 counter No. 4 and determine its counting rate. You should have to wait at least two minutes to observe the transition from 0 to 1. Remember to reset to 0000 the four-decade counter by pressing the reset pulser before you start each experiment.

Step 7

Vary the timing capacitance on the clock Outboard module, and fill in the following chart with your results.

Save this circuit and continue to Experiment No. 6.

Questions

1. Explain how the two 74153 data selectors/multiplexers actually "multiplex" the outputs from the four 7490 decade counters. Be detailed in your answer to this question.

Timing Capacitance, μF	Frequency, Counts/min	Counts/sec	Frequency^{-1}, Seconds/count	Frequency × Capacitance 10^{-8} Farads/sec
0.00005				
0.0001				
0.0002				
0.0005				
0.001				
0.002				
0.005				
0.01				
0.02				
0.05				
0.10				
0.22				
0.47				
1.0				
2.2				

2. What would happen if the STROBE inputs at pins 1 and 15 on the 74153 chips would be all set to a "logic 1" state? You may wish to try this experiment while the counter is counting.

3. Would a pair of pulsers or a pair of logic switches be more appropriate for the select inputs to the 74153 chips? It may be a matter of preference, but what do you think, and why?

EXPERIMENT NO. 6

Purpose

The purpose of this experiment is to demonstrate the phenomenon of contact bounce and how it can be eliminated. You will wire several simple circuits that produce one or more bounces.

Pin Configuration of Integrated-Circuit Chip (Fig. 4-33)

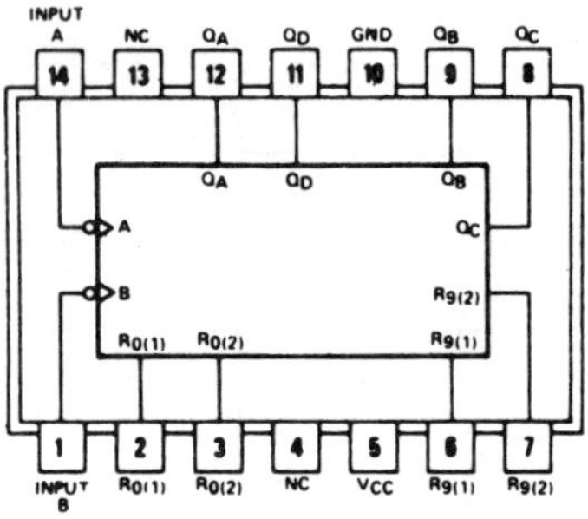

7490

Fig. 4-33.

Schematic Diagrams of Circuits (Figs. 4-34 through 4-36)

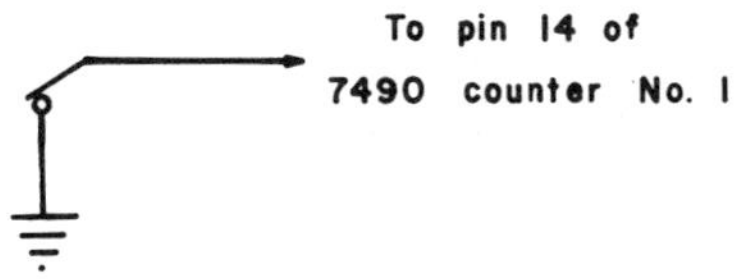

Fig. 4-34.

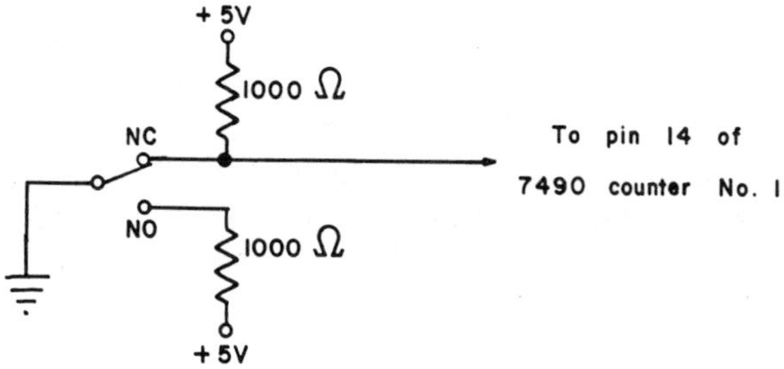

Fig. 4-35.

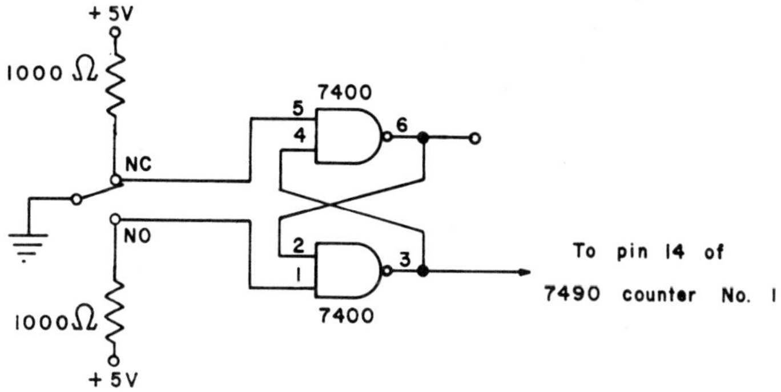

Fig. 4-36.

Step 1

The phenomenon of contact bounce can be represented schematically by the following timing diagram:

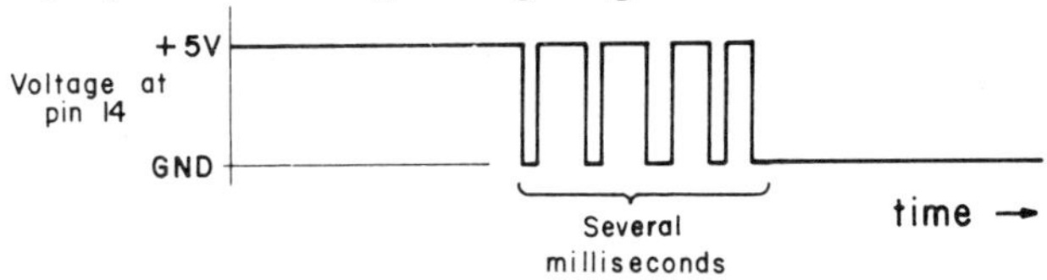

The logic state oscillates between logic 0 and logic 1 until it finally settles down, as in this case, to a logic 0. If you attempt to ground a wire initially at +5 volts, you will observe such a phenomenon if you were able to follow it with a storage oscilloscope.

Step 2

Your multidecade counting circuit should already be wired. It is either the circuit of Experiment No. 2 or the circuit of Experiment No. 4. You will need no more than three decades. Wire the 7490 chips so that you can clear all of them together.

Step 3

Connect a long wire to the input of the multidecade counter and ground it quickly. How many bounces do you detect? Clear the counter and then remove the wire from ground. Do you detect any additional bounces?

Step 4

Repeat the grounding/clearing/ungrounding process until you satisfy yourself that it is difficult to generate one and only one clock pulse through the action of mechanically grounding a wire.

Step 5

Wire the second circuit shown in Fig. 4-35. Use either an spdt switch, or else several wires and resistors to simulate an spdt switch.

Operate the spdt switch, or its equivalent, and demonstrate the fact that there exists contact bounce. Some switches are more "bouncy" than others. We found one that bounced 563 times before settling down to a logic 0 state.

Step 6

Now wire the third circuit shown in Fig. 4-36, which consists of a pair of 2-input NAND gates wired as a simple latch.

Operate the debounced switch circuit and demonstrate that the switch produces only a single pulse when it is pressed and released. Try it many times. In each case, only a single pulse will be generated.

Discussion

Another way to debounce a mechanical switch is to (a) detect when the first contact is made, (b) wait a predetermined period of time, such as 10 ms, and (c) detect whether or not contact continues. A 10-millisecond waiting period is sufficient to permit the bouncing switch to settle down.

Questions

1. Explain what must occur for you to observe a single bounce. HINT: A "bounce" is somehow related to a clock pulse.

2. Draw the appropriate digital waveform for five quick bounces when you ground the CLOCK INPUT A, pin 14, to 7490 decade counter No. 1.

3. It is common for us to state that a pulser or microswitch is "undebounced." What, in your opinion, do we mean by the term, "undebounced?" Try to reason this question through to a correct answer.

4. What other types of mechanical switches exist? Do you expect such switches to exhibit contact bounces when pressed and re-

leased or when switched from position to position? What is a mercury switch and how does it behave?

5. Explain, in full detail, why the pair of 7400 2-input NAND gates succeed in debouncing the spdt microswitch discussed and used in this experiment.

6. Will the following circuit "debounce" an spdt microswitch? If the answer is yes, please explain in full detail how the circuit works.

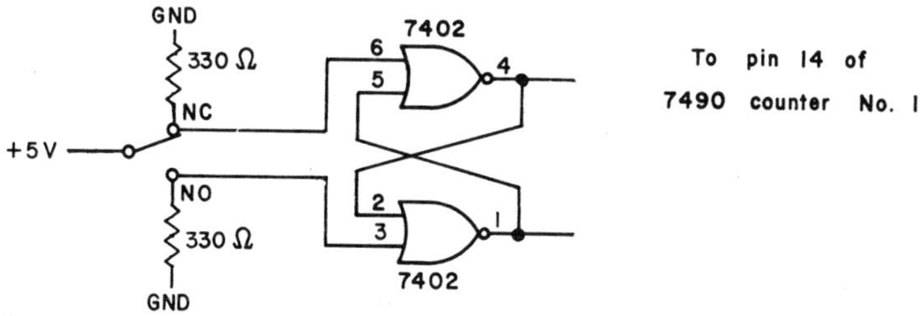

EXPERIMENT NO. 7

Purpose

The purpose of this experiment is to measure the pulse width produced by a 74121 monostable chip.

Pin Configuration of Integrated-Circuit Chip (Fig. 4-37)

Schematic Diagram of Circuit (Fig. 4-38)

Step 1

You will require a multidecade counter for this experiment. One that you probably have already wired is the four-decade counter circuit shown in Fig. 4-39. Wire the clear inputs at pin 2 together and attach the connecting wire either to logic 0 or to a pulser. You will be making repeated measurements, so you will need a convenient counter clearing mechanism.

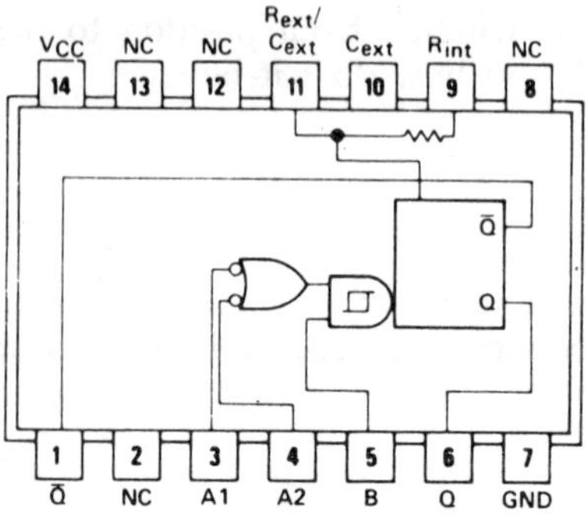

74121

Fig. 4-37.

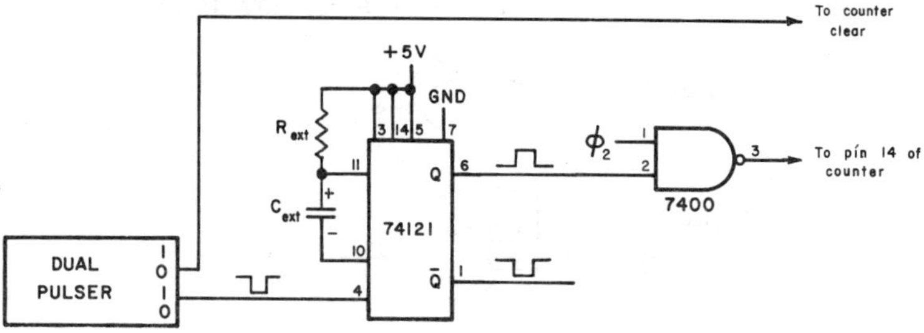

Fig. 4-38.

Step 2

Wire the 74121 monostable circuit shown in the schematic diagram. The clock frequency, Φ_2, can be that of your LR-5 Clock module with no timing capacitor present. For this experiment, our measured frequency was 62.31 kHz, which we set equal to Φ_2.

Keep in mind that you will probably be using ±20% resistors and capacitors. Any pulse width value within 10% of the predicted value is acceptable. Actually, the error in your pulse width could be greater than 20% since both the resistor and the capacitor have such a tolerance. With electrolytic capacitors, the error may be greater, as much as ±50%.

Step 3

Choose $R_{ext} = 10\ k\Omega$ and $C_{ext} = 0.33\ \mu F$. The predicted pulse width can be calculated according to the following equation:

$$t_w = R_{ext}\, C_{ext} \ln 2$$

where,

t_w is the pulse width in seconds,

R_{ext} is the external timing resistance in ohms,

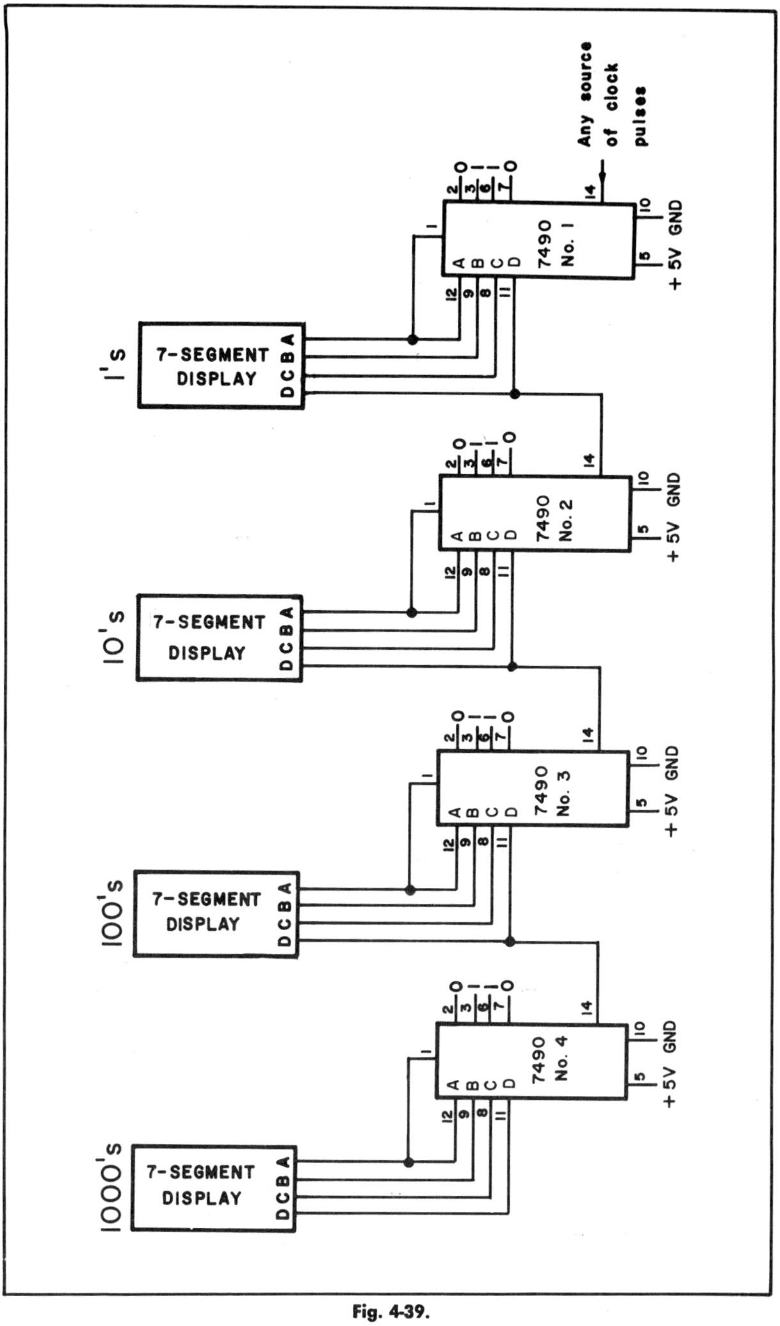
1's
10's
100's
1000's
7-SEGMENT DISPLAY
D C B A
7490 No. 1
7490 No. 2
7490 No. 3
7490 No. 4
Any source of clock pulses
+ 5V GND

Fig. 4-39.

C_{ext} is the external timing capacitance in farads,
ln 2 is the natural logarithm of 2, or 0.693.

The repeatability of the 74121 chip is probably no better than ±0.5%. With the values of R_{ext} and C_{ext} given above, the theoretical value for the pulse width is:

$$t_w = 2.29 \text{ ms}$$

For a clock input of 62.31 kHz to the 2-input NAND gate, the total number of counts that should appear on the counter is:

$$\begin{aligned} \text{Counts} &= 62{,}310 \cdot 0.00229 \\ &= 142 \end{aligned}$$

Step 4

Measure the number of counts for the values of the timing resistance and capacitance given in the preceding step. How many counts do you observe? Repeat the measurement about ten times over a period of five minutes.

We measured 152 counts, although this number varied from 152 to 153 over a period of five minutes. The error in our measurement, when based upon the theoretical value of 142 counts, was 7.0%, within the tolerance of our passive components.

Step 5

Measure the number of counts for a variety of different capacitors, and fill in the table below. Retain R_{ext} at 10 kilohms.

Capacitance, μF	Theoretical counts	Measured counts	Percentage error
0.02			
0.05			
0.10			
0.33			
1.0			
2.2			
3.3			

We obtained the following results for a clock frequency of 62.31 kHz. We show several measurements for each timing capacitor to indicate the variations observed.

Theoretical Counts	Capacitance, μF	Measured counts
8	0.02	8, 8, 9, 8, 8, 8, 9, 9, 8
21	0.05	16, 16, 15, 16, 16, 16, 16, 16
43	0.10 (Mylar®)	43, 43, 43, 44, 43, 44, 43
142	0.33	152, 152, 152, 152, 152, 152, 152
215	0.50 (Mylar®)	213, 212, 212, 212, 212, 212, 212
1424	3.3	1523, 1549, 1551, 1569, 1554, 1554, 1551

When using the 74121 integrated circuit, make certain that your power supply provides at least +5.5 volts to the LR-1 Power module.

Step 6

What is your measured pulse width for the 0.10- and 0.33-μF capacitors?

We calculated pulse widths of 693 μs and 2.29 ms for a clock frequency of 62.31 kHz.

Step 7

What would be the theoretical pulse width if C_{ext} = 50 pF and R_{ext} = 10 kilohms?

Approximately 347 ns.

If you used such a gating pulse to gate a 750-kHz source of clock pulses into your counter, how many counts would you observe?

Either no counts or one count, depending upon whether or not a negative edge appeared at the output of the 2-input NAND gate. With a trial run of fifty pulses, we observed eleven no counts and thirty-nine single counts.

Save this circuit and continue to the following experiment.

Questions

1. Based on your experimental observations, please explain what we mean by a "monostable multivibrator." What does it do? If possible, give your answer in terms of a digital waveform.

2. According to the manufacturer of the 74121 monostable multivibrator, the output pulse length can be as low as 40 nanoseconds. What is a nanosecond? How many nanoseconds are there in a microsecond? In a millisecond? In a second?

3. Calculate the predicted pulse width for $R_{ext} = 2000$ ohms and $C_{ext} = 500$ picofarads.

4. Calculate the predicted pulse width for $R_{ext} = 30{,}000$ ohms and $C_{ext} = 0.01$ microfarad.

EXPERIMENT NO. 8

Purpose

The purpose of this experiment is to demonstrate the use of the 555 Timer chip as a monostable multivibrator. a one shot

Schematic Diagrams of Circuits (Figs. 4-40 and 4-41)

Pin Configurations of Integrated-Circuit Chips (Figs. 4-42 and 4-43)

Step 1

The 555 monostable multivibrator requires an input trigger pulse that is shorter than the pulse that it outputs. Valid pulse widths

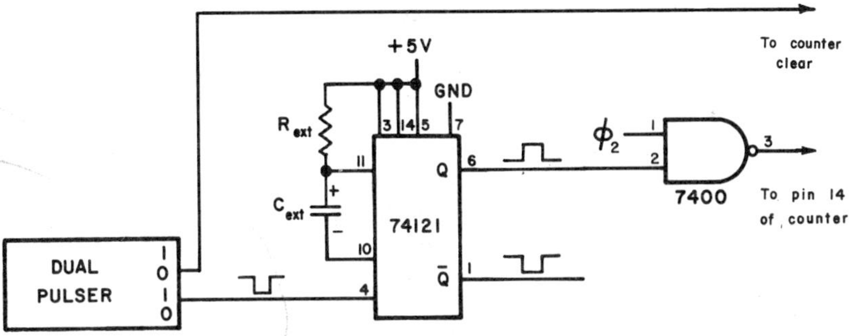

Fig. 4-40.

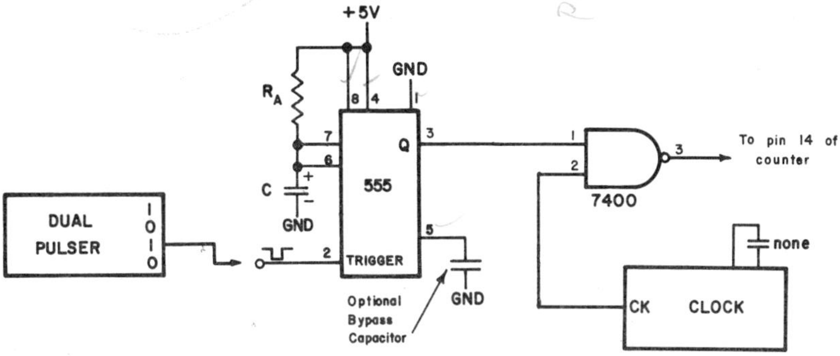

Fig. 4-41.

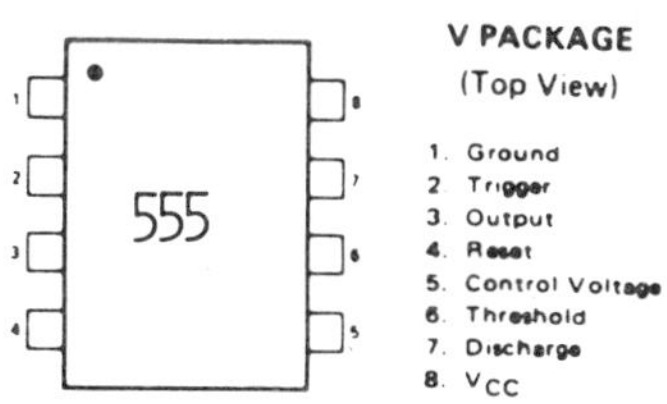

Fig. 4-42.

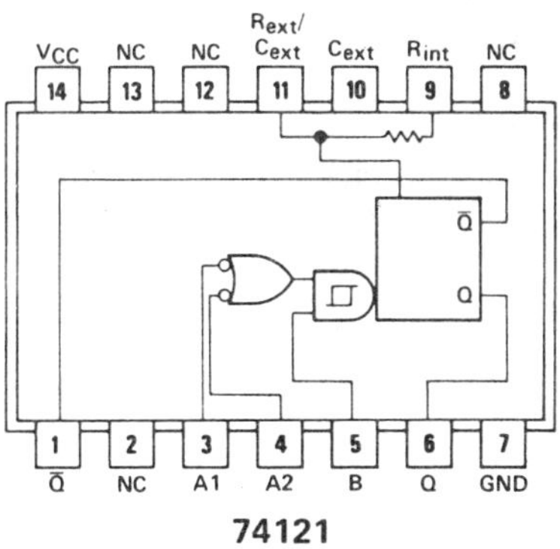

Fig. 4-43.

range from approximately 2 μs to several seconds. In this experiment, you will use a 74121 monostable to trigger the 555 monostable.

Step 2

Wire the circuits shown. Use the following passive components:

74121 monostable: C_{ext} = none
R_{ext} = none
555 monostable: C = 0.33 μF
R_A = 100 kilohms

The clock Outboard module should be used as a source of clock pulses. Connect pin 1 of the 74121 chip to pin 2 of the 555 chip. Do not use the pulser indicated in the diagram for the 555 monostable circuit. You will use the pulser later.

Step 3

The interesting aspect of this experiment is that you are using a 555 monostable to gate clock pulses from a 555 astable (the clock Outboard module) into the multidecade counter. Both the 555 monostable and astable have a very high degree of stability, so you should observe better results than you did in preceding experiments.

Measure the number of counts when you trigger the 74121 monostable. Write your results in the following space. Repeat this experiment several times. 2788, 2787, 2788, 2788, 2789

When we performed this experiment, we measured the following: 2789 counts, 2788, 2789, 2789, 2785, 2789, 2788, 2788, 2788, 2788, 2781, 2788, and 2788.

Step 4

Use the clock module frequency that you determined in Experiment No. 2 to calculate the pulse width of your 555 monostable. Show your calculations in the space below.

Our clock module frequency was 62.31 kHz. The pulse width is therefore:

$$t_w = \frac{\text{Number of counts}}{\text{Frequency}}$$
$$= 2788 \text{ counts}/62{,}310 \text{ Hz}$$
$$= 44.7 \text{ ms}$$

The theoretical pulse width is:

$$t_w = 1.1\, R_A\, C$$
$$= (1.1)(100{,}000 \text{ ohms})(0.33\ \mu F)$$
$$= 36.3 \text{ ms}$$

The agreement is fair. The theoretical formula was obtained from the Signetics Corporation digital/linear/MOS Application Manual.

Step 5

Now eliminate the 74121 circuit and use a pulser to trigger the 555 monostable. Clear the counter. Now press the trigger pulser and keep it pressed. You should observe that the counter continues to count. The reason? The trigger pulse width is greater than the desired output pulse width. This phenomenon will always occur under such conditions.

EXPERIMENT NO. 9

Purpose

The purpose of this experiment is to wire a 555 Timer chip as an astable multivibrator and to measure its frequency.

Schematic Diagram of Circuit (Fig. 4-44)

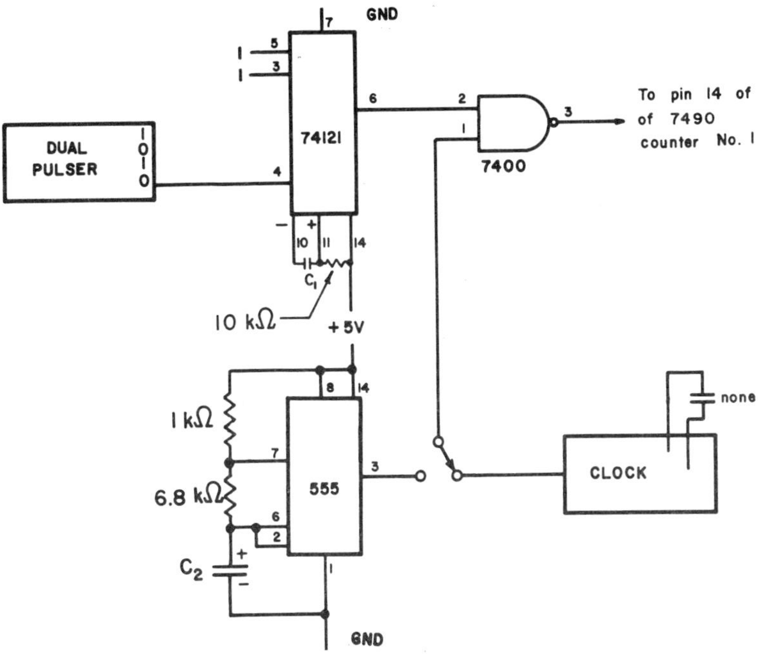

Fig. 4-44.

Pin Configurations of Integrated-Circuit Chips (Figs. 4-45 through 4-47)

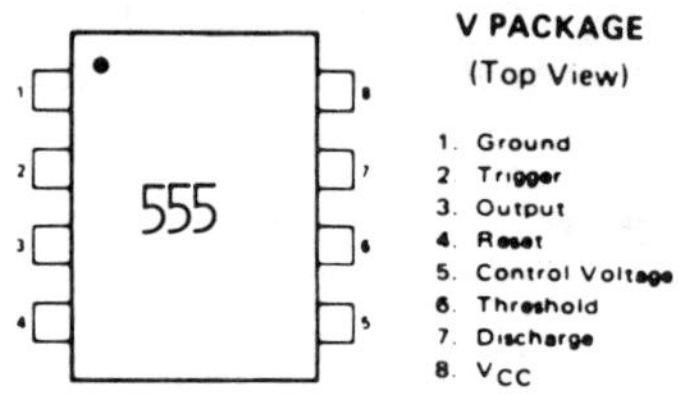

Fig. 4-45.

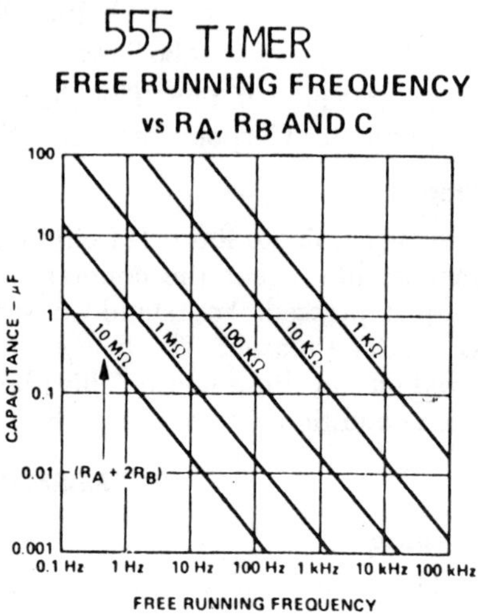

Fig. 4-46.

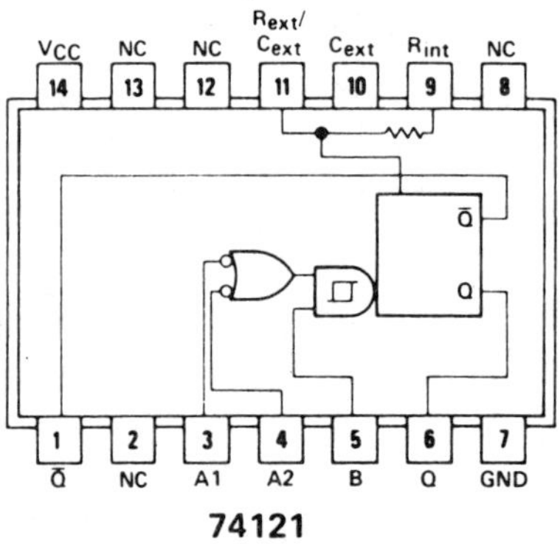

Fig. 4-47.

Step 1

In this experiment, you will wire a clock directly from a 555 Timer chip. The timing resistors are identical to those present on the clock Outboard module.

Wire the circuit and apply power to the breadboard. Set $C_2 = 0$ (no capacitor).

Step 2

Use one of the pulse widths for the 74121 chip that you used in Experiment No. 7. With this pulse width, determine the number of counts with your 555 astable multivibrator as a source of clock pulses.

We used a pulse width of 3.48 ms and observed the following counts: 224, 225, 224, 220, 226, 225, 228.

Step 3

Calculate the frequency of the 555 astable multivibrator.

We calculated a frequency of 64.6 kHz. *Save this circuit.*

TEST

This test probes your understanding of the digital electronic concepts that we have described in this chapter. Please write your answers on a separate piece of paper.

1. Define the term, digital waveform, and give an example of same.
2. In your own words, distinguish between the following sets of terms:
   a. positive clock pulse vs. negative clock pulse
   b. positive edge transition vs. negative edge transition
   c. positive edge vs. negative edge
   d. positive edge vs. leading edge
   e. positive edge vs. trailing edge
   f. positive edge vs. positive leading edge
   g. negative edge vs. negative trailing edge
   h. positive leading edge vs. negative leading edge
   i. negative trailing edge vs. positive trailing edge
   j. rising edge vs. falling edge
   k. positive edge vs. rising edge
   l. negative edge vs. falling edge
3. In your own words, explain what happens when a mechanical switch "bounces."
4. Draw the digital waveforms for the A CLOCK INPUT and the four outputs, A, B, C, and D, from a 7490 decade counter.
5. List four different sources of clock pulses that can be measured by a four-decade counter.
6. Write two different sources of gating pulses that can be used to gate a four-decade counter.
7. Suggest at least one set of values for R_A, R_B, and C that will lead to a frequency of 60 cycles/second (*i.e.*, 60 counts/second) on the 555 Timer.

8. Explain the significance of the graph (Fig. 4-48) for the 555 Timer entitled "Free Running Frequency vs. R_A, and R_B, and C" that appears in the pin configurations section of Experiment No. 8.

Your performance on this test will be acceptable if you can answer all of the above questions correctly in a two-hour closed book examination. You have encountered all of the above concepts in this chapter.

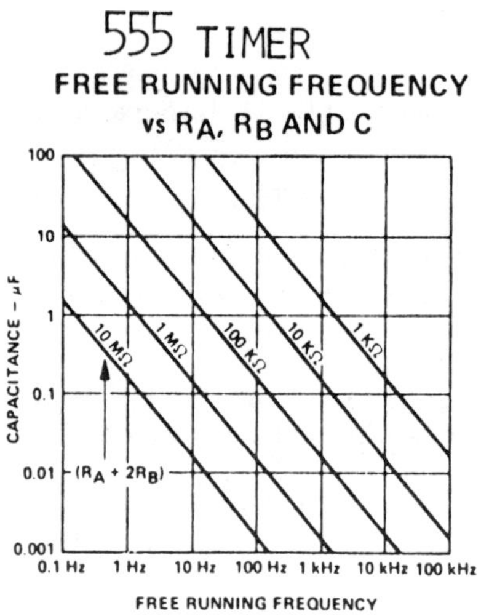

Courtesy Signetics Corp.

Fig. 4-48.

WHAT HAVE YOU ACCOMPLISHED IN THIS CHAPTER?

Review of Objectives

We have stated in the Introduction to this chapter that at the end you should be able to do the following:

1. Construct and operate a four-decade counter consisting of four cascaded 7490 counters.
   The four-decade counter was constructed in Experiment No. 2. You operated it in Experiments Nos. 2 through 4 and 6 through 9.
2. Employ a four-decade counter to measure both frequency and capacitance over a 50,000-fold range.
   You accomplished this task in Experiment No. 2.
3. Count contact bounces when you attempt to ground a wire.
   You observed, in Experiment No. 6, as many as several hundred bounces when you attempted to ground pin 14 of 7490 counter No. 1.

4. Count contact bounces in a single-pole double-throw (spdt) switch.
   You demonstrated this phenomenon in Experiment No. 6.
5. Construct a circuit that eliminates contact bounces in an spdt switch.
   You succeeded in doing so in Experiment No. 6, where you employed a pair of 7400 2-input positive NAND gates.
6. Employ a 555 Timer as a source of clock pulses at frequencies up to 70 kHz.
   You did this in Experiment No. 9.
7. Employ a 74121 monostable multivibrator integrated circuit as a source of single gating pulses.
   You did this in Experiment Nos. 7 and 9.
8. Pass a test in which you are asked to distinguish between different types of clock pulses and clock pulse transitions.
   You have been asked to do this in question No. 2 on the test.
9. Explain, in your own words, what is meant by the term, digital waveform, and give an example.
   You have been asked to do this in question No. 1 on the test.

5

Decoders, Demultiplexers, Multiplexers, and Sequencers

INTRODUCTION

Another important concept in digital electronics is that of a *digital code*. Digital codes are the languages of digital electronic systems, without which it might be difficult, if not impossible, for one device to communicate with another. In this chapter, we shall explore the basis for digital codes, describe five or six of the most important codes used today, and then explain the use of special integrated circuits called *decoders*, which convert one code into another. While on the subject of decoders, we have found it appropriate to also treat the subjects of *demultiplexers, multiplexers,* and *sequencers.* The experiments are slightly more advanced than those given in Chapter 3, but less complex and difficult to wire than those given in Chapter 4.

OBJECTIVES

At the completion of this chapter, you will be able to do the following:

- Pass a test in which you describe the five commonly used digital codes, including *binary, octal, hexadecimal, binary, binary coded decimal,* and *half-ASCII.*

- Pass a test in which you distinguish between the following types of devices:

  multiplexer
  demultiplexer
  sequencer
  programmable sequencer
  decoder

- Pass a test in which you explain the difference between a *decoder* and a *decoder/driver*.
- Construct decoders from 7442, 74154, and 74155 integrated-circuit chips.
- Construct sequencers and programmable sequencers from 7442 and 74154 integrated-circuit chips using 7490 and 7493 counters, respectively.
- Construct demultiplexers from 7442, 74154, and 74155 integrated-circuit chips.
- Construct multiplexers from 74150 and 74153 integrated-circuit chips.

DEFINITIONS

alphanumeric—Pertaining to a character set that contains both letters and numerals, and usually other characteristics.[1]

alphanumeric characters—Used to include numeric digits, alphabetic characters, and special characters.[1]

alphanumeric code—A code used to express numerically the letters of the alphabet.[1]

ASCII code—The American Standard Code for Information Interchange. A 7-bit character code without the parity bit, or an 8-bit character code with the parity bit.

bcd-to-decimal decoder—A decoder that converts binary coded decimal code to a one-of-ten code.

binary code—A code in which each code element is one of two distinct kinds of values, for example, the presence and absence of a pulse.[1]

binary-coded—Expressed by a series of binary digits (0's and 1's).[1]

binary-coded character—A decimal digit, alphabetic letter, punctuation mark, etc., represented by a fixed number of consecutive binary digits.[1]

binary coded decimal—Abbreviated bcd. A system of number representation in which each decimal digit of a number is expressed by binary numbers.[1] Also known as the 8 4 2 1 code.

binary-coded digit—On element of a notation system for representing a decimal digit by a fixed number of binary positions.[1]

binary-coded octal—An octal numbering system in which each octal digit is represented by a three-place binary number.[1]

bit—Abbreviation for binary digit. A unit of information equal to one binary decision, or the designation of one of two possible and equally likely values or states (such as 0 or 1) of anything used to store or convey information. It may also mean "yes" or "no."[1]

block diagram—A diagram in which the essential units of any system are drawn in the form of blocks, and their relationship to each other is indicated by appropriately connected lines.[1]

channel—The path along which information, particularly a series of digits or characters, may flow.[1]

channel sampling rate—The number of times per second that individual channels are sampled.[1]

channel selector—A switch or dial used for selecting a desired channel.[1]

channel sequence number—A number assigned for reference purposes to a channel.[1]

code—A system of symbols that represent data values and make up a special language that a computer or a digital circuit can understand and use.[1]

code conversion—The changing of the bit grouping for a character in one code into the corresponding bit grouping in another code.[1]

code converter—A code conversion device.

decoder—A code conversion device.

decoder/driver—A code conversion device that can also have sufficient voltage or current output to drive an external device such as a display or a lamp monitor.

to decode—To use a code to reverse a previous encoding.[1]

demultiplexer—A digital device that directs information from a single input to one of several outputs. Information for output-channel selection usually is presented to the device in binary weighted form and is decoded internally. The device also acts as a single-pole multiposition switch that passes digital information in a direction opposite to that of a multiplexer.[1]

digital code—See code.

digital waveform—A graphical representation of a digital signal, showing the variations in logic state as a function of time. Also known as a timing diagram.

EBCDIC code—The Extended Binary Coded Decimal Interchange Code, a digital code primarily used by IBM equipment. It closely resembles the half-ASCII code.

to encode—To use a code, frequently one composed of binary numbers, to represent individual characters or groups of characters in a message. To change from one digital code to another. If the codes are greatly different, the process is called code conversion.[1]

excess-three bcd—Excess-three binary-coded decimal. Pertaining to a code based on adding 3 to a decimal digit and then converting the result directly to binary form. Use of this code simplifies the execution of certain mathematical operations in a binary computer that must handle decimal numbers.[1]

Gray code—A positional binary number notation in which any two numbers whose difference is one are represented by expressions that are the same except in one place or column and differ by only one unit in that place or column.[1]

half-ASCII—A 64-character code that contains the code words for numeric digits, alphabetic characters, and symbols but not keyboard operations.

hexadecimal code—A digital code based upon the radix 16, in which the decimal numbers 0 through 9 and the letters A through F represent the sixteen distinct states in the code.

least significant bit (lsb)—The digit with the lowest weighting in a binary number. Abbreviated lsb.[1]

most significant bit (msb)—The digit with the highest weighting in a binary number. Abbreviated msb.[1]

multiplexer, data selector—A digital device that can select one of a number of inputs and pass the logic level of that input on to the output. Information for input-channel selection usually is presented to the device in binary weighted form and is decoded internally. The device acts as a single-pole multiposition switch that passes digital information in one direction only.[1]

octal code—Pertaining to a binary-coded number system with the radix 8, in which the natural binary values of 0 through 7 are used to represent octal digits with values from 0 to 7.[1]

radix—Also called the base. The total number of distinct marks or symbols used in a numbering system. For example, since the decimal numbering system uses ten symbols, the radix is 10. In the binary numbering system, the radix is 2, because there are only two marks or symbols (0 and 1).[1] In the octal numbering system, radix is 8, in the hexadecimal system, the radix is 16.

Selectric code—An alphanumeric code that is used with IBM Selectric typewriters.

sequence (noun)—The order in which objects or items are arranged.[1]

to sequence—To place in order.[1]

sequencer—An electronic device that may be set to initiate a series of events and to make the events follow in sequence, i.e., in order.[1]

sequential operation—The carrying out of operations one after the other.[1]

timing diagram—See digital waveform.

waveform—The graphic representation of an electromagnetic wave, showing the variations in amplitude with time.[1]

DIGITAL CODES

One of the most important characteristics that all animals must possess is the ability to communicate with other animals of the same species. Such communication gives an animal a definite survival advantage, in the Darwinian sense of the term, and is certainly found in most multicellular creatures starting with insects on up to man himself. With insects, there are several modes of communications, including the dance of a bee and forms of chemical communication through remarkable chemical agents known as pheromones. Man can communicate with his kind through all of his five senses, as amply illustrated by the performance of handicapped individuals who have lost one or more of their senses but are nevertheless highly sensitive with the remaining ones.

Assuming that an individual wishes to communicate with another through the sense of hearing and the use of speech, it is clear that there must be some general agreement concerning how a specific spoken sound will be interpreted by the one who hears it. Over the centuries, different regions around the world have separately developed their own consensus regarding the meaning of specific sounds and their transcription onto paper. We call such consensus a "language." Hundreds of different languages exist, although only a relatively small number of them are used by significant percentages of the world's population. In addition, the popularity of a specific language waxes and wanes through the course of civilization. Latin, once a dominant language in Europe, is now considered a "dead" language, although it clearly has influenced most of the European languages in very profound ways. English may well be at the peak of its popularity at the present time, and may conceivably decline several centuries from now.

The object of this brief digression is to develop an analogy between spoken languages and digital languages, more commonly known as *digital codes.* Just as there are numerous spoken languages, there are also a variety of digital codes, including binary, binary coded decimal (bcd), gray, 4 2 2 1, excess 3, 2 out of 5, American Standard Code for Information Interchange (ASCII), and Extended Binary Coded Decimal Interchange Code (EBCDIC). These codes are, in effect, the true "languages" of digital electronics. Such codes can be sub-divided into three important categories:

Category I. Codes employed by electronic circuitry to perform various digital tasks. Examples: binary

Category II. Codes employed to convert the decimal integers 0, 1, 2, 3, 4, 5, 6, 7, 8, and 9 into digital form. Examples: binary coded decimal (bcd), gray, 4 2 2 1, 2 out of 5, 4 of 8, and biquinary 5 1 1 1 1.

Category III. Codes employed to convert decimal integers, the 26-letter English alphabet, symbols, and keyboard operations into digital form. Examples: ASCII, EBCDIC, and Baudot

In this chapter, we shall pay particular attention to three codes: binary, binary coded decimal (bcd) or 8 4 2 1, and ASCII.

BINARY CODE

The simplest digital code is a two-state, or *binary*, code that consists of a "logic 0" and a "logic 1" state. From discussions given in previous chapters, it is clear that our breadboarding system is designed to operate in binary code, where a "logic 0" state is at GROUND potential and a "logic 1" state is at +5 Volts.

In binary code, the decimal number 0 is represented by a "logic 0" and the decimal number 1 is represented by a "logic 1." This should be quite clear. How, on the other hand, do we represent higher decimal numbers, such as 3, 17, 568, etc., using binary code? The answer is that we generate numbers larger than decimal 1 through the use of a binary counting system that is based upon a *base*, or *radix*, of two. For example, the binary number 11101_2, where the subscript "$_2$" represents the binary system, is equivalent to, reading from left to right,

$$11101_2 = (1 \times 2^4) + (1 \times 2^3) + (1 \times 2^2) + (0 \times 2^1) + (1 \times 2^0) = 29_{10}$$

where you should keep in mind that

$$2^4 = 16 \text{ in decimal notation} = 16_{10}$$
$$2^3 = 8 \text{ in decimal notation} = 8_{10}$$
$$2^2 = 4 \text{ in decimal notation} = 4_{10}$$
$$2^1 = 2 \text{ in decimal notation} = 2_{10}$$
$$2^0 = 1 \text{ in decimal notation} = 1_{10}$$

Thus, $11101_2 = 16_{10} + 8_{10} + 4_{10} + 0 + 1_{10} = 29_{10}$, where the subscript "$_{10}$" associated with these numbers represents the decimal system, a system that is based upon a base, or radix, of 10. Let us make a brief table that would allow us to convert simple decimal numbers into binary form (Table 5-1).

Thus, a sequence of four binary digits can represent any of sixteen different decimal numbers ranging from zero to fifteen. Decimal numbers larger than fifteen require more binary digits, as shown in Table 5-2. It is also useful to note the implications of Table 5-3.

Thus, a 12-*bit* binary number can *encode* four thousand and ninety-six different decimal numbers, ranging from 0 to 4095_{10}, into

Table 5-1.

Decimal Number	Binary Number
0	0000
1	0001
2	0010
3	0011
4	0100
5	0101
6	0110
7	0111
8	1000
9	1001
10	1010
11	1011
12	1100
13	1101
14	1110
15	1111
16	10000
17	10001
18	10010
19	10011
20	10100
21	10101
22	10110
23	10111
24	11000

binary form. A 3-bit binary number can only encode eight decimal numbers ranging from 0 to 7_{10}, whereas a 4-bit binary number can encode sixteen decimal numbers ranging from 0 to 15_{10}.

Table 5-2.

Decimal Number	Binary Number
0	0
1	1
2	10
4	100
8	1000
16	10000
32	100000
64	1000000
128	10000000
256	100000000
512	1000000000
1024	10000000000
2048	100000000000

Table 5-3.

Decimal Number	Binary Number
1	1
3	11
7	111
15	1111
31	11111
63	111111
127	1111111
255	11111111
511	111111111
1023	1111111111
2047	11111111111
4095	111111111111

Table 5-4.

Octal Number	Binary Number
0	000
1	001
2	010
3	011
4	100
5	101
6	110
7	111

Table 5-5.

Hexadecimal Number or Letter	Binary Number
0	0000
1	0001
2	0010
3	0011
4	0100
5	0101
6	0110
7	0111
8	1000
9	1001
A	1010
B	1011
C	1100
D	1101
E	1110
F	1111

OCTAL AND HEXADECIMAL CODE

Two commonly used binary codes are the *octal code* and the *hexadecimal code*. The complete octal code is given in Table 5-4.

Clearly, octal code is a 3-bit binary code. Hexidecimal code, which is becoming ever more popular with the advent of 16-bit minicomputers and other digital electronic systems, is given by Table 5-5.

Thus, hexadecimal code is a 4-bit binary code. We employ the letters A through F instead of the decimal numbers 10 through 15 so that we can use LED displays that read only a single letter or digit. On the commonly used seven-segment LED display, the decimal numbers between 10_{10} and 15_{10} are represented by unusual symbols, as shown in Table 5-6.

Table 5-6.

Decimal Number	LED Symbol
10	
11	
12	
13	
14	
15	blank

BINARY CODED DECIMAL

One of the most popular digital codes for encoding decimal numbers is the 8 4 2 1 code, otherwise known as *binary coded decimal*

Table 5-7.

Decimal Number	Binary Coded Decimal
0	0000
1	0001
2	0010
3	0011
4	0100
5	0101
6	0110
7	0111
8	1000
9	1001

(*bcd*). Table 5-7 relates decimal numbers to the binary coded decimal system. Clearly, binary coded decimal is a simple 4-bit binary code that starts at 0000_2 and terminates at 1001_2. Each decimal digit is encoded by four binary bits, as shown for the decimal numeral nine:

	2^3	2^2	2^1	2^0
decimal 9 =	1	0	0	1
Notation on IC chip =	D	C	B	A
Other notation =	msb			lsb

We have given two notation systems for binary coded decimal that merit discussion. First, we have the DCBA system, where D corresponds to the logic state for the binary bit in the 2^3 position, C corresponds to the logic state for the binary bit in the 2^2 position, B corresponds to the logic state for the binary bit in the 2^1 position, and A corresponds to the logic state for the binary bit in the 2^0 position. The second notation system is the lsb/msb system, where lsb stands for *least significant bit* and *msb* stands for *most significant bit*. Clearly, the lsb is A and the msb is D in a binary coded decimal number.

Binary coded decimal is a convenient but rather wasteful code for representing decimal numbers. Too many binary bits are required to represent a decimal number as, for example, the decimal number 4095_{10}, which requires not 12 bits as is the case for the binary system, but rather 16 bits to be encoded in binary coded decimal,

$$4095_{10} = 111111111111_2 = \underbrace{0100}_{4}\ \underbrace{0000}_{0}\ \underbrace{1001}_{9}\ \underbrace{0101}_{5} \text{ in bcd}$$

THE VERBS: TO ENCODE AND TO DECODE

Although almost all electronic circuits operate with two logic states and, therefore, perform operations using binary arithmetic, it is clear that individuals, for historical and other reasons, prefer to work with decimal arithmetic, in which the integers 0, 1, 2, 3, 4, 5, 6, 7, 8, and 9 are used, along with powers of the base 10, to represent any number desired. In view of this fact, it is quite advantageous to *encode* a decimal number into binary form, have the electronic circuitry operate upon the resulting binary number and, finally, when a readout or output is desired, *decode* the binary numbers generated back into decimal form. While a computer both encodes decimal numbers and decodes binary numbers stored in core, a laboratory instrument needs only to decode binary information.

The terms, *to encode* and *to decode*, can be defined in the following way:

to encode—To use a code, frequently one composed of binary numbers, to represent individual characters or groups of characters in a message. To change from one digital code to another. If the codes are greatly different, the process is called conversion.[1]

to decode—To use a code to reverse a previous encoding. To determine the meaning of a set of pulses that describes an instruction, a command, or an operation to be carried out.[1]

ALPHANUMERIC CODES

You need not restrict the encoding and decoding process simply to decimal numbers. It is equally important to be able to encode the entire English alphabet into digital form, as is done, for example by a teletypewriter. When one encodes the English alphabet, it is always the case that spare positions in the *alphanumeric digital code* exist. These are used to encode such symbols as !, @, #, $, %, ¢, &, *, (,), −, +, =, ", ', etc., as well as certain keyboard operations such as "tab return," "back space," "shift," "carriage return," "space," etc. *The object of a digital code, therefore, is to encode decimal numbers, alphabets, control functions, keyboard operations, and/or symbols in digital form,* a task which is certainly feasible in the United States, Europe, or Latin America, but considerably more difficult in China.

ASCII CODE

The most popular alphanumeric code for the transmission of the twenty-six letters of the alphabet, the ten decimal numerals, twenty-eight symbols, and a variety of keyboard functions is known as the *ASCII code,* otherwise known as the *American Standard Code for Information Interchange.* The ASCII code is a 7-bit binary code that has the potential of encoding $2^7 = 128$ different pieces of information.

Given in Table 5-8 is a half-ASCII table, *i.e.,* a listing of the 64 7-bit ASCII code words for the letters of the English alphabet, numerals, and symbols. The remaining 64 ASCII code words are keyboard operations or the lower case alphabet.

Note that the ASCII code is even more inefficient than binary coded decimal for encoding decimal numerals. For example, the decimal numeral nine in the ASCII code is 011 1001, which compares unfavorably with 1001_2 in either binary or binary coded decimal. Maximum efficiency, however, is not always the most iimportant aspect of a digital code. Properties such as compatibility, utility, and readability can be more important in many situations.

In Table 5-8, the least significant bit (lsb) is on the far right and the most significant bit (msb) is on the far left. For example, in the ASCII word for the letter T, the msb is logic 1 and the lsb is logic 0.

Table 5-8.

Character	ASCII code	Character	ASCII Code
@	100 0000	space	010 0000
A	100 0001	!	010 0001
B	100 0010	"	010 0010
C	100 0011	#	010 0011
D	100 0100	$	010 0100
E	100 0101	%	010 0101
F	100 0110	&	010 0110
G	100 0111	'	010 0111
H	100 1000	(	010 1000
I	100 1001	)	010 1001
J	100 1010	*	010 1010
K	100 1011	+	010 1011
L	100 1100	,	010 1100
M	100 1101	—	010 1101
N	100 1110	.	010 1110
O	100 1111	/	010 1111
P	101 0000	0	011 0000
Q	101 0001	1	011 0001
R	101 0010	2	011 0010
S	101 0011	3	011 0011
T	101 0100	4	011 0100
U	101 0101	5	011 0101
V	101 0110	6	011 0110
W	101 0111	7	011 0111
X	101 1000	8	011 1000
Y	101 1001	9	011 1001
Z	101 1010	:	011 1010
[	101 1011	;	011 1011
\	101 1100	<	011 1100
]	101 1101	=	011 1101
^	101 1110	>	011 1110
←	101 1111	?	011 1111

Other popular alphanumeric codes include the *Selectric code* and the *Extended Binary Coded Decimal Interchange Code* (*EBCDIC* code). Both are used predominantly by IBM machines. Fairchild Semiconductor manufactures MOS chips that convert ASCII to either Selectric or EBCDIC, or vice versa (Fairchild chip numbers 3512A and 3514A/3514B).

CODE CONVERSION

In the United Nations, we have human "interpreters" who "translate" one language into another and thus permit the representatives of the more than one hundred member countries to communicate with each other. In the field of digital electronics, we have a similar requirement to "translate" one digital code into another. The process is called *code conversion,* which has been defined by Graf as the changing of the bit grouping for a character in one code into the corresponding bit grouping in another code.[1] A code conversion device can be represented as shown in Fig. 5-1, and has the following truth table (Table 5-9), where the rows and columns contain the

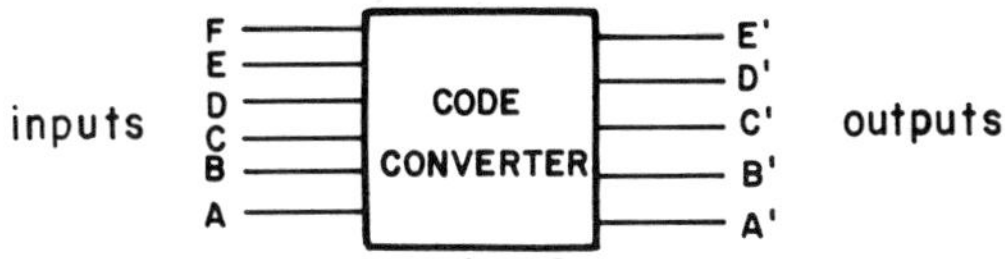

Fig. 5-1.

Table 5-9.

Inputs						Outputs				
F	E	D	C	B	A	E′	D′	C′	B′	A′

appropriate "logic 0" and "logic 1" entries. In the above truth table, there are 64 distinctly different input conditions, not all of which need to be used in a specific code converter.

The 7400 series of integrated-circuit chips contains a variety of very popular code converters, otherwise known as *decoders* or *decoder/drivers.* These include the

- 4-line-to-10-line decoders and decoder/drivers such as the 7441, 7442, 7443, 7444, 7445, 74141, and 74145

- 4-line-to-7-line decoder/drivers such as the 7446, 7447, and 7448
- 2-line-to-4-line decoders such as the 74139, 74155, and 74156
- 3-line-to-8-line decoders such the 74138
- 4-line-to-16-line decoders such as the 74154 and 74159

We shall now in turn describe each of these different types of decoders.

4-Line-to-10-Line Decoders

The specific purpose of a 4-line-to-10-line decoder is to convert a 4-bit binary word into one of ten outputs. The most common such decoder is one that converts 8 4 2 1 binary-coded decimal (bcd) into 1-of-10 output lines. The truth table for such a decoder is given in Table 5-10.

Table 5-10.

Inputs				Outputs									
D	C	B	A	0	1	2	3	4	5	6	7	8	9
0	0	0	0	0	1	1	1	1	1	1	1	1	1
0	0	0	1	1	0	1	1	1	1	1	1	1	1
0	0	1	0	1	1	0	1	1	1	1	1	1	1
0	0	1	1	1	1	1	0	1	1	1	1	1	1
0	1	0	0	1	1	1	1	0	1	1	1	1	1
0	1	0	1	1	1	1	1	1	0	1	1	1	1
0	1	1	0	1	1	1	1	1	1	0	1	1	1
0	1	1	1	1	1	1	1	1	1	1	0	1	1
1	0	0	0	1	1	1	1	1	1	1	1	0	1
1	0	0	1	1	1	1	1	1	1	1	1	1	0
1	0	1	0	1	1	1	1	1	1	1	1	1	1
1	0	1	1	1	1	1	1	1	1	1	1	1	1
1	1	0	0	1	1	1	1	1	1	1	1	1	1
1	1	0	1	1	1	1	1	1	1	1	1	1	1
1	1	1	0	1	1	1	1	1	1	1	1	1	1
1	1	1	1	1	1	1	1	1	1	1	1	1	1

It should be quite clear why this type of decoder is called a "1-of-10" decoder. For any of the ten decimal numerals between 0 and 9, only one of the outputs is at a "logic 0" state; the remaining nine outputs are all at "logic 1." The 7441, 7442, 7445, 74141, and 74145 decoders all fall into this category. The main difference between such decoders is simply the ability of the ten outputs to tolerate either low (+5 volts) or high (+5, +30, or +70 volts) output voltages. These decoders are frequently called *bcd-to-decimal decoders* or *bcd-to-decimal decoder/drivers*, the term "decoder/driver" being reserved for those decoders whose outputs can tolerate high voltages.

The two remaining 4-line-to-10-line decoders are known as the *excess-3-to-decimal decoder* and the *excess-3-gray-to-decimal decoder*. The "excess-3" and "excess-3-gray" codes are other forms of binary coded decimal that are used in special situations. A combined truth table for these two decoders is given in Table 5-11.

Table 5-11.

Inputs								Outputs									
excess-3				excess-3-gray													
D	C	B	A	D	C	B	A	0	1	2	3	4	5	6	7	8	9
0	0	1	1	0	0	1	0	0	1	1	1	1	1	1	1	1	1
0	1	0	0	0	1	1	0	1	0	1	1	1	1	1	1	1	1
0	1	0	1	0	1	1	1	1	1	0	1	1	1	1	1	1	1
0	1	1	0	0	1	0	1	1	1	1	0	1	1	1	1	1	1
0	1	1	1	0	1	0	0	1	1	1	1	0	1	1	1	1	1
1	0	0	0	1	1	0	0	1	1	1	1	1	0	1	1	1	1
1	0	0	1	1	1	0	1	1	1	1	1	1	1	0	1	1	1
1	0	1	0	1	1	1	1	1	1	1	1	1	1	1	0	1	1
1	0	1	1	1	1	1	0	1	1	1	1	1	1	1	1	0	1
1	1	0	0	1	0	1	0	1	1	1	1	1	1	1	1	1	0
1	1	0	1	1	0	1	1	1	1	1	1	1	1	1	1	1	1
1	1	1	0	1	0	0	1	1	1	1	1	1	1	1	1	1	1
1	1	1	1	1	0	0	0	1	1	1	1	1	1	1	1	1	1
0	0	0	0	0	0	0	0	1	1	1	1	1	1	1	1	1	1
0	0	0	1	0	0	0	1	1	1	1	1	1	1	1	1	1	1
0	0	1	0	0	0	1	1	1	1	1	1	1	1	1	1	1	1

Notice that in all three truth tables, bcd, excess-3, and excess-3-gray, the outputs are identical. The "unique" output state, in this number corresponds to the binary code. In binary coded decimal, output line 9 is at a "logic 0" state only when the DCBA inputs are 1001_2. Output line 9 is also at a "logic 0" state when DCBA is 1100_2 in the excess-3 code or when DCBA is 1010_2 in the excess-3-gray code. Whichever decoder is used—and the 7442 is the most popular—it is clear that you can use a 4-bit binary code to choose one of ten distinct output lines. In the Experiments section, you shall explore various uses for this capability.

Owing to its low cost, widespread availability, and utility in simple TTL circuits, we shall only consider the 7442 bcd-to-decimal decoder, the diagram and pin configuration of which are shown in Figs. 5-2 and 5-3.

4-Line-to-16-Line Decoders

Since it is possible to decode a 4-bit binary input into any one of sixteen different output lines, it is not surprising that there is a chip that performs this type of operation. The truth table for the

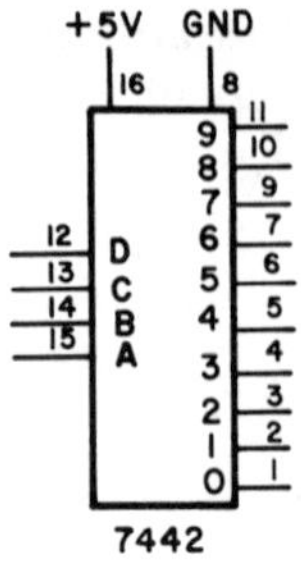

Fig. 5-2.

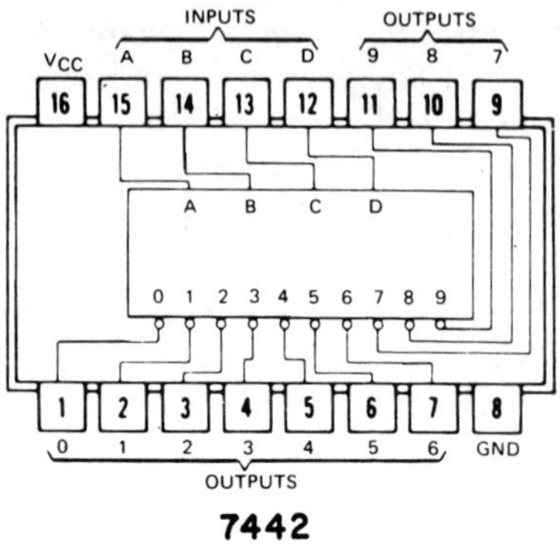

Fig. 5-3.

74154 4-line-to-16-line decoder/demultiplexer is given in Table 5-12. Again, it should be clear why we call this chip a "1-of-16" decoder.

Table 5-12.

Inputs				Outputs															
D	**C**	**B**	**A**	**0**	**1**	**2**	**3**	**4**	**5**	**6**	**7**	**8**	**9**	**10**	**11**	**12**	**13**	**14**	**15**
0	0	0	0	0	1	1	1	1	1	1	1	1	1	1	1	1	1	1	1
0	0	0	1	1	0	1	1	1	1	1	1	1	1	1	1	1	1	1	1
0	0	1	0	1	1	0	1	1	1	1	1	1	1	1	1	1	1	1	1
0	0	1	1	1	1	1	0	1	1	1	1	1	1	1	1	1	1	1	1
0	1	0	0	1	1	1	1	0	1	1	1	1	1	1	1	1	1	1	1
0	1	0	1	1	1	1	1	1	0	1	1	1	1	1	1	1	1	1	1
0	1	1	0	1	1	1	1	1	1	0	1	1	1	1	1	1	1	1	1
0	1	1	1	1	1	1	1	1	1	1	0	1	1	1	1	1	1	1	1
1	0	0	0	1	1	1	1	1	1	1	1	0	1	1	1	1	1	1	1
1	0	0	1	1	1	1	1	1	1	1	1	1	0	1	1	1	1	1	1
1	0	1	0	1	1	1	1	1	1	1	1	1	1	0	1	1	1	1	1
1	0	1	1	1	1	1	1	1	1	1	1	1	1	1	0	1	1	1	1
1	1	0	0	1	1	1	1	1	1	1	1	1	1	1	1	0	1	1	1
1	1	0	1	1	1	1	1	1	1	1	1	1	1	1	1	1	0	1	1
1	1	1	0	1	1	1	1	1	1	1	1	1	1	1	1	1	1	0	1
1	1	1	1	1	1	1	1	1	1	1	1	1	1	1	1	1	1	1	0

The diagram and pin configuration are, respectively, shown in Figs. 5-4 and 5-5. The two "strobe" inputs G1 and G2 must be at a "logic 0" state for the 74154 chip to properly function as a 4-line-to-16-line decoder. These two strobe inputs are also important whenever the chip is employed as a "demultiplexer," a subject that we shall discuss shortly.

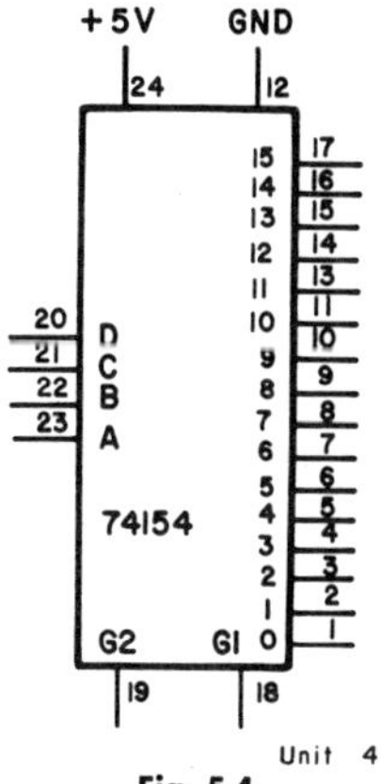

Fig. 5-4.

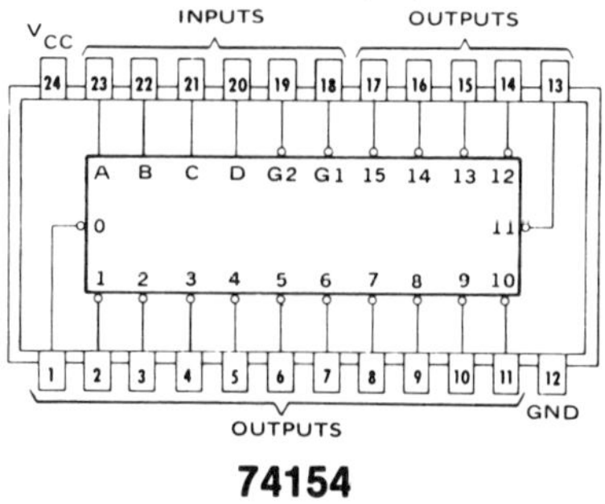

Fig. 5-5.

3-Line-to-8-Line Decoders

Although the 74138 integrated circuit is listed as a 3-line-to-8-line decoder/demultiplexer, it is rarely used. Instead, the 7442 bcd-to-decimal decoder is converted into a 3-line-to-8-line decoder by the simple expedient of grounding the input D at pin 12. The representation for this specific use of the 7442 chip is shown in Fig. 5-6.

The truth table for the 7442 integrated circuit reduces, for D = 0, to that for a simple 3-line-to-8-line decoder (Table 5-13). It is also possible to convert the 74154 integrated circuit to a 3-line-to-8-line decoder by grounding the D input at pin 20.

Fig. 5-6.

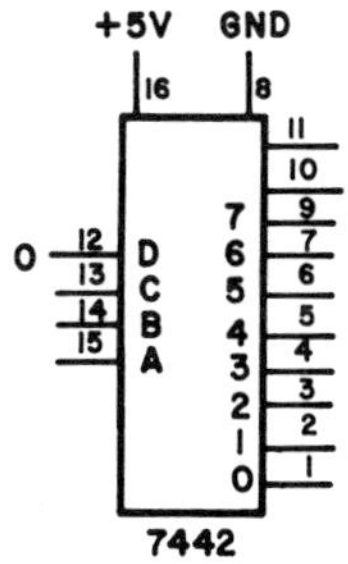

2-Line-to-4-Line Decoders

A simple 2-line-to-4-line decoder can be constructed from a 7442 chip by grounding both the C and D inputs, as shown in Fig. 5-7.

The corresponding truth table is given in Table 5-14. Of greater interest are the 74155/74156 dual 2-line-to-4-line decoder/demulti-

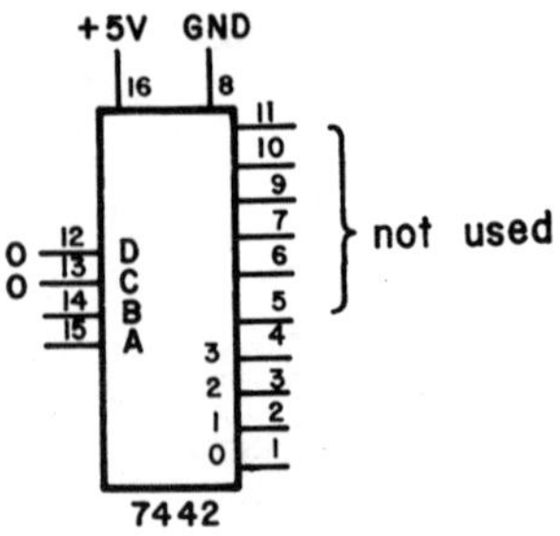

Fig. 5-7.

Table 5-13.

Inputs			Outputs							
C	B	A	0	1	2	3	4	5	6	7
0	0	0	0	1	1	1	1	1	1	1
0	0	1	1	0	1	1	1	1	1	1
0	1	0	1	1	0	1	1	1	1	1
0	1	1	1	1	1	0	1	1	1	1
1	0	0	1	1	1	1	0	1	1	1
1	0	1	1	1	1	1	1	0	1	1
1	1	0	1	1	1	1	1	1	0	1
1	1	1	1	1	1	1	1	1	1	0

Table 5-14.

B	A	0	1	2	3
0	0	0	1	1	1
0	1	1	0	1	1
1	0	1	1	0	1
1	1	1	1	1	0

plexer integrated circuits, which have the pin configuration shown in Fig. 5-8, and can be diagrammed in the manner shown in Fig. 5-9.

Note in Fig. 5-9 that only a single set of inputs, A and B, are used in both 2-line-to-4-line decoders. The inputs 1C, 1G, 2C, and 2G must all be at a "logic 0" state for you to observe simple decoder behavior. Otherwise, each decoder functions as a 1-line-to-4-line demultiplexer.

4-Line-to-7-Line Decoders

Other than the 7442 integrated circuit, the most important decoder chip is the 4-line-to-7-line decoder/driver, otherwise known as the *bcd-to-seven-segment decoder/driver.* Such an integrated circuit converts binary coded decimal information to seven-segment coded

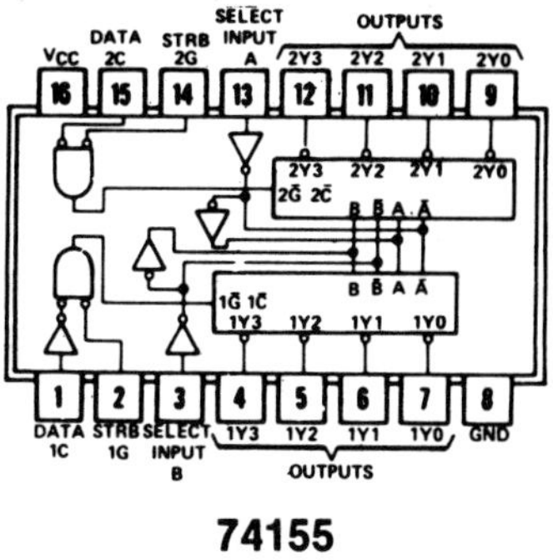

Fig. 5-8.

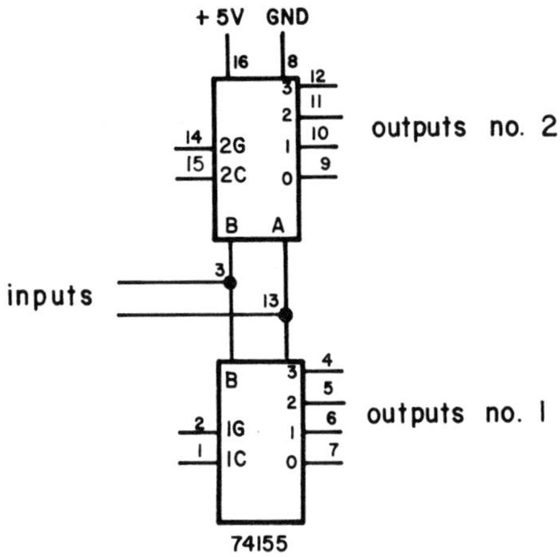

Fig. 5-9.

information that is suitable for driving a seven-segment LED display to produce the decimal numerals 0, 1, 2, 3, 4, 5, 6, 7, 8, and 9 directly.

We should first demonstrate that it is possible to "simulate" a decimal numeral between 0 and 9 using only seven straight segments. Let us consider the following seven-segment symbol, where the individual segments are labeled a, b, c, d, e, f, and g, as shown in Fig. 5-10. The numeral zero, 0, should be easy: only segments a, b, c, d, e, and f are required (Fig. 5-11). The nunmeral one, 1, is even simpler, requiring only segments b and c (Fig. 5-12). The number two requires segments a, b, g, e, and d (Fig. 5-13).

In a similar manner, we can write the numbers 3 through 9 as shown in Fig. 5-14.

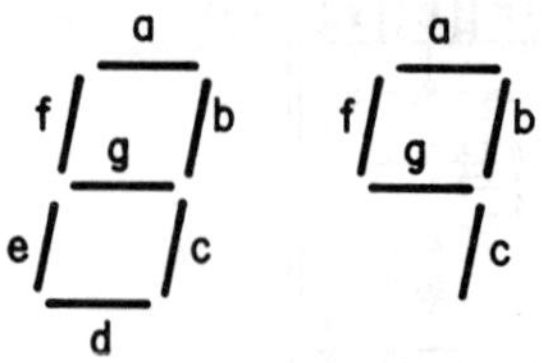

Fig. 5-10.

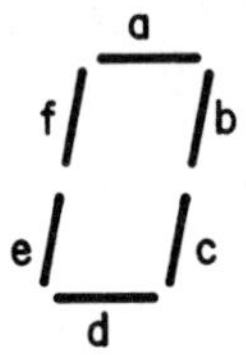

Fig. 5-11.

Fig. 5-12.

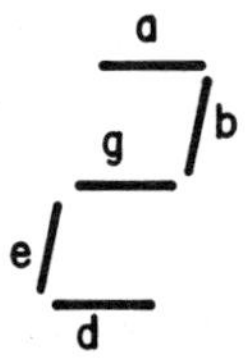

Fig. 5-13.

Let us now write a "truth table" for this type of display. For reasons that we shall not discuss here, we shall write a "logic 0" each time we use a segment, and a "logic 1" each time the segment is absent from the representation of the decimal numeral. The truth table is given in Table 5-15. Note that we have also listed the bcd form for the decimal numeral in the truth table.

The unusual characteristic of a typical 4-line-to-7-segment decoder/driver, the 7447 integrated circuit, can be immediately seen in the truth table that it is not a 1-of-7 decoder! One set of input conditions may have only a single unlit segment, whereas another

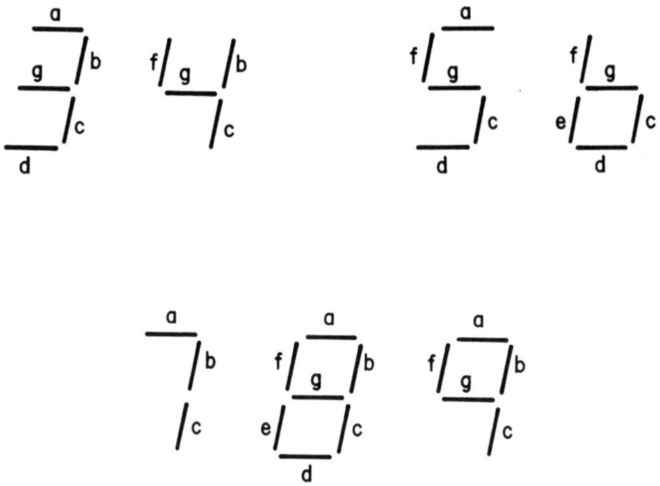

Fig. 5-14.

Table 5-15.

Decimal Numeral	Inputs				Outputs						
	D	C	B	A	a	b	c	d	e	f	g
0	0	0	0	0	0	0	0	0	0	0	1
1	0	0	0	1	1	0	0	1	1	1	1
2	0	0	1	0	0	0	1	0	0	1	0
3	0	0	1	1	0	0	0	0	1	1	0
4	0	1	0	0	1	0	0	1	1	0	0
5	0	1	0	1	0	1	0	0	1	0	0
6*	0	1	1	0	1*	1	0	0	0	0	0
7	0	1	1	1	0	0	0	1	1	1	1
8	1	0	0	0	0	0	0	0	0	0	0
9*	1	0	0	1	0	0	0	1*	1	0	0

* There are two different ways to represent the numerals 6 and 9. Segment **a** in numeral 6 and segment **d** in numeral 9 could be lit.

set of input conditions, e.g., DCBA = 0001_2, may have as many as five unlit segments.

The pin configuration and diagram of the 7447 bcd-to-seven-segment decoder/driver are shown in Figs. 5-15 and 5-16. The pin

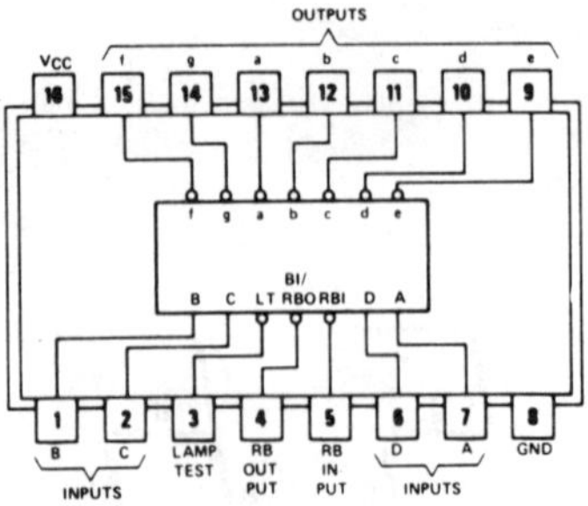

7447

Fig. 5-15.

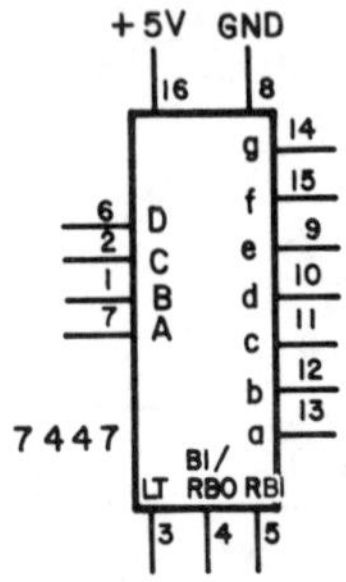

Fig. 5-16.

configuration for a related chip, the 7448 bcd-to-seven-segment decoder/driver, is given in Fig. 5-17.

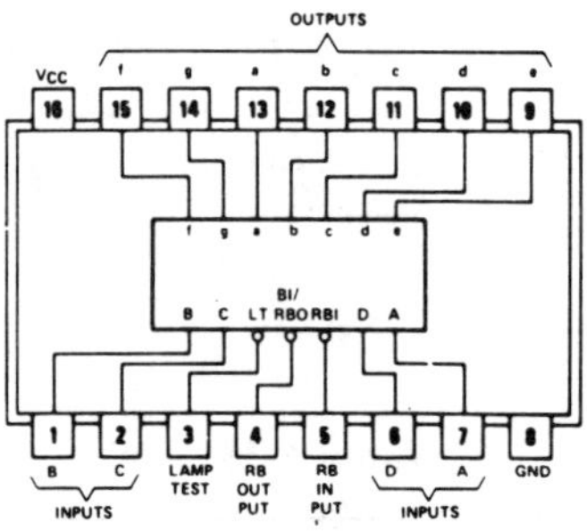

7448

Fig. 5-17.

In Figs. 5-15 through 5-17, you should be able to understand all input and output pins with the exception of pins 3, 4, and 5, the LAMP TEST, BLANKING INPUT/RIPPLE BLANKING OUTPUT, and RIPPLE BLANKING INPUT pins, abbreviated LT, BI/RBO, and RBI, respectively. These three inputs perform, not collectively but individually, the following operations:

1. Test all seven segments to determine that they are capable of being lit; this occurs if BI = 1 and LT = 0.
2. Force all seven output segments to go to a "logic 1" state if a "logic 0" state is present at RBI; concurrently, force a "logic 0" state to appear at the BI/RBO pin.
3. Force all seven output segments to go to a "logic 1" state regardless of the logic state of any other input condition; this occurs only if BI is at a "logic 0" state,(BI = 0.)

4. If the pins are not connected, all three remain at a "logic 1" state and are considered to be disabled, — they do not perform their intended function.

At present, we are not interested in pins 3 through 5 and will not discuss them further.

The 7446 integrated circuit is quite similar to the 7447 chip; the only difference is that the 7446 outputs become damaged only if the output voltage for any segment exceeds 30 volts. For the 7447 chip, the outputs become damaged if the output voltage for any segment exceeds 15 volts.

The truth table, shown in Table 5-16, for the 7448 bcd-to-seven-segment decoder/driver is the complement of the truth table for the 7447 chip.

Thus, for the 7448 chip, a lit segment corresponds to a "logic 1" state, and an unlit segment corresponds to a "logic 0" state. Either the 7447 or 7448 integrated circuit is quite acceptable as a driver and decoder for a seven-segment LED display. Some seven-segment LED displays require a logic 0 to light a segment (the OPCOA SLA-1, for example), whereas other seven-segment LED displays require a logic 1 to light a segment (the Litronix Data-Lit 704, for example). The SLA-1 seven-segment displays are more common.

We can ask one remaining question about the 7447 integrated circuit: What happens if we apply the following bcd inputs to the chip? These six 4-bit binary words are the binary equivalents of decimal 10, 11, 12, 13, 14, and 15. The 7447 chip cannot provide outputs for two decimal digits, so it instead gives the following outputs for the seven segments (Table 5-18):

The final entry, the one for DCBA = 1111_2, corresponds to a completely blank display. The following symbols are observed for the remaining five input conditions (Table 5-19):

Table 5-16.

Decimal Numeral	Inputs				Outputs						
	D	C	B	A	a	b	c	d	e	f	g
0	0	0	0	0	1	1	1	1	1	1	0
1	0	0	0	1	0	1	1	0	0	0	0
2	0	0	1	0	1	1	0	1	1	0	1
3	0	0	1	1	1	1	1	1	0	0	1
4	0	1	0	0	0	1	1	0	0	1	1
5	0	1	0	1	1	0	1	1	0	1	1
6	0	1	1	0	1	0	1	1	1	1	1
7	0	1	1	1	1	1	1	0	0	0	0
8	1	0	0	0	1	1	1	1	1	1	1
9	1	0	0	1	1	1	1	1	0	1	1

Table 5-17.

D	C	B	A	a	b	c	d	e	f	g
1	0	1	0							
1	0	1	1							
1	1	0	0							
1	1	0	1							
1	1	1	0							
1	1	1	1							

Table 5-18.

D	C	B	A	a	b	c	d	e	f	g
1	0	1	0	1	1	1	0	0	1	0
1	0	1	1	1	1	0	0	1	1	0
1	1	0	0	1	0	1	1	1	0	0
1	1	0	1	0	1	1	0	1	0	0
1	1	1	0	1	1	1	0	0	0	0
1	1	1	1	1	1	1	1	1	1	1

Table 5-19.

D	C	B	A	Symbol
1	0	1	0	
1	0	1	1	
1	1	0	0	
1	1	0	1	
1	1	1	0	
1	1	1	1	blank

WHAT IS THE DIFFERENCE BETWEEN A DECODER AND A DECODER/DRIVER?

Graf[1] has defined a *decoder* as a device that converts coded information into a more usable form as, for example, a bcd-to-decimal decoder. In considering this definition, we can inquire, In what way is the form more usable? Some possible answers to this question are as follows:

- To select a single line among many for the transmission of a negative pulse.
- To convert one type of digital code into another, the latter being more appropriate for the scientific instrument or computer being used.
- To drive an alphanumeric or simple numeric display.

Let us further consider the last of these three applications, the driving of a display. Shown in Fig. 5-18 are five different displays that are driven by decoder/drivers:

(a) A Nixie* tube display driven by a 74141 integrated circuit
(b) A series of ten lamps driven by a 7445 integrated circuit
(c) A seven-segment incandescent display such as the RCA Numitron
(d) A seven-segment common-anode LED display such as the MAN-1
(e) A seven-segment common-cathode LED display

Whereas the 7442, 7443, 7444, and 74154 decoders only have outputs that can drive other 7400-series integrated-circuit chips, the 741, 74141, 7445, 74145, 7446, 7447, and 7448 can drive external devices such as displays. In order to do so, these chips must have higher current-carrying outputs as well as outputs that can tolerate larger voltage drops. The 74141 decoder/driver is particularly remarkable in that it can withstand voltages as high as 55 volts. In view of a previously given definition for a *driver* (an element coupled to the output stage of a circuit to increase the power- or current-handling capability of the stage) we can define a *decoder/driver* as a decoder whose outputs are capable of handling increased currents and voltages when compared to the outputs from simple transistor-transistor logic (TTL) integrated-circuit chips. If used in such high-current high-voltage applications, a simple TTL decoder would burn out.

MULTIPLEXERS AND DEMULTIPLEXERS

The terms *multiplexer* and *demultiplexer* can be defined as follows:

multiplexer—A digital device that can select one of a number of inputs and pass the logic level of that input on to the output. Information for input-channel selection usually is presented to the device in binary weighted form and decoded internally. The device acts as a single-pole multiposition switch that passes digital information in one direction only.[1]

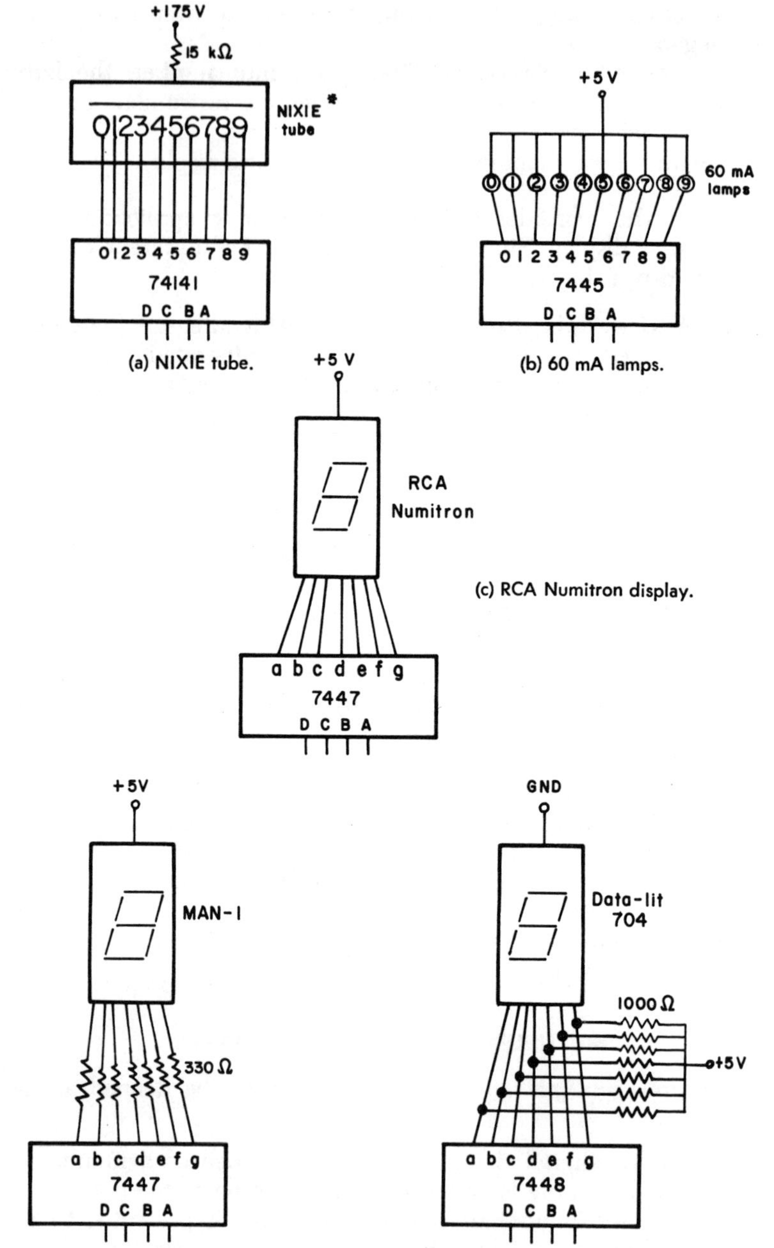

(a) NIXIE tube.

(b) 60 mA lamps.

(c) RCA Numitron display.

(d) Seven-segment common anode display.

(e) Seven-segment common cathode display.

Fig. 5-18. Five displays driven by decoder/driver ICs.

demultiplexer—A digital device that directs information from a single input to one of several outputs. Information for output-channel selection usually is presented to the device in binary weighted form and decoded internally. The device passes digital information in a direction opposite to that of a multiplexer.

Rather than discuss these two concepts in words, it is more appropriate to illustrate the function of both a multiplexer and a demultiplexer in schematic form. Consider the 74150 multiplexer chip and the 74154 decoder/demultiplexer chip shown in the schematic diagram in Fig. 5-19. We have indicated that input data, in this case a train of clock pulses, enters at input channel 8 on the multiplexer and leaves at output channel 3 on the demultiplexer. How are these channels selected? On the 74150 chip, we have applied the 4-bit binary word 1000 at the input-channel select pins DCBA; on the 74154 chip, we have applied the 4-bit word 0011 at the output-channel select pins

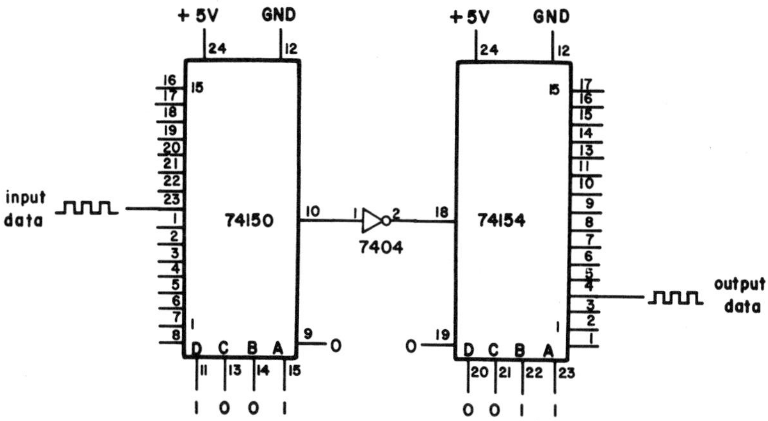

Fig. 5-19.

DCBA. We can thus independently select any one of sixteen input channels and any one of sixteen output channels using these two sets of 4-bit binary inputs! The 74150 multiplexer is certainly selecting one of a number of inputs, and the 74154 demultiplexer is certainly directing the output from the 74150 multiplexer to one of a number of outputs.

One possible use for a multiplexer/demultiplexer pair is to minimize the number of wires that are required to send data over relatively "long" TTL distances, such as several feet. As shown in Fig. 5-20, you can employ a 16-wire cable between the two points.

Alternatively, you can connect the DCBA input- and output-channel select pins together and sequence the 4-bit binary select in-

Fig. 5-20.

formation so that the appropriate input information is transmitted to the demultiplexer located several feet away. As a further capability, you have the ability to eliminate data transmission altogether by applying a "logic 1" state to the STROBE input at pin 9 of the multiplexer as well as at the G2 input at pin 19 on the demultiplexer (Fig. 5-21). A total of six lines can convey the information required by sixteen parallel lines! The only penalty you pay in this case is the fact that the digital information is not transmitted from one point to another simultaneously, i.e., in parallel. The best that you can do is to use a 4-bit binary counter, such as the 7493 or 74193 integrated circuit, and sequence the input/outputs. The input information appears one at a time at the output of the 74150 integrated circuit,

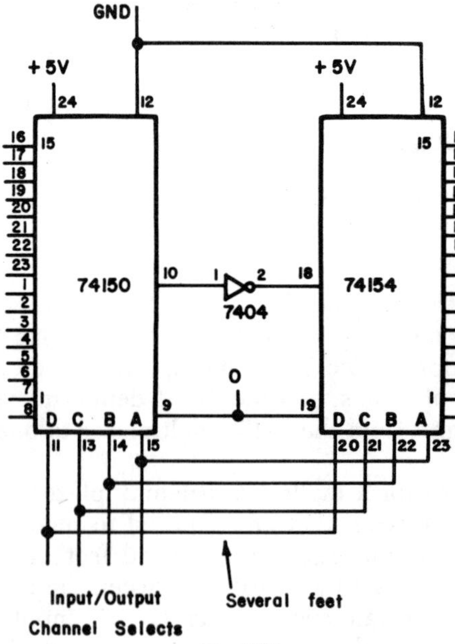

Fig. 5-21.

and is transmitted to the input of the 74154 integrated circuit, which then decodes the information to the appropriate output pin. With such a system, it can take sixteen times as long to transmit the sixteen lines of input information as it would have in the case of sixteen parallel lines each several feet long. In many cases, the decrease of transmission speed is quite acceptable. Keep in mind that it is difficult to transmit TTL signals reliably at high speeds over distances greater than one foot. To transmit digital data over several hundred feet, current loops are required.

We shall now demonstrate how it is possible to operate the 7442 bcd-to-decimal decoder as a 1-of-8 demultiplexer. First, we draw the schematic diagram (Fig. 5-22) for the 7442 integrated circuit in a way that is slightly different from that shown previously in Fig. 5-6.

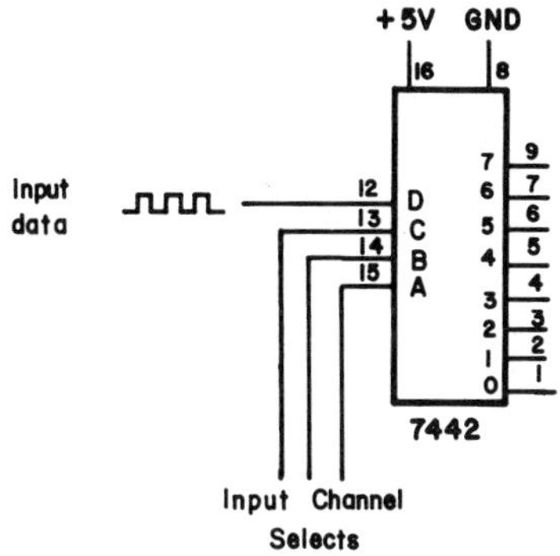

Fig. 5-22.

Next, we determine the truth table for two situations: input data = 0, and input data = 1. The tables are given in Tables 5-20 and 5-21, respectively.

Note that output lines 8 and 9 are not used; we cannot select such lines using only a 3-bit binary word. Clearly, the 7442 integrated

Table 5-20.

C	B	A	0	1	2	3	4	5	6	7
0	0	0	0	1	1	1	1	1	1	7
0	0	1	1	0	1	1	1	1	1	1
0	1	0	1	1	0	1	1	1	1	1
0	1	1	1	1	1	0	1	1	1	1
1	0	0	1	1	1	1	0	1	1	1
1	0	1	1	1	1	1	1	0	1	1
1	1	0	1	1	1	1	1	1	0	1
1	1	1	1	1	1	1	1	1	1	0

Table 5-21.

C	B	A	0	1	2	3	4	5	6	7
0	0	0	1	1	1	1	1	1	1	1
0	0	1	1	1	1	1	1	1	1	1
0	1	0	1	1	1	1	1	1	1	1
0	1	1	1	1	1	1	1	1	1	1
1	0	0	1	1	1	1	1	1	1	1
1	0	1	1	1	1	1	1	1	1	1
1	1	0	1	1	1	1	1	1	1	1

circuit can be employed to demultiplex input data to any one of eight different output lines. The data is transmitted to the output lines directly, without inversion. Another way that we can indicate this behavior is shown in Table 5-22.

Table 5-22.

Select Inputs			Output Channel at Which Data Appears
C	B	A	Number
0	0	0	0
0	0	1	1
0	1	0	2
0	1	1	3
1	0	0	4
1	0	1	5
1	1	0	6
1	1	1	7

We can also use the 7442 integrated circuit as a 1-of-4 demultiplexer by setting D = 0 and having the input data enter at C. The diagram and truth table for such an application are given in Fig. 5-23 and Tables 5-23 and 5-24.

Table 5-23. Input Data = 0

B	A	0	1	2	3
0	0	0	1	1	1
0	1	1	0	1	1
1	0	1	1	0	1
1	1	1	1	1	0

We shall stop our discussion of demultiplexers, with these examples. Keep in mind that the 74139, 74155, 74156, and 74138 chips can also be used as demultiplexers.

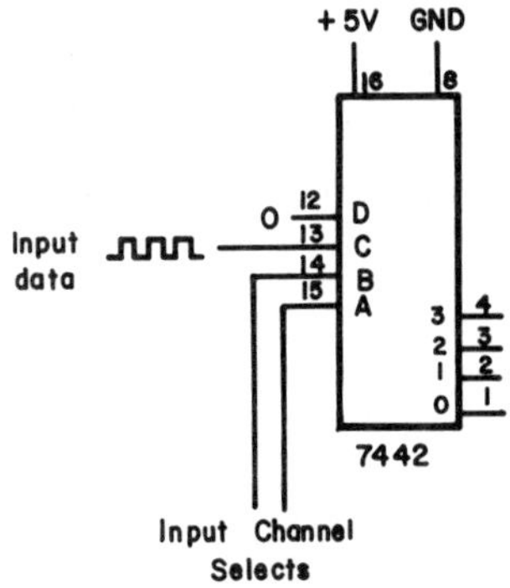

Fig. 5-23.

Table 5-24. Input Data = 1

B	A	0	1	2	3
0	0	1	1	1	1
0	1	1	1	1	1
1	0	1	1	1	1
1	1	1	1	1	1

SEQUENCERS

In the preceding section, we mentioned the fact that we could sequence a multiplexer/demultiplexer pair and thus transmit sixteen channels of data over a distance of several hundred feet using only six wires. It is appropriate to define the term *sequence* as well as the term *sequencer* and to discuss in more detail what these two concepts mean.

sequence (noun)—The order in which objects or items are arranged.[1]
to sequence (verb)—To place in order.[1]
sequencer—An electronic device that may be set to initiate a series of events and to make the events follow in sequence, *i.e.*, in order.[1]
sequential operation—The carrying out of operations one after the other.[1]

Let us consider the 74150/74154 multiplexer/demultiplexer pair discussed in the preceding section. If we add a 7493 binary counter and drive the input/output channel select inputs DCBA sequentially from 0000 to 1111, we cause the 74150/74154 pair to sequence the sixteen inputs from input 0 to input 15 and back again to input 0. We thus have "ordered" our selection of both inputs and outputs; each integrated-circuit chip—both the 74150 and the 74154—acts as a sequencer:

- The 74150 chip sequences the sixteen inputs and causes input data to appear at the output pin 10 in the order, channel 0, 1, 2, 3, . . 15.
- The 74154 chip takes input data appearing at input pin 18 and sequences it to the sixteen outputs. Data appears at the outputs in the order, channel 0, 1, 2, 3, . . 15.

The circuit required for the 7493 chip is shown in Fig. 5-24. There are other ways to sequence the multiplexer/demultiplexer combination, but the use of the 7493 is the most convenient and is the preferred technique wherever possible.

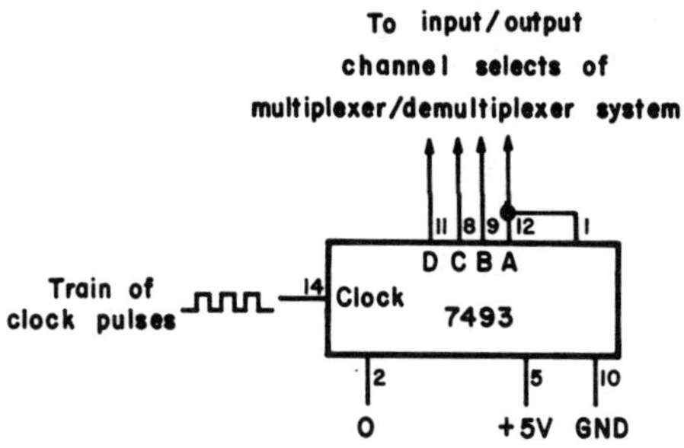

Fig. 5-24.

To sequence the 7442 bcd-to-decimal decoder, we employ the 7490 circuit shown in Fig. 5-25 in which all ten channels are sequenced. We have used this circuit in an instrument that acquires digital data, an "automated digital data logger" (Analytical Chemistry, *45*, 219, 1973).

As the 7490 decade counter counts from 0 to 9, the ten output channels 0, 1, 2, . . . 9 on the 7442, in sequence, exhibit a "logic 0" state for a full clock pulse. This behavior can best be seen with the aid of digital waveforms for the clock input to the 7490 chip and the ten outputs from the 7442 chip (Fig. 5-26). To save space, we have only shown a few representative output channels on the 7442 chip. Clearly, the negative pulses coming from the outputs of the 7442 chip appear in order from channel 0 through 9, i.e., they are sequenced.

It is possible to construct other types of sequencers that sequence between a total of 2, 3, 4, 5, 6, 7, 8, 9, 11, 12, 13, 14, 15, or 16 output channels. The 7490, 7492, 7493, 74192, and 74193 counters can all be wired to accomplish such a purpose. We shall discuss such counters briefly in Chapter 10 of Book 2. For now, we shall be content simply with a decade sequencer that is based on the 7490 decade counter or a hexadecimal sequencer that is based on the 7493

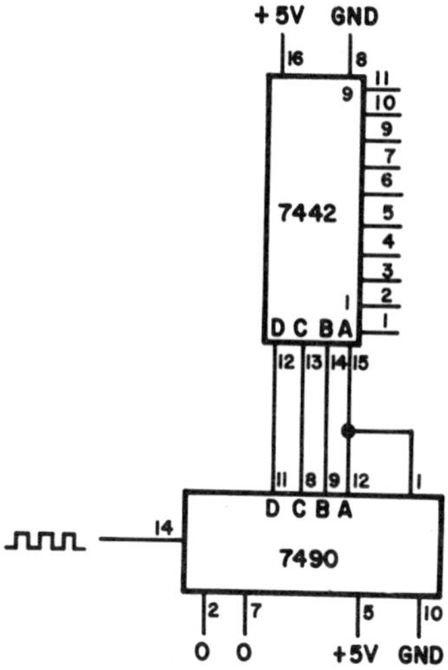

Fig. 5-25. Ten-state sequencer circuit consisting of 7490 decade counter and 7442 bcd-to-decimal decoder.

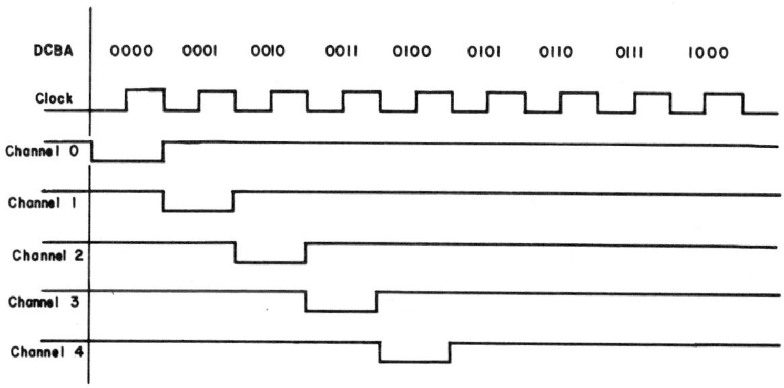

Fig. 5-26. Typical waveforms for sequencer circuit in Fig. 5-25; only outputs from channels 0 through 4 are shown.

binary counter. These two sequencers will be described in the experimental section.

PIN CONFIGURATIONS FOR DATA SELECTORS/MULTIPLEXERS

The three 7400-series data selector/multiplexer chips that you will most commonly encounter are the

- 74150 16-line-to-1-line data selector/multiplexer
- 74151 8-line-to-1-line data selector/multiplexer
- 74153 dual 4-line-to-1-line data selector/multiplexer

It is useful to represent the above multiplexers in a way that would make subsequent circuit diagrams that incorporate such chips easier to read. In this book, we shall place the input lines, and usually the strobe or enable input, on the left-hand side of the rectangle, the single output on the right-hand side, the two power connections at the top, and the data select inputs at the bottom of the rectangle. This is illustrated in Fig. 5-27 and will now be applied to the three multiplexer chips.

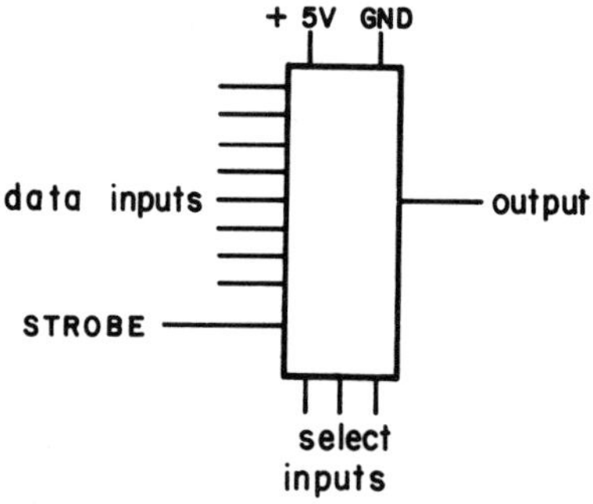

Fig. 5-27.

The 74150 16-Line-to-1-Line Data Selector/Multiplexer

The 74150 integrated circuit is a 16-line-to-1-line data selector/multiplexer whose pin configuration and schematic representation is shown in Figs. 5-28 and 5-29, respectively. The output, W, at pin 10 on the 74150 chip is the inverted value of the data input, so you

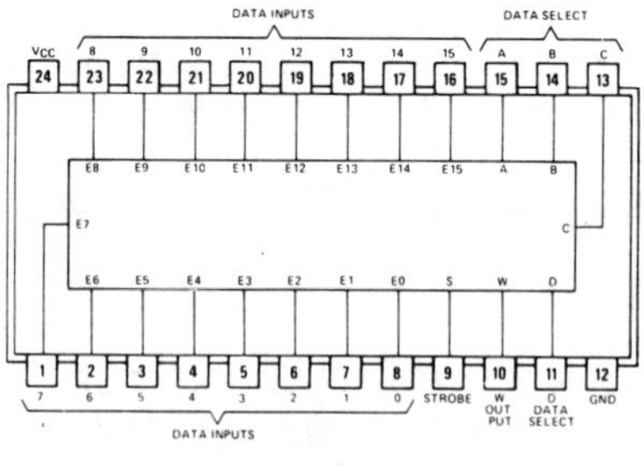

74150

Fig. 5-28.

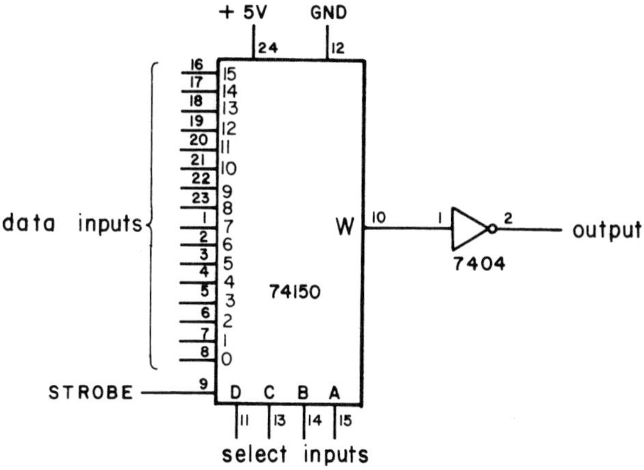

Fig. 5-29.

must employ a 7404 inverter, as has been done in the above diagram, to produce a normal output.

The 74153 Dual 4-Line-to-1-Line Data Selector/Multiplexer

The pin configuration for the 74153 dual 4-line-to-1-line data selector/multiplexer integrated circuit is shown in Fig. 5-30, and the schematic diagram in Fig. 5-31. Note the adjective, "dual," which

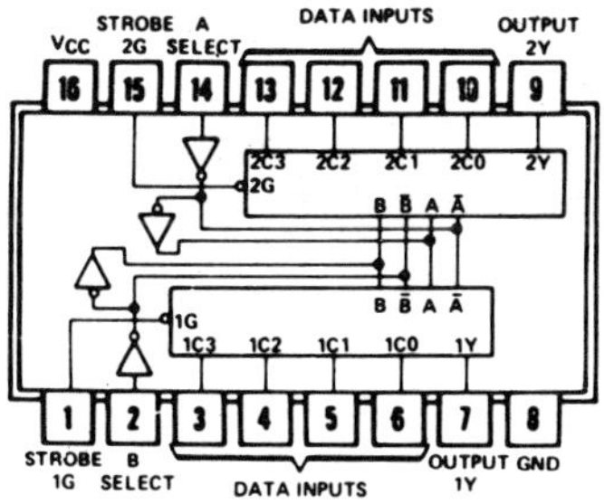

74153

Fig. 5-30.

means you have two different data selectors/multiplexers on the chip. Unfortunately, the two data selectors are *not independent;* there is only a single pair of select inputs, A and B, for the two selectors. Data selector/multiplexer No. 1 has data inputs at pins 3, 4, 5, and 6; a strobe input at pin 1; and a single normal output at

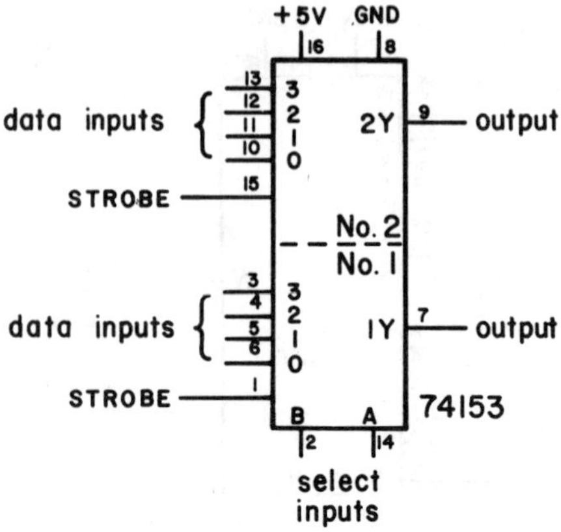

Fig. 5-31.

pin 7. Data selector/multiplexer No. 2 has data inputs at pins 10, 11, 12, and 13; a strobe input at pin 15; and a single normal output at pin 9. You can independently select either of the two data selectors by applying the appropriate logic states to the strobe inputs. For example, when STROBE 1G = 0 and STROBE 2G = 1, you select data selector No. 1; when STROBE 1G = 1 and STROBE 2G = 0, you select data selector No. 2.

The 74151 8-Line-to-1-Line Data Selector/Multiplexer

The 74151 8-line-to-1-line data selector/multiplexer has the pin configuration and schematic representation shown in Figs. 5-32 and 5-33, respectively.

The binary system is employed with the data select inputs CBA to choose the desired data input channel. Thus, if CBA = 000, chan-

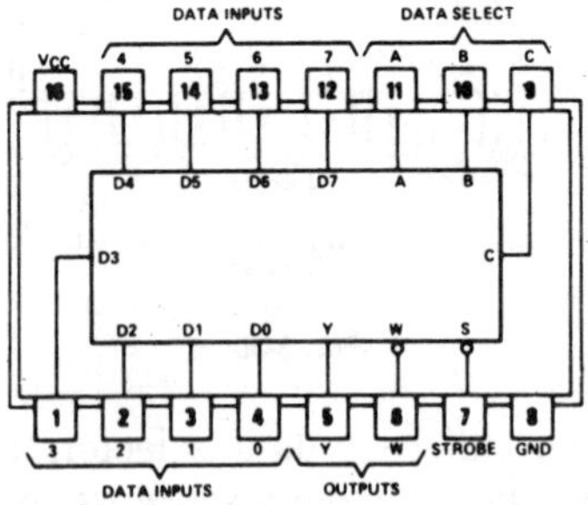

74151

Fig. 5-32.

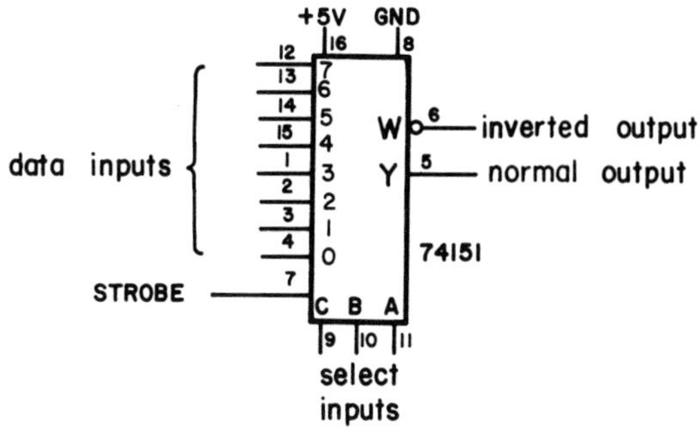

Fig. 5-33.

nel 0 at pin 4 is selected. If CBA = 111, channel 7 at pin 12 is selected. As a convenience to the user, the manufacturers of this chip have provided both a normal output, Y, at pin 5 and an inverted output, W, at pin 6. W is the complement of Y, and vice versa.

If the STROBE input is at logic 1, the entire 74151 chip is inactivated. The consequence is that output Y becomes logic 0 and output W becomes logic 1, independent of the information that appears at either the data or select inputs. If the STROBE input is at logic 0, the chip is considered to be "strobed" or "enabled" and functions normally. Thus, the STROBE input allows you to select one or more among many independent 74151 chips; the STROBE input acts as a "chip select" input.

ADVANCED LOGIC TECHNIQUES

The repertoire of available 7400-series gates is quite limited. For example, 2-input AND, NAND, OR, NOR, and Exclusive-OR gates are all available, but only 3-input AND, NAND, and NOR gates are conveniently provided as integrated circuits 7410, 7411, and 7427, respectively. With 4-input gates, only the 7420 NAND and 7421 AND are available.

With this background in mind, you might ask, Is it possible to "create" other types of 2-, 3-, and 4-input gates using existing 7400-series chips? The answer is yes. Using data selector/multiplexer integrated circuits, you can create any type of 2-input, 3-input, 4-input, or even 5-input logic gate. The basic idea, or "gimmick," is to provide the truth table for the desired gate as the inputs to the data selector/multiplexer. The A, B, C, and D inputs in the desired truth table function as the select inputs on the data selector/multiplexer, and the outputs Q in the desired truth table function as the data inputs to the data selector/multiplexer. For further details, consider Experiments Nos. 14 and 15 in this chapter. Question No. 3 in Experiment No. 15 demonstrates the use of the "folding technique" to create a 4-input gate using an 8-line-to-1-line data selector/multiplexer.

INTRODUCTION TO THE EXPERIMENTS

In this chapter, you will construct decoders, sequencers, programmable sequencers, demultiplexers, multiplexers and gates using two types of counters, the 7490 and 7493 chips; three multiplexers, the 74150, 74151, and 74153 chips; and two decoder/demultiplexers, the 7442 and 74154 chips. This is the first time that you will use a 24-pin integrated-circuit chip.

You will require two seven-segment LED displays for Experiment No. 5. Team up with a friend. The experiment is somewhat complicated, but is quite worthwhile. You should also plan to spend considerable time on Experiment No. 4; it is not difficult, but it is important.

The experiments in this chapter can best be summarized by listing the type of device that you will construct. Thus,

Experiment No.	*Type of Device*
1	decoder
2	demultiplexer
3	sequencer
4	programmable sequencer
5	programmable sequencer
6	decoder
7	programmable sequencer
8	demultiplexer
9	multiplexer plus demultiplexer
10	decoder
11	demultiplexer
12	decoder
13	demultiplexer
14	2-input logic gates
15	AND-OR-INVERT gate

EXPERIMENT NO. 1

Purpose

The purpose of this experiment is to demonstrate the operation of the 7442 chip as a simple bcd-to-decimal decoder.

Pin Configuration of Integrated-Circuit Chip (Fig. 5-34)

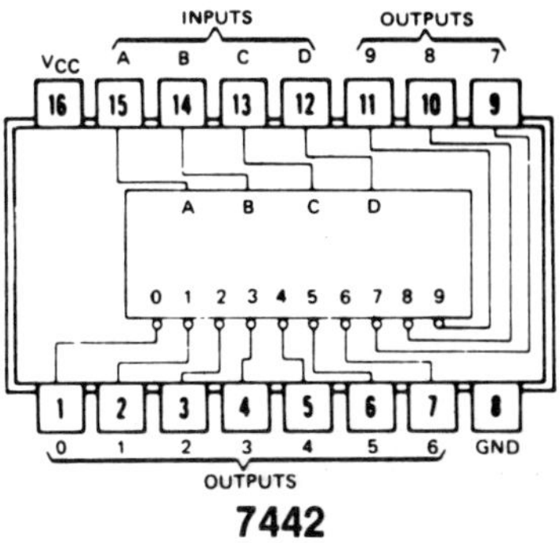

Fig. 5-34.

Schematic Diagram of Circuit (Fig. 5-35)

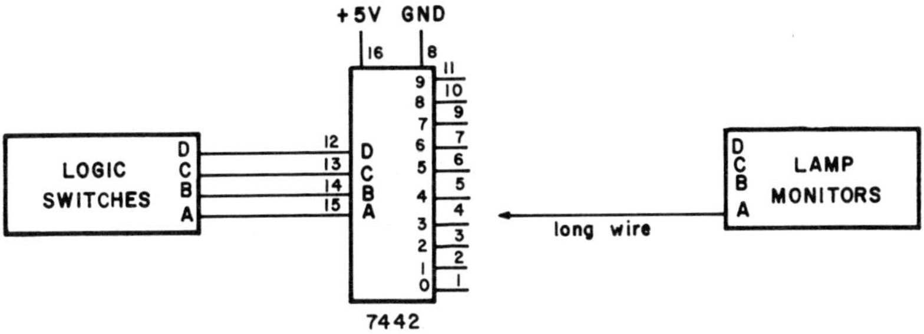

Fig. 5-35.

Step 1

Carefully study the schematic diagram in Fig. 5-35. Note that the pin numbers are given outside the box, whereas the channel numbers of the ten outputs from the 7442 chip are given on the inside of the box. Only a single lamp monitor is required, and this should be connected to a long wire that can be employed to test the logic state of all ten channels.

Siep 2

Wire the circuit as shown. It should be rather easy to do so, since only seven connections need to be made. Leave the long wire attached to lamp monitor A unconnected.

Step 3

Apply power to the breadboard and set the four logic switches to binary 0000. Connect the long wire from lamp monitor A to pin 1 of the 7442 chip. What do you observe?

If you did this experiment correctly, you should observe that the lamp monitor is unlit. Now, touch the long wire to each of the remaining nine output channels at pins 2 through 7 and 9 through 11 on the 7442 chip. What do you observe with the lamp monitor?

If the circuit is wired up correctly, you should observe that the lamp monitor is lit for all remaining nine channels on the decoder. This means that each of these channels is at a "logic 1" state, whereas channel 0 at pin 1 is at a "logic 0" state.

Step 4

Set the four logic switches to DCBA = 0001. Which single output channel among the ten output channels is at a "logic 0" state?

If you have done this experiment correctly, you should observe that channel 1 at pin 2 is at a "logic 0" state, and all remaining channels, including channel 0, are at a "logic 1" state.

Step 5

Set the four logic switches to DCBA = 0010 and determine which single output channel among the ten output channels is now at a "logic 0" state.

If your answer is that channel 2 at pin 3 is at a "logic 0" state, you are correct. Test the remaining channels and demonstrate to yourself that they are all at a "logic 1" state.

Step 6

By now, a pattern has emerged and you should be able to write a truth table that summarizes your experimental observations. Fill in the following truth table.

Channel Number	D	C	B	A	Channel Number 0	1	2	3	4	5	6	7	8	9
0	0	0	0	0	0	1	1	1	1	1	1	1	1	1
1	0	0	0	1	1	0	1	1	1	1	1	1	1	1
2	0	0	1	0	1	1	0	1	1	1	1	1	1	1
3	0	0	1	1										
4	0	1	0	0										
5	0	1	0	1										
6	0	1	1	0										
7	0	1	1	1										
8	1	0	0	0										
9	1	0	0	1										

Remember, you will have to change the logic switch settings and then test all output channels to determine which single channel is at a "logic 0" state. We are aware of the fact that we have given you this truth table in the discussion prior to this experiment. We request that you demonstrate to yourself that this truth table is a correct representation of the actual behavior of the 7442 chip.

Step 7

Using four logic switches, it is possible to generate six additional binary (DCBA) states. These states correspond to the binary words 1010, 1011, 1100, 1101, 1110, and 1111. Set the logic switches to each of these states in turn and demonstrate that there no longer is a single output channel that is at a "logic 0" state. This is the reason that we call the 7442 chip a bcd-to-decimal decoder. There are only ten valid output conditions, hence the word, "decimal."

Questions

1. Why do you use the term, bcd, in the description of the 7442 chip as a bcd-to-decimal decoder?

2. What output pin would be at a "logic 0" state if you applied the following binary words to the four input pins on the 7442 chip?

DCBA	Pin Number
0100	
0111	
0011	
0001	
1001	

EXPERIMENT NO. 2

Purpose

The purpose of this experiment is to demonstrate the operation of the 7442 chip as a 1-of-8 demultiplexer.

Pin Configuration of Integrated-Circuit Chip (Fig. 5-36)

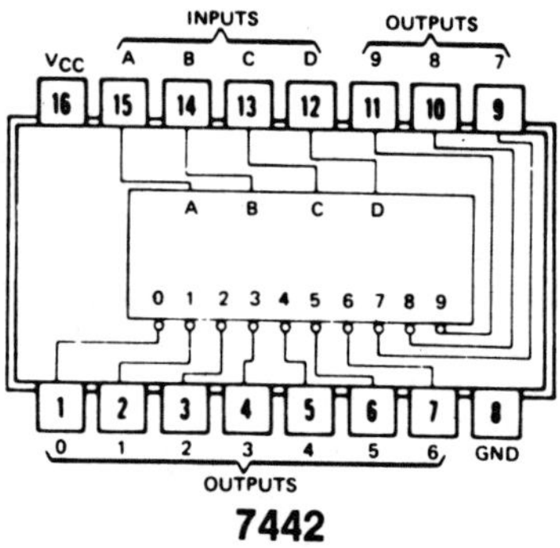

Fig. 5-36.

Schematic Diagram of Circuit (Fig. 5-37)

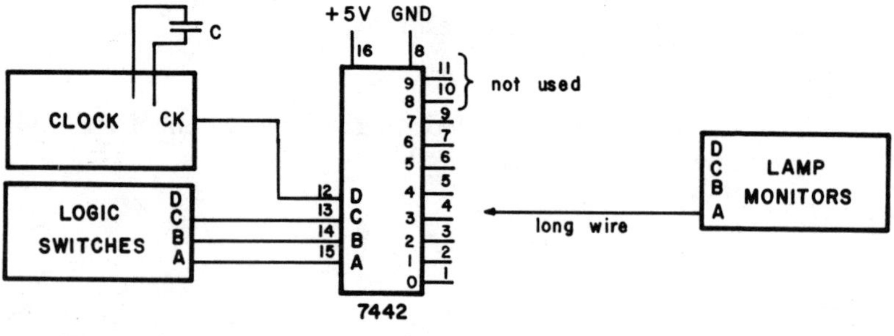

Fig. 5-37.

Step 1

Carefully study the schematic diagram and note that the only change from Experiment No. 1 is the substitution of a clock input at pin 12 of the 7442 chip. The clock supplies a train of clock pulses, which act as the "input data" to the chip. The experimental question that you will answer is: At what output channel will this train of clock pulses appear?

Step 2

Wire the circuit and apply power to the breadboard. Set the three remaining logic switches, CBA, all to a "logic 0" state, corresponding to the 3-bit binary word CBA = 000. At what output channel would you predict the train of clock pulses to appear (and be detected by a lamp monitor connected to that channel)?

If you picked channel 0 at pin 1, you are entirely correct. Connect the lamp monitor to this channel and observe that the light goes on and off, indicating a train of clock pulses. (NOTE: Channel 8 at pin 10 is an equally correct answer.)

Step 3

What output channel would you predict if the three logic switches were set at CBA = 001? Try it.

When we tried this experiment, the "input data," a train of clock pulses, appeared at pin 2, which corresponds to channel 1. Hopefully, you observed an identical result. (NOTE: Channel 9 at pin 11 is an equally correct answer.)

Step 4

Set the three logic switches to CBA = 110 and determine at which channel the input data now appears.

The correct answer here is channel 6 at pin 7. What is the logic state of the remaining seven channels, i.e., channels 0 through 5 and channel 7?

We observed a "logic 1" state for all of the remaining channels. Hopefully, you did also.

Step 5

Is the pattern becoming clear again? The results you observe here should bear a strong resemblance to the results you observed in the preceding experiment. You should now be able to confirm a table that we gave you in the discussion section in this chapter. In the following table, fill in the output channel at which the data appears. *Note that we have scrambled the select input states; an answer such as 0, 1, 2, 3, 4, 5, 6, and 7 is incorrect!*

Select Inputs			Output Channel at Which Data Appears
C	B	A	Number
1	1	0	
0	0	1	
0	1	0	
1	0	1	
0	1	1	
1	1	1	
0	0	0	0
1	0	0	

In conclusion, you have demonstrated how you can transmit input data to any of eight different output channels. The transmitted data is not inverted. Any channel among the eight output channels that does not contain data is at a "logic 1" state. We call this process, demultiplexing, and the specific circuit, a demultiplexer.

Questions

1. Define in your own words the term, demultiplexer.

2. Using the 7442 chip, how can you construct a 1-of-4 demultiplexer? Draw the circuit.

3. Complete the truth table for a 1-of-5 decoder.

C	B	A	0 1 2 3 4
0	0	0	
0	0	1	
0	1	0	
0	1	1	
1	0	0	
1	0	1	

EXPERIMENT NO. 3

Purpose

The purpose of this experiment is to demonstrate the operation of the 7442 chip as a decade sequencer.

Pin Configurations of Integrated-Circuit Chips (Figs. 5-38 and 5-39)

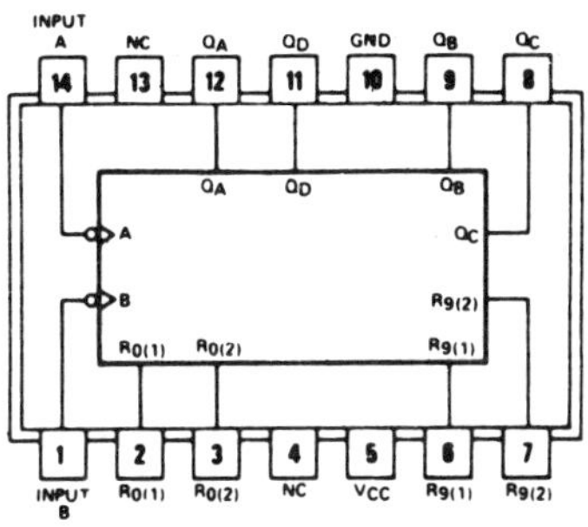

Fig. 5-38.

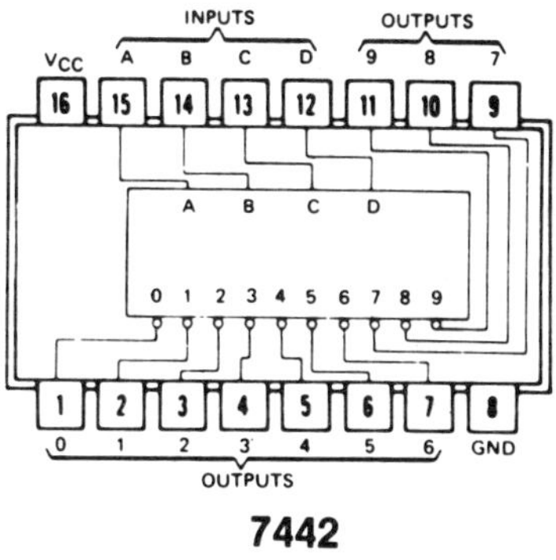

Fig. 5-39.

Schematic Diagram of the Circuit (Fig. 5-40)

Step 1

Wire the circuit shown in the schematic diagram. The object here is to demonstrate that you are able to sequence the ten output channels on the 7442 chip with the aid of the 7490 decade counter applied at the inputs DCBA to the 7442 chip. Rather than use a

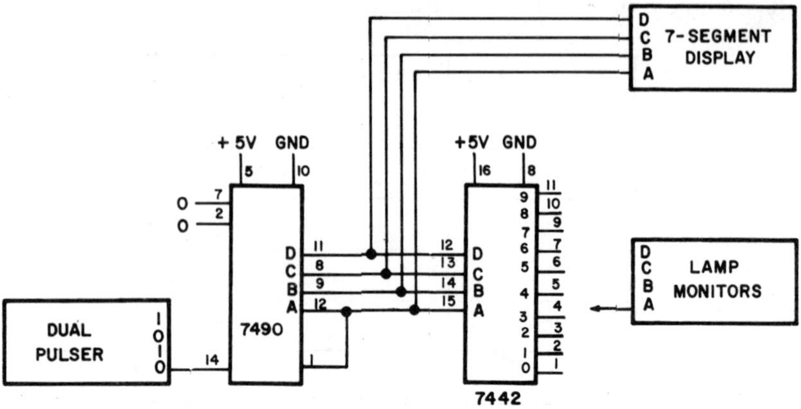

Fig. 5-40.

clock input at pin 1 of the 7490 chip, you instead use a pulser. This allows you to slow down the sequencing process so that you can observe what is happening at all ten output channels, should you so desire.

Step 2

Apply power to the breadboard and connect the lamp monitor to channel 0 of the 7442 chip. Repeatedly press and release the pulser, thus changing the decimal numeral on the seven-segment display, until the lamp monitor goes off. When this happens, what is the numeral that you read on the LED display?

Our lamp monitor became unlit, corresponding to a "logic 0" state, when a decimal zero appeared on the display. How about you? Did you observe the same thing?

Step 3

Now connect the lamp monitor to channel 1 of the 7442 chip (pin 2). Cycle the counter once more and stop at the point when the lamp monitor goes off once again. What numeral is displayed on the LED readout?

It should be a decimal one.

Step 4

Connect the lamp monitor to channel 7 at pin 9. Cycle the counter once again and stop when the lamp monitor goes off. What decimal numeral now appears on the display?

It should be decimal seven, 7. What is the logic state of the lamp monitor for the remaining nine decimal numerals? Your answer here should be a logic 1.

Step 5

Since you have four lamp monitors, connect all four to channels 0, 1, 2, and 3 on the 7442 chip. This is shown in the following partial diagram of Fig. 5-41. Make the connections indicated and apply power to the breadboard once again. Cycle through the counter

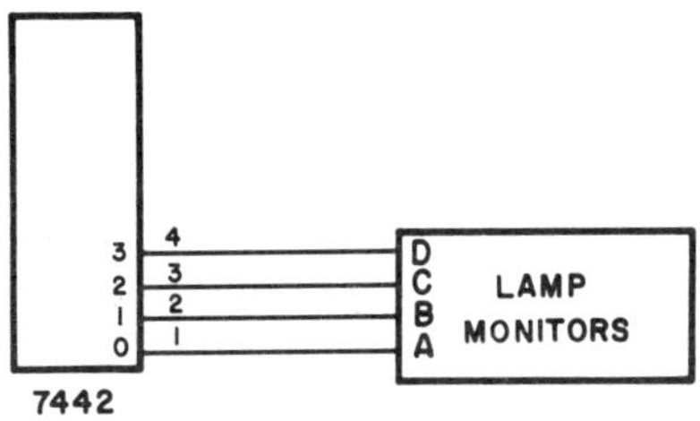

Fig. 5-41.

from 0 to 9 several times. Do the four lamp monitors go out "in order," *i.e.*, do they go out sequentially? Your answer should be yes.

Step 6

Explain what would happen in the partial diagram in Fig. 5-42 (the remaining part of the diagram is the same as before). We are

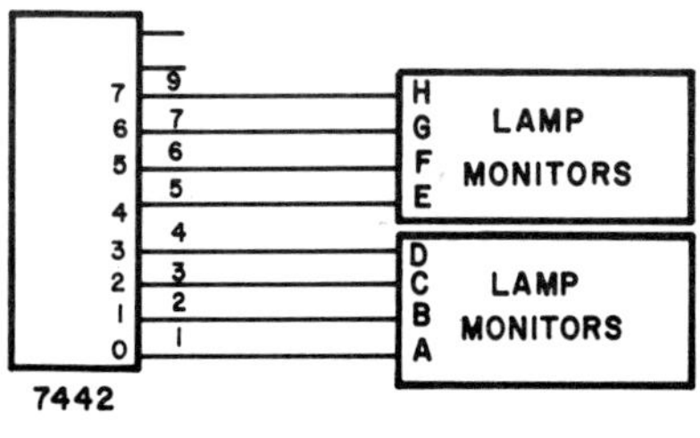

Fig. 5-42.

interested in whether the eight lamp monitors go out sequentially or not.

Your answer should be that a "logic 0" state (unlit LED) sequences through the eight output channels, but that it must also sequence through channels 8 and 9 before it again appears at channel 0.

Questions

1. Explain how the 7442 chip in the above experiment acts as a sequencer. What is sequenced? How do you experimentally detect the fact that the chip is acting as a sequencer?

2. What type of counter would be required if you wanted to make an octal sequencer from the 7442 chip? In other words, what must the counter do for you to observe an octal sequencer?

3. What type of counter would be required if you wanted to sequence between channels 0, 1, 2, 3, and 4?

4. What type of counter would be required if you wanted to sequence between channels 0, 1, 2, 3, 4, 5, 6, 7, and 8?

EXPERIMENT NO. 4

Purpose

The purpose of this experiment is to demonstrate how the 7442 chip can be used to sequence among any number of states between two and nine.

Pin Configurations of Integrated-Circuit Chips (Figs. 5-43 through 5-45)

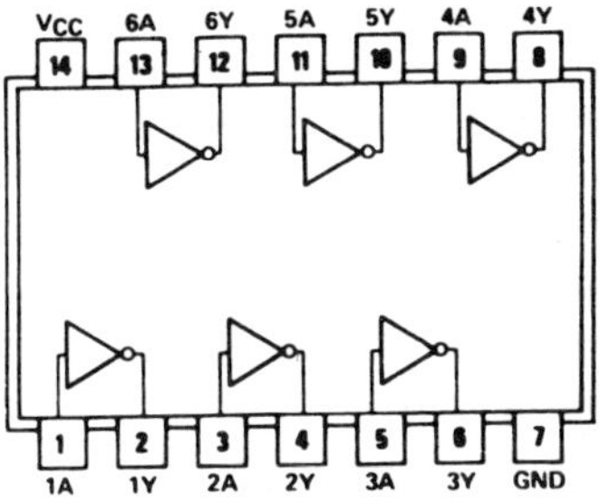

7404

Fig. 5-43.

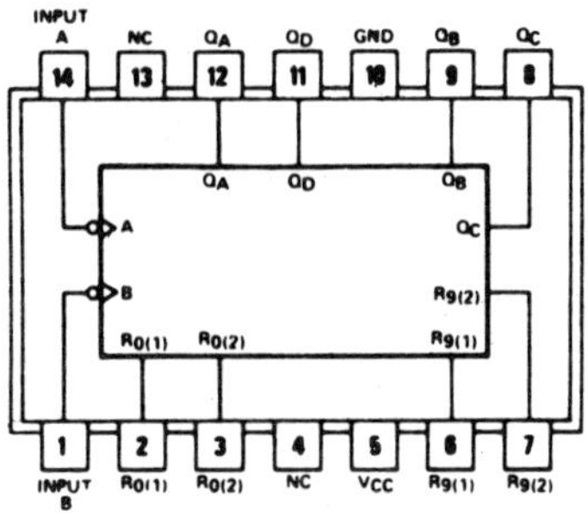

7490

Fig. 5-44.

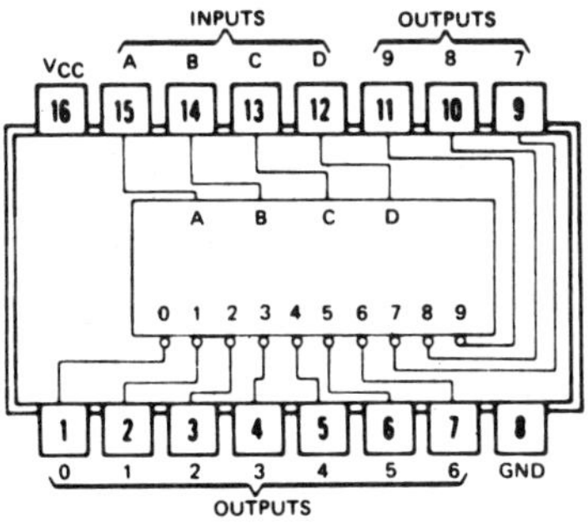

7442

Fig. 5-45.

Schematic Diagram of Circuit (Fig. 5-46)

Step 1

This is a very interesting circuit that you may use many times in the future. It permits you to construct a sequencer that sequences between any number of channels between two channels up to and including nine different channels. For example, a three-channel or three-state sequencer will sequence in the following manner: 0, 1, 2, 0, 1, 2, 0, 1, 2, 0, 1, 2, . . . etc. A five-channel sequencer will sequence as follows: 0, 1, 2, 3, 4, 0, 1, 2, 3, 4, 0, 1, 2, 3, 4, . . . etc.

Step 2

Wire the circuit as shown and connect the wire from pin 1 of the 7404 inverter counter to output pin 4 of the 7442 decoder. Apply

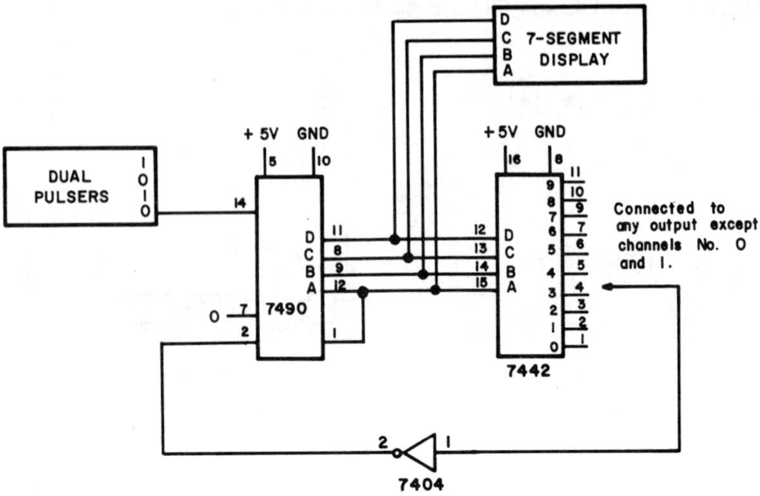

Fig. 5-46.

power to the breadboard and write down the counting sequence that you observe on the seven-segment LED display. Remember, you will have to repeatedly press and release the pulser in order to clock the 7490 counter.

We observed the following sequence: 0, 1, 2, 0, 1, 2, 0, 1, 2, etc. Actually, the number 3 also appeared on the display, but it appeared for such a short period of time (perhaps 40 or 50 nanoseconds) that we could not see it.

Step 3

Now connect the wire from pin 1 of the inverter to output pin 5 of the 7442 decoder. The counting sequence now should be 0, 1, 2, 3, 0, 1, 2, 3, 0, 1, 2, 3, etc. Do you observe such a result?

Connect the wire from pin 1 of the 7404 inverter to pin 7 of the 7442 chip. What is the counting sequence now?

The sequence should be 0, 1, 2, 3, 4, 5, 0, 1, 2, 3, . . . etc.

Step 4

Fill in the following table.

Pin 1 of 7404 Inverter Connected to Following Channel of 7442 Chip	Counting Sequence Observed
0	0
1	0
2	0, 1, 0, 1, etc.
3	0, 1, 2, 0, etc.
4	
5	
6	
7	
8	
9	0, 1, 2, 3, 4, 5, 6, 7, 8, 0, etc.

Step 5

A subtle aspect of this experiment, already mentioned in Step 2, is that there may be a hidden digit that you cannot observe because it appears for such a short time. The following truth table summarizes the digit that you are not able to see:

Pin 1 of 7404 Inverter Connected to Following Channel of 7442 Chip	Digit That Cannot Be Seen
0	—
1	1
2	2
3	3
4	4
5	5
6	6
7	7
8	8
9	9

How would you determine whether or not the channel that corresponds to a digit that cannot be seen really was at logic 0 for a brief instant of time (approximately 40 nanoseconds)?

You should attach a counter to the channel in question. If the channel goes from logic 1 to logic 0 and back to logic 1, even for 40 nanoseconds, a negative edge is generated and can be counted by the counter. Try it!

Questions

1. Would you have any suggestions concerning how you would construct a sequencer that would count up to decimal 37 and then cycle back to decimal 0 again? You will require more than one 7442 chip and more than one 7490 counter. Give your suggested circuit in the space below.

2. Explain how you would have the sequencer sequence through eight channels and yet have the lamp monitor light only on channel 5. Show the output from the 7442 chip that is required for such a circuit.

EXPERIMENT NO. 5

Purpose

The purpose of this experiment is to demonstrate how a pair of 7442 chips can be used (a) to sequence among any number of channels up to ninety-one different channels, and (b) to decode any single channel between channel 0 and up to and including channel 99. This experiment is somewhat more difficult than the average experiment in this chapter.

Pin Configurations of Integrated-Circuit Chips (Figs. 5-47 through 5-49)

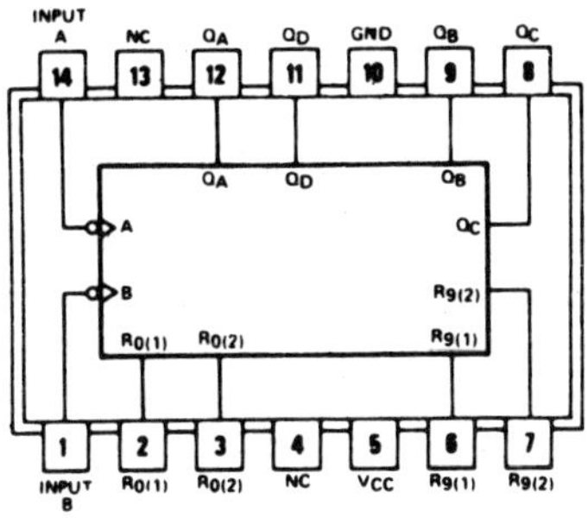

7490

Fig. 5-47.

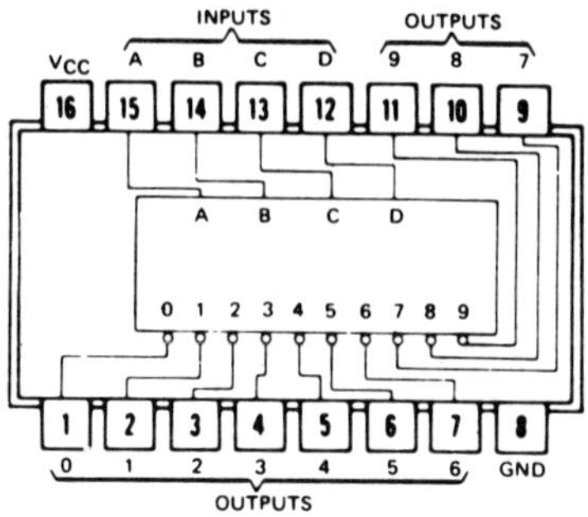

7442

Fig. 5-48.

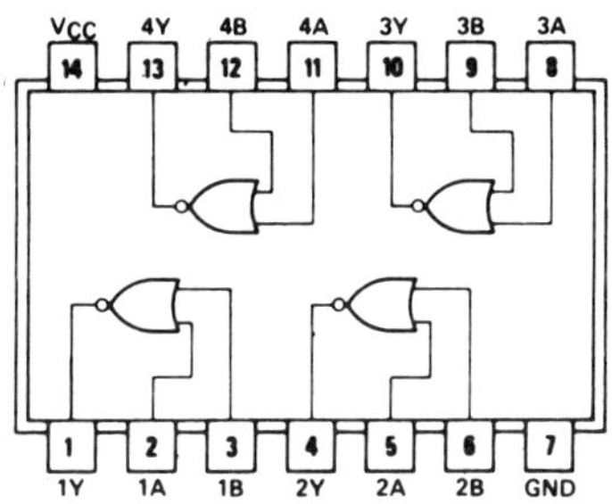

7402

Fig. 5-49.

Schematic Diagram of the Circuit (Fig. 5-50)

Step 1

Carefully study the schematic diagram in Fig. 5-50 and note that the two pairs of 7442 and 7490 chips are linked together by a 7402 2-input positive NOR gate. When the output at pin 1 of gate G1 goes to a "logic 1" state, both the 7490 counters are reset to decimal 0. In order for this to occur, the two inputs to G1 at pins 2 and 3 must both be at a "logic 0" state.

As an example, assume that you want to sequence between channels 0, 1, 2, 3, . . . 37, 0, 1, 2, . . . 37, 0, etc. Only at decimal 38 will the outputs of the two 7442 chips both be at a "logic 0" state, which causes the output from the NOR gate G1 to be at logic 1, which

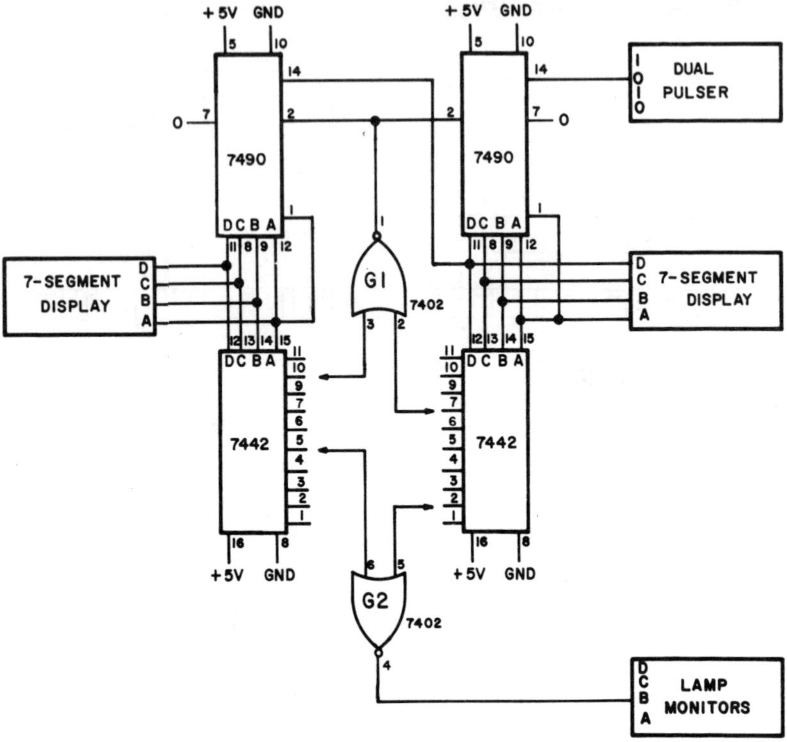

Fig. 5-50.

resets both 7490 decade counters at pin 2 of the 7490 chip. The resetting of the counters occurs extremely quickly, so fast that you will not be able to see the number 38 on the display.

Gate G2, which is also a 7402 2-input positive NOR gate, functions as a simple 1-of-N decoder, where N is the number of channels among which we are sequencing. When a channel is decoded, the lamp monitor will light up; for all other nondecoded channels, the lamp monitor will remain unlit.

Step 2

Wire the circuit shown in Fig. 5-51. Borrow an extra seven-segment LED display so that you can observe decimal numbers between 00 and 99. Connect pin 2 of gate G1 to pin 2 on the right-hand 7442 chip in the schematic diagram; connect pin 3 of gate G1 to pin 2 on the left-hand 7442 chip. Finally, connect pins 5 and 6 of gate G2 to pins 1 on the respective 7442 chips.

Step 3

Apply power to the breadboard. Both displays will probably be at decimal zero, corresponding to 00_{10}, and the lamp monitor should be lit. Press and release the pulser until you obtain a reading of

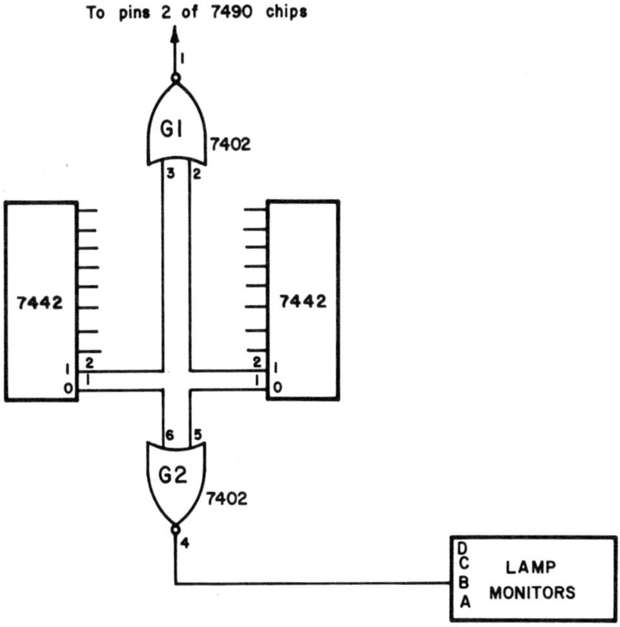

Fig. 5-51.

decimal 10 on the pair of LED displays. Stop for a moment and observe that the lamp monitor is unlit. Press and release the pulser one more time. What reading do you observe on the LED displays, and is the lamp monitor lit or unlit?

We shall not give you the correct answer here. The circuit that you have wired is sufficiently simple that you should have the capability of answering the above questions correctly. You should be able to predict what should happen from the schematic diagram and a knowledge that a logic 1 state at pin 2 of a 7490 chip will, if pin 3 on the 7490 chip if left unconnected, reset the decade counter to decimal 0.

Step 4

Now change the connections to gates G1 and G2 to those shown in the following schematic diagram (Fig. 5-52). Remember to turn the power off while you are making the changes.

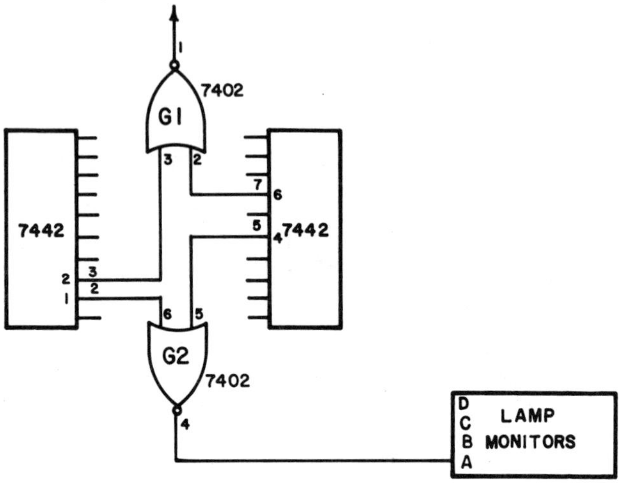

Fig. 5-52.

Step 5

Apply power to the breadboard. The two LED displays should be at 00. Repeatedly press and release the pulser until the lamp monitor becomes lit. What is the display reading when this occurs?

In our case, it was decimal 14. How about you? Continue pressing and releasing the pulser and note the largest LED display reading before both displays become reset to 00. What is this largest display reading?

We observed a reading of 25 before the two 7490 counters were reset.

Let us explain the above, and your, observations. When the LED display is at 14, pin 2 on the left-hand 7442 chip in the schematic diagram and pin 5 on the right-hand 7442 chip are both at a logic 0 state. Two logic 0 inputs to gate G2 produce a logic 1 output at pin 4 that lights up the lamp monitor. In the second case, when the counters reach 26, pin 3 on the left-hand 7442 chip and pin 7 on the right-hand 7442 chip are both at a logic 0 state. As with gate G2, two logic 0 inputs to gate G1 produce a logic 1 at output at pin 1, which is applied at pins 2 on the pair of 7490 counters and resets them to 0. The resetting of the counters occurs extremely quickly; you never have a chance to see decimal 26 on the displays.

In summary, for both the lighting of the lamp monitor and the resetting of the 7490 counters, a 2-input positive NOR gate is required. The basic logic of the decoding process in each case is the same, only two different actions are produced.

Questions

1. Write the partial schematic diagram, as you did in Steps 2 and 5 above, for a sequencer that sequences to 83 on the displays and a decoder that causes a lamp monitor to light at channel 57. You may wish to wire your suggested circuit and demonstrate that it operates as you predicted.

2. Would it be possible to substitute a 7400 2-input positive NAND gate for gates G1 or G2? Why or why not?

3. Would it be possible to substitute a 7432 2-input positive OR gate for gates G1 and G2? Why or why not?

4. If you had to substitute a gate for G1 or G2, which would you prefer, a 7400 2-input positive NAND gate or a 7432 2-input posi-

tive OR gate? Why? Would it help if we allowed you the use of a single 7404 inverter with either gate? Why?

5. What type of gate would you require if you wanted to sequence between 000 and N, where N is less than or equal to 999?

EXPERIMENT NO. 6

Purpose

The purpose of this experiment is to demonstrate the operation of the 74154 chip as a simple 4-line-to-16-line decoder.

Pin Configurations of Integrated-Circuit Chips (Figs. 5-53 and 5-54)

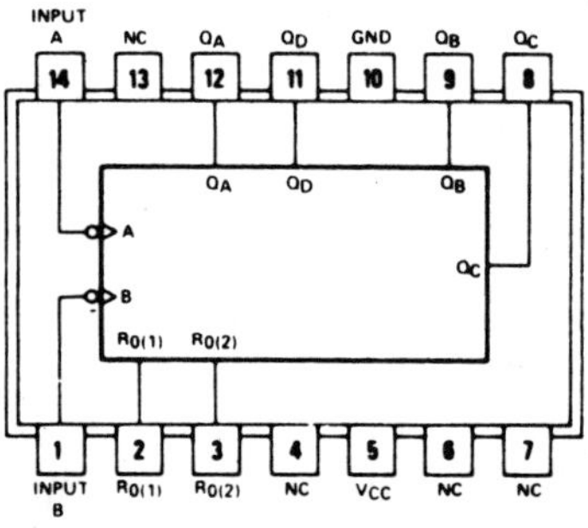

7493

Fig. 5-53.

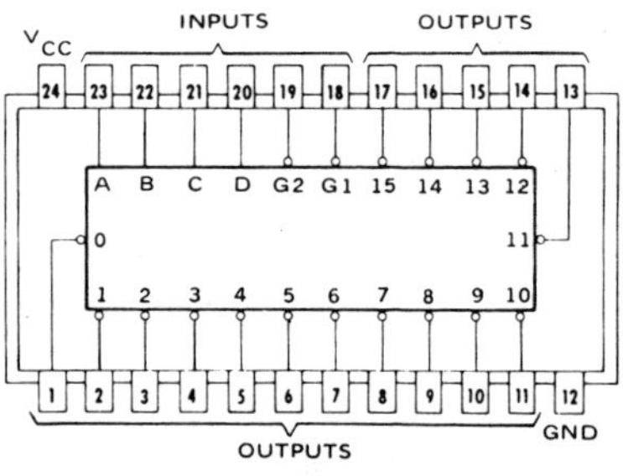

74154

Fig. 5-54.

Schematic Diagram of Circuit (Fig. 5-55)

Step 1

Carefully study the schematic diagram given for this experiment. For the 74154 chip, the output channel numbers are within the box, whereas the pin numbers are outside the box. You only need a single lamp monitor to perform this experiment.

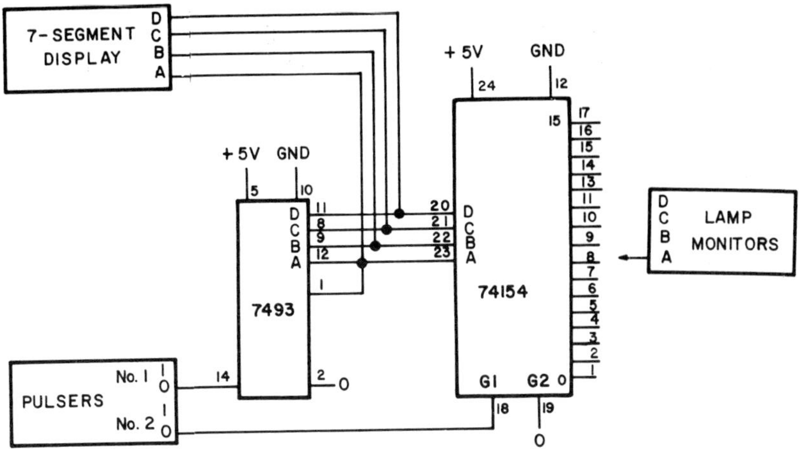

Fig. 5-55.

Step 2

Wire the circuit. Leave unconnected a long wire to the lamp monitor; you will touch it to the output channels of the 74154 chip to determine their logic states.

Step 3

Apply power to the breadboard and connect the lamp monitor to pin 1 of the 74154 chip. Is the lamp monitor lit? What is the reading on seven-segment LED display?

In our case, the lamp monitor was lit and the reading on the seven-segment display was decimal 1. You may observe a result that is slightly different than this.

Step 4

Repeatedly press and release the pulser until a decimal 0 appears on the display. Is the lamp monitor lit or unlit?

The lamp monitor should be unlit!

Step 5

Repeatedly press and release the pulser and determine if the lamp monitor is unlit for any other decimal numeral or display symbol between binary 0000 and binary 1111.

In our case, the lamp monitor was unlit only when the display read decimal 0. For all other display readings, the lamp monitor was lit! We concluded the following: channel 0 is at a logic 0 state only when DCBA = 0000, which corresponds to a seven-segment display reading of decimal 0.

Step 6

Now connect the lamp monitor to pin 2 on the 74154 chip. For what LED display reading is the lamp monitor unlit?

We observed that the lamp monitor was unlit only when the display reading was decimal 1, corresponding to DCBA = 0001.

Step 7

Connect the lamp monitor to pin 3 on the 74154 chip. For what LED display reading is the lamp monitor unlit?

For a display reading of decimal 2, corresponding to DCBA = 0010, pin 3 was at logic 0.

Step 8

Fill in the truth table for the 74154 chip (Table 5-25). To obtain the correct logic states, proceed as you did in Steps 4 through 7. We have helped you with the truth table by filling the first three rows.

Step 9

To complete your understanding of the 74154 chip, you must determine the function of the inputs G1 and G2 at pins 18 and 19. These are a pair of gating inputs, and they determine whether or not the decoder is enabled.

Connect the lamp monitor to pin 1 and set G1 = G2 = logic 0. The lamp monitor should go off only when the decimal numeral 0 appears on the display. Now vary the logic states of both G1 and G2 and determine if the lamp monitor connected to pin 1 will become unlit for any other condition beside DCBA = 0000 and G1 = G2 = 0. You should observe that the decoder is disabled, i.e., inoperative, when G1 and G2 have values other than both being at a "logic 0" state. We can write the truth table as given in Table 5-26.

Table 5-25. Truth Table for 74154

Inputs				Output Channels																Numeral or Symbol
D	C	B	A	0	1	2	3	4	5	6	7	8	9	10	11	12	13	14	15	
0	0	0	0	0	1	1	1	1	1	1	1	1	1	1	1	1	1	1	1	0
0	0	0	1	1	0	1	1	1	1	1	1	1	1	1	1	1	1	1	1	1
0	0	1	0	1	1	0	1	1	1	1	1	1	1	1	1	1	1	1	1	2
0	0	1	1																	3
0	1	0	0																	4
0	1	0	1																	5
0	1	1	0																	6
0	1	1	1																	7
1	0	0	0																	8
1	0	0	1																	9
1	0	1	0																	
1	0	1	1																	
1	1	0	0																	
1	1	0	1																	
1	1	1	0																	
1	1	1	1																	blank

Table 5-26.

G1	G2	Condition of Decoder
0	0	enabled
0	1	disabled
1	0	disabled
0	1	disabled

By "disabled," we mean that there are no values of DCBA between 0000 and 1111 for which the lamp monitor from any outlet channel is at a "logic 0" state.

Questions

1. Define the terms, "enable" and "disabled."

2. What logic states have a unique output state when the inputs are at a "logic 0" state? There are two such gates that have this characteristic.

3. How would you make a programmable sequencer using the 74154 chip and the 7493 binary counter?

EXPERIMENT NO. 7

Purpose

The purpose of this experiment is to demonstrate how the 74154 chip can be used to sequence among any number of states between two and fifteen. Such a sequencer is called a *programmable sequencer*. This experiment is similar to Experiment No. 4.

Pin Configurations of Integrated-Circuit Chips (Figs. 5-56 through 5-58)

Schematic Diagram of Circuit (Fig. 5-59)

Step 1

Note that the circuit diagram is essentially identical to that given in Experiment No. 6. The only change is that a lamp monitor is no longer required and that pin 2 on the 7493 counter is connected to one of the output channels from the 74154 chip.

Step 2

Wire the circuit. Connect pin 2 on the 7493 chip to pin 2 on the 7404 inverter and pin 1 on the inverter to pin 3 on the 74154 chip.

Step 3

Apply power to the breadboard and repeatedly press and release the pulser. In the following space, explain what happens.

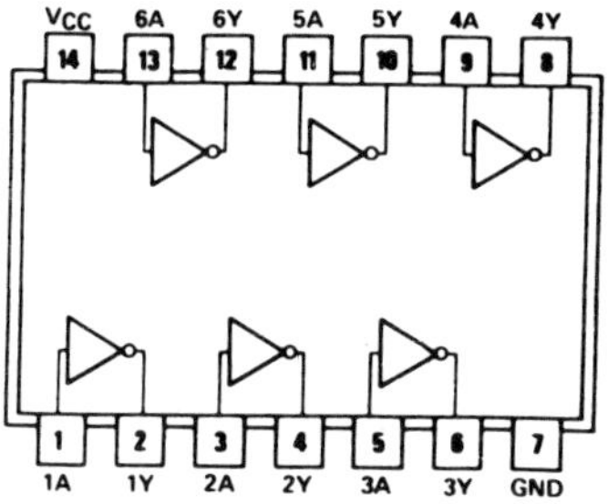

7404

Fig. 5-56.

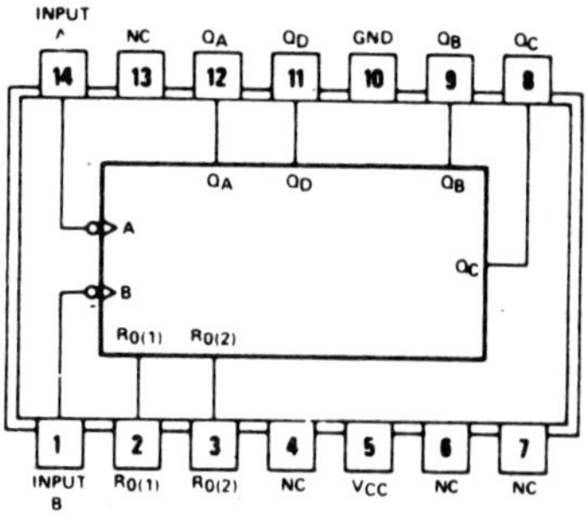

7493

Fig. 5-57.

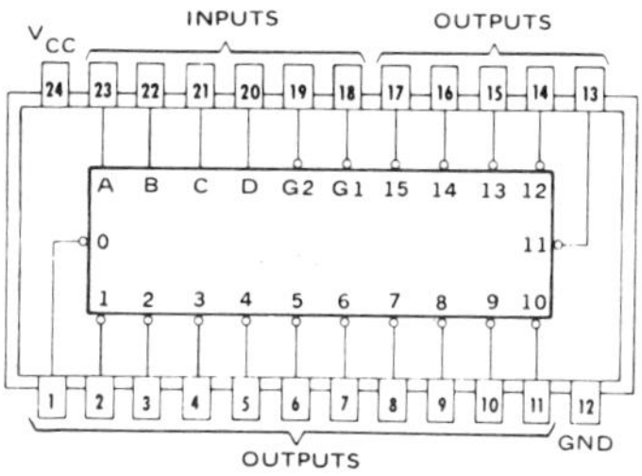

74154

Fig. 5-58.

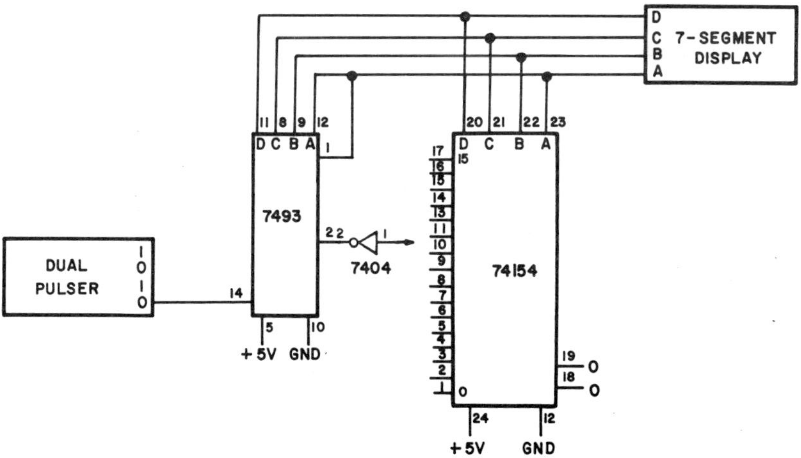

Fig. 5-59.

The number sequence 0,1,0,1,0 . . . appears on the display. The number 2 also appears on the display, but for such a short time (40 nanoseconds) that it cannot be seen visually.

Step 4

Connect pin 1 on the 7404 chip to pin 5 on the 74154 chip. Repeatedly press and release the pulser and explain what you observe on the LED display in the following space.

The number sequence 0,1,2,3,0,1,2,3,0,1 . . . appears on the display. The number 4 also appears, but it cannot be seen visually. Among how many different states are you sequencing?

Your answer should be 5 states.

Step 5

Connect pin 1 on the 7404 chip to pin 13 on the 74154 chip. Write down the number of different states in this new sequencer that you have constructed. Remember that decimal 0 is a separate state.

Your answer should be 12 states: 0, 1, 2, 3, 4, 5, 6, 7, 8, 9,⊏ , ⊐, and ⊔, the last symbol not being seen on the display.

Step 6

It should be clear that you can program the number of states among which you sequence. You can sequence among as few as two different states, or among as many as sixteen different states!

Questions

1. What is meant by a "programmable" sequencer? How do you program the sequencer?

2. Why did you use the 7493 binary counter instead of the 7490 decade counter in this experiment?

3. Would it be possible to use the 7490 decade counter in this experiment? What limitations would you observe with the programmable sequencer if you would do so?

EXPERIMENT NO. 8

Purpose

The purpose of this experiment is to demonstrate the operation of the 74154 chip as a 1-of-16 demultiplexer.

Pin Configurations of Integrated-Circuit Chips (Figs. 5-60 and 5-61)

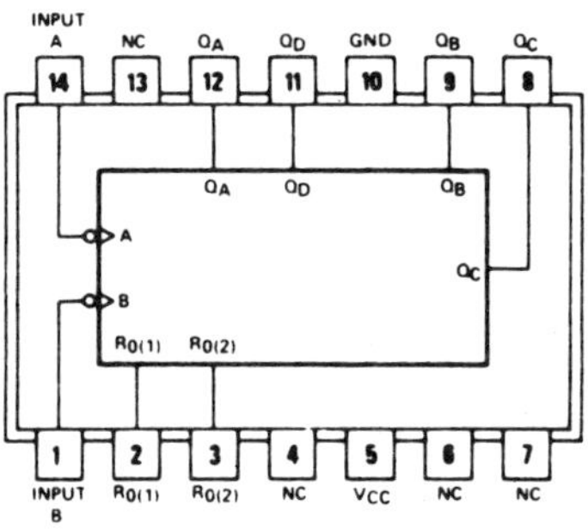

7493

Fig. 5-60.

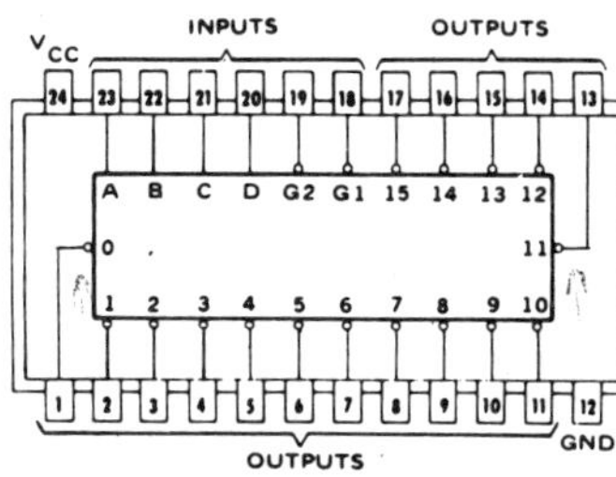

74154

Fig. 5-61.

Schematic Diagram of Circuit (Fig. 5-62)

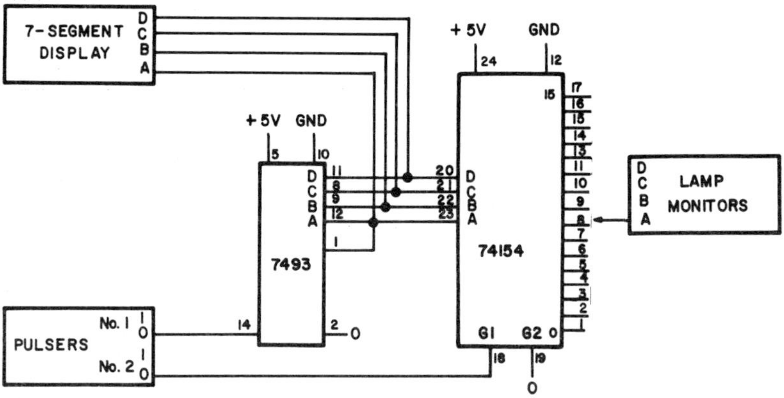

Fig. 5-62.

Step 1

Study the schematic diagram and note the following:

- Pulser No. 1 selects the output channel on the 74154 chip at which the input data at G1 (pin 18) appears.
- Pulser No. 2 represents the input data, a variable sequence of "logic 0" and "logic 1" states. A clock would be an appropriate substitute for this pulser. Note that the input data enters the circuit at G1.
- The output channel at which the input data appears is shown on the seven-segment LED display.
- You have a choice of sixteen different output channels to which you can direct the input data at G1. This is why we call such a circuit, a 1-of-16 demultiplexer.
- Input data only appears at that output channel that you select using the 7493 binary counter. The output will be a "logic 1" state for all other channels.
- The input data appears at the chosen output channel without inversion.

Step 2

Wire the circuit with the power to the breadboard disconnected. Note that the circuit is almost identical to the circuits given in Experiments Nos. 6 and 7.

Step 3

Apply power to the breadboard. Connect the lamp monitor to output channel 0 at pin 1 on the 74154 chip. Sequence through the sixteen output channels until channel 0 appears on the seven-segment LED display. Use pulser No. 1 to sequence through the channels.

Now repeatedly press and release pulser No. 2. If you have wired the circuit correctly, you should observe that the lamp monitor follows the logic state of the pulser, which is the "input data" to the demultiplexer. In other words, input data at G1 is now appearing at output channel 0.

Step 4

You should test whether or not the input data is appearing at any other channel. Disconnect the lamp monitor wire from pin 1 and touch it to pins 2 through 11 and 13 through 17 in succession while you repeatedly press and release pulser No. 2. You should observe that the lamp monitor remains illuminated for all channels except channel 0.

Step 5

Now connect the lamp monitor to output channel 4 at pin 5 on the 74154 chip. Repeatedly press and release pulser No. 2 while you sequence the output channels from channel 00 to channel 15. At which output channel on the 74154 chip do you now observe the input data at G1?

Output channel 4.

Step 6

Connect the lamp monitor to pin 15 on the 74154 chip. Repeatedly press and release pulser No. 2 while you sequence the output channels from channel 00 to channel 15 using pulser No. 1. At which output channel on the 74154 chip do you observe the input data. Write the numeral or symbol on the LED display in the following space, then convert it to the correct channel number.

The symbol observed is , which corresponds to output channel 13.

Step 7

A demultiplexer is a digital device that directs information from a single input to one of several outputs. Is it clear that the circuit that you have wired is acting as a 1-of-16 demultiplexer? If the answer to this question is no, then you should carefully repeat the experiment.

Questions

1. Based on the circuit diagrams given in experiments 6 through 8, draw a circuit diagram for a 1-of-13 demultiplexer. Use the following space for your circuit. HINT: You will require the ability to sequence among thirteen different states. Show the correct pin numbers in your circuit, and make it as simple as possible.

EXPERIMENT NO. 9

Purpose

The purpose of this experiment is to demonstrate the operation of both a 1-of-16 multiplexer and also a 1-of-16 demultiplexer. The 74150 and 74154 chips are used in this circuit; both are wired to the same 7493 binary counter, thus producing a sequenced multiplexer/demultiplexer.

Schematic Diagram of the Circuit (Fig. 5-63)

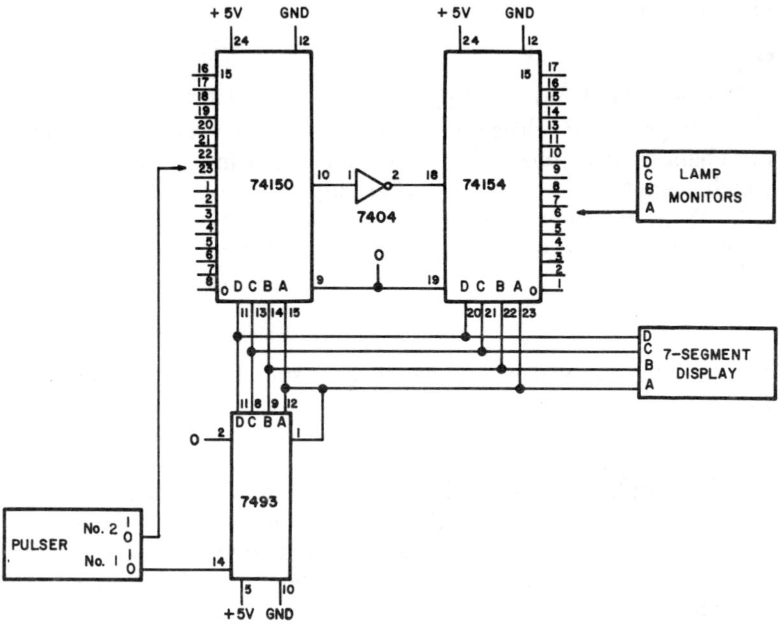

Fig. 5-63.

Pin Configurations of Integrated-Circuit Chips (Figs. 5-64 through 5-66)

Step 1

In the above schematic diagram, note the following:

- The 74150 multiplexer and 74154 demultiplexer are *synchronized*, – the channels are sequenced by the same 7493 binary counter.
- Pulser No. 2 is the “input data” to the multiplexer/demultiplexer system.

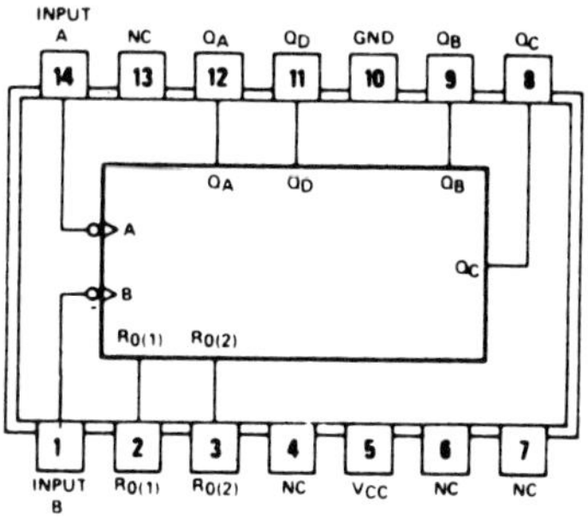

7493

Fig. 5-64.

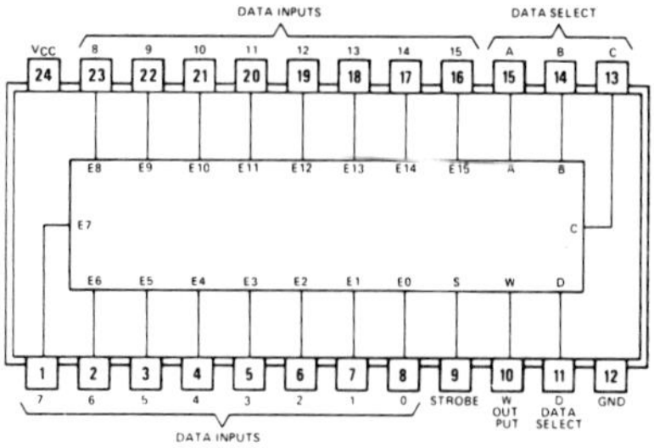

74150

Fig. 5-65.

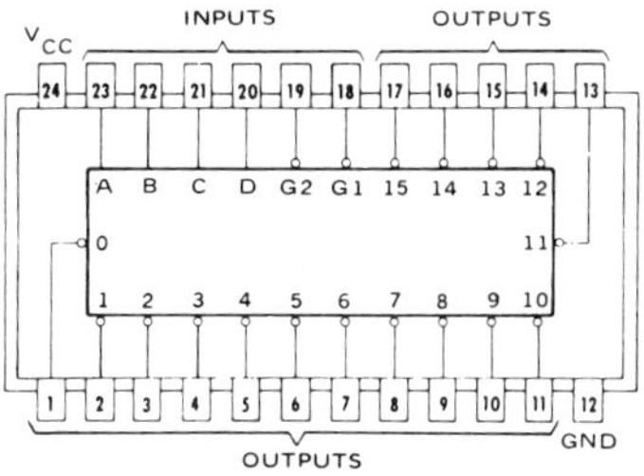

74154

Fig. 5-66.

- Pulser No. 1 selects the specific pair of channels over which data is transmitted. If, for example, input data enters at channel 5 on the multiplexer, it leaves at channel 5 on the demultiplexer. Sixteen channels between channel 00 and channel 15 can be selected.
- The 7404 inverter is required to ensure that the input data leaves the demultiplexer as a normal, as opposed to inverted, output.

Step 2

Wire the circuit as shown. Apply power to the breadboard. Connect pulser No. 2 to pin 8 of the 74150 chip. Connect the lamp monitor to pin 1 of the 74154 chip. Sequence the 7493 counter until decimal 0 appears on the LED display. Repeatedly press and release pulser No. 2 and observe that the lamp monitor "follows" the logic states of the pulser. In other words, you are transmitting input data through the 74150 multiplexer, the 7404 inverter, and the 74154 demultiplexer to the output at channel 0. Check your wiring if you do not observe this result.

Step 3

Now connect pulser No. 2 to pin 7 of the 74150 chip and the lamp monitor to pin 2 of the 74154 chip. Sequence the 7493 counter until decimal 1 appears on the display. Repeatedly press and release pulser No. 2 and observe again that lamp monitor A is blinking on and off. You have now selected channel 1 as your data transmission channel.

Step 4

You should now know how to select any one of the sixteen data transmission channels in the circuit. You simply sequence the 7493 counter until you reach the correct channel number, connect the pulser No. 2 and lamp monitor A to the appropriate pins on the 74150 and 74154 chips, respectively, and transmit your data.

Questions

1. You desire to send data over channel 11 in the multiplexer/demultiplexer system. To which input pin on the 74150 chip do you connect pulser No. 2? To which output pin on the 74154 chip do you connect lamp monitor A?

2. Repeat question No. 1, this time for channel 6.

3. Repeat question No. 1, this time for channel 15.

4. Repeat question No. 1, this time for channel 3.

EXPERIMENT NO. 10

Purpose

The purpose of this experiment is to demonstrate the operation of a 74155 chip as a 2-line-to-4-line decoder.

Pin Configuration of Integrated-Circuit Chip (Fig. 5-67)

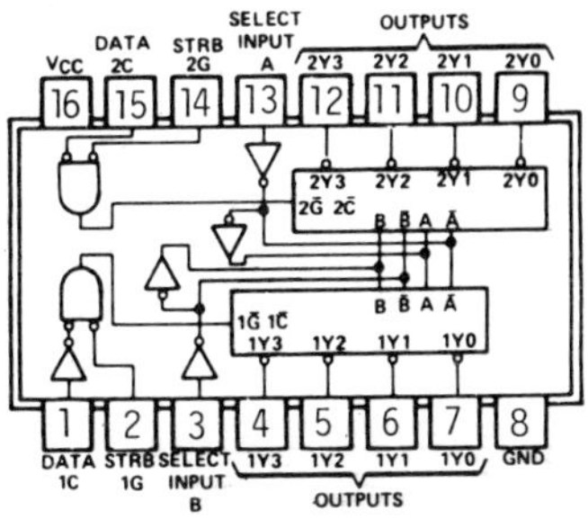

74155

Fig. 5-67.

Schematic Diagram of Circuit (Fig. 5-68)

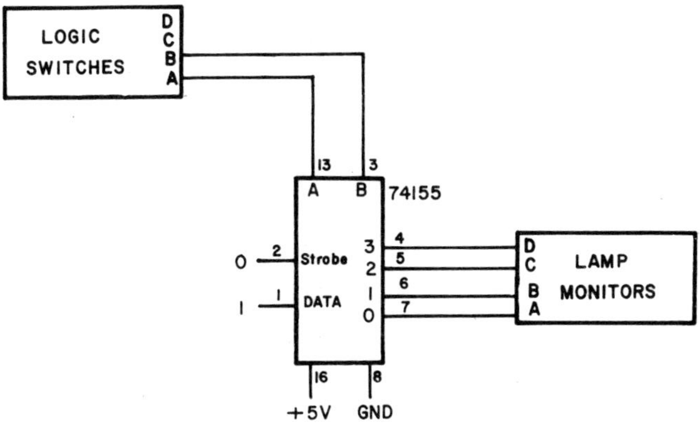

Fig. 5-68.

Step 1

Wire the circuit as shown. Your objective is to produce in succession, a "logic 0" state at each of the four output channels on the 74155 chip.

Step 2

Apply power to the breadboard. By varying the logic states of the A and B select inputs, determine the logic states of the four output channels and fill in the following truth table. Note that this time we have used the 0123 output channel notation, rather than DCBA. Channel 0 is shown as 1Y0 on the pin configuration for the chip.

B	A	0	1	2	3
0	0				
0	1				
1	0				
1	1				

Questions

1. Assume that the 74155 decoder is connected to a seven-segment LED display rather than to four lamp monitors, as shown in the diagram of Fig. 5-69. In the following truth table, write down

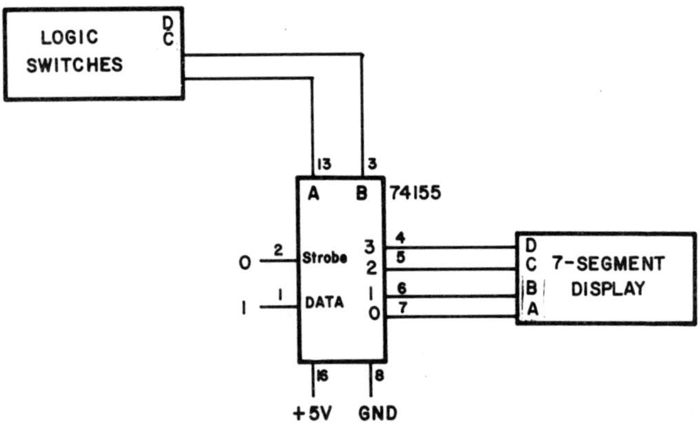

Fig. 5-69.

the numerals or symbols that you would expect to appear on the display. You may wish to try this experiment to confirm your predictions.

B	A	Numeral or Symbol
0	0	
0	1	
1	0	
1	1	

EXPERIMENT NO. 11

Purpose

The purpose of this experiment is to demonstrate the operation of the 74155 chip as a 1-of-4 demultiplexer.

Pin Configuration of Integrated-Circuit Chip (Fig. 5-70)

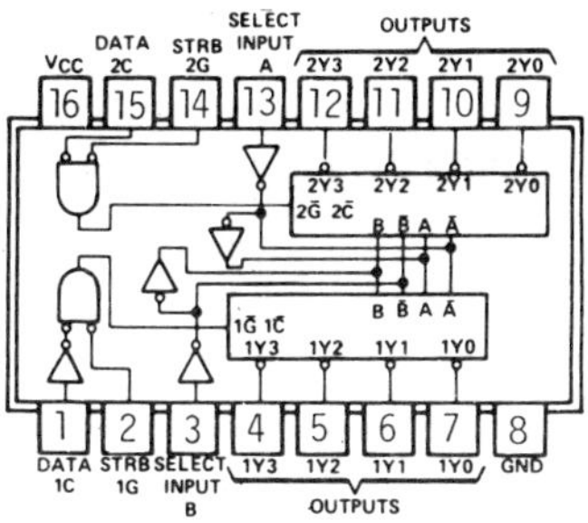

74155

Fig. 5-70.

Schematic Diagram of Circuit (Fig. 5-71)

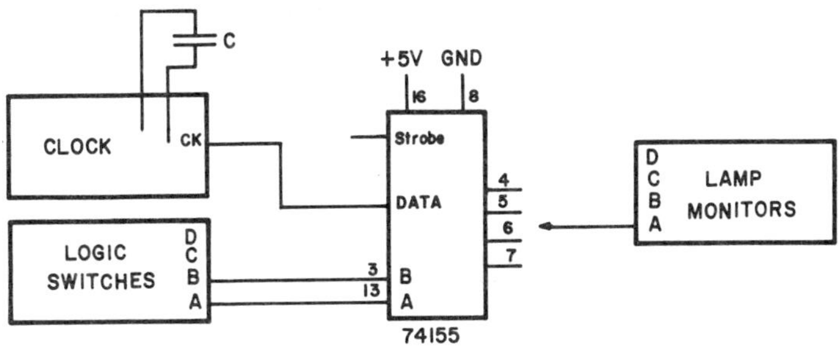

Fig. 5-71.

Step 1

The object of this experiment is to detect the clock pulses using the lamp monitor, which is connected to one of four different output channels on the 74155 chip. The logic switches A and B determine at which channel the data appears.

Step 2

Wire the circuit as shown. Leave a long wire connected to the lamp monitor unconnected. It will be inserted into pins 4, 5, 6, or 7.

Step 3

Set B = A = 0 and determine at which output channel the "data," in this case a train of clock pulses, appears.

In our circuit, the data appeared at pin 7, corresponding to channel 0.

Step 4

Now set B = A = 1 and again determine at which output channel the data appears.

We observed that the data appeared at pin 4, corresponding to channel 3.

Step 5

Fill in the truth table below. Enter the channel number at which the data appears in the far right column.

B	A	Channel Number
0	0	0
0	1	
1	0	
1	1	3

Step 6

Slow down the clock pulses so that you can determine whether the "data" is appearing at the output channel in normal or in inverted form. Is the data inverted or normal? Your answer should be, inverted.

Step 7

Now rewire the connections to pins 1 and 2 on the 74155 integrated circuit. The new schematic diagram is shown in Fig. 5-72.

Step 8

Fill in the following truth table. Again, enter the channel number at which the data appears in the far right column.

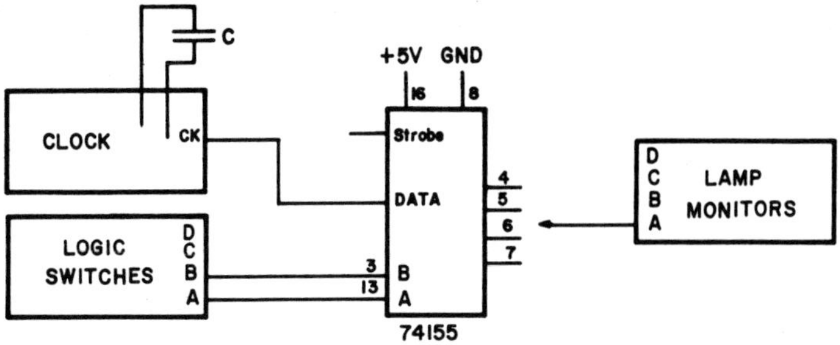

Fig. 5-72.

B	A	Channel Number
0	0	
0	1	
1	0	
1	1	

Step 9

Slow down the clock pulses and determine whether the data appears at the output channel in normal or inverted form. Is the data normal or inverted? Your answer should be, normal.

Step 10

Do both circuits above function in the same manner, or are they different? If they are different, what is the difference that you observe?

In one case, the output data was inverted; in the other, the output data was the same as the input data.

Step 11

Explain the function of the STROBE input at pin 2. Do so in the following space.

A logic 0 at the STROBE input at pin 2 enables the 74155 integrated circuit.

Questions

1. Fill in the following truth table. The entry, "X" means that either a "logic 0" or a "logic 1" state is acceptable.

Inputs				Outputs			
Select		STROBE	DATA				
B	A	1G	1C	0	1	2	3
X	X	1	X				
0	0	0	1				
0	1	0	1				
1	0	0	1				
1	1	0	1				
X	X	X	0				

EXPERIMENT NO. 12

Purpose

The purpose of this experiment is to demonstrate the operation of the 74155 chip as a 3-line-to-8-line decoder.

Pin Configuration of Integrated-Circuit Chip (Fig. 5-73)

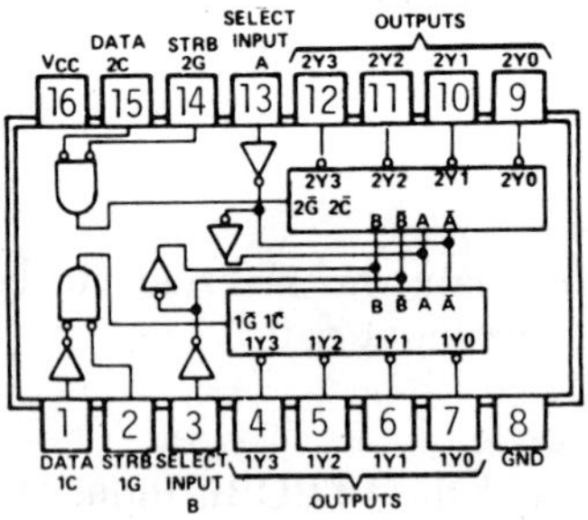

74155

Fig. 5-73.

Schematic Diagram of Circuit (Fig. 5-74)

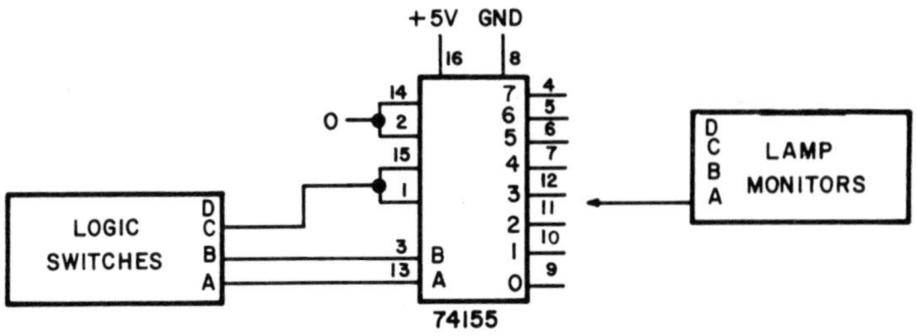

Fig. 5-74.

Step 1

Study the circuit carefully. Note that you connect the two DATA inputs at pins 1 and 15 together, and then connect the pair to logic switch C. The two STROBE inputs are tied together and then grounded.

Step 2

Wire the circuit. Leave a long wire dangling from the lamp monitor. You will test the logic state of the eight output channels using this wire.

Step 3

Set CBA = 000 and determine which output channel goes to a "logic 0" state. Pin 9 should; it corresponds to channel 0. Now set CBA = 001 and again determine which output channel goes to a "logic 0" state. You should observe that pin 10 goes to logic 0, and, thus, must be output channel 1.

Step 4

Fill in the following truth table. You will have to determine the logic states of the remaining output channels besides the specific channel that goes to a "logic 0" state.

Make sure that the pins given in the schematic diagram correspond to the correct output channels.

Questions

1. Can you explain why it was necessary to connect the two DATA inputs at pins 1 and 15 together? Why can they thus function as the third select input, logic switch C? This is not an easy question to answer.

C	B	A	0	1	2	3	4	5	6	7
0	0	0	0							
0	0	1		0						
0	1	0								
0	1	1								
1	0	0								
1	0	1								
1	1	0								
1	1	1								

EXPERIMENT NO. 13

Purpose

The purpose of this experiment is to demonstrate the operation of the 74155 chip as a 1-of-8 demultiplexer.

Pin Configuration of Integrated-Circuit Chip (Fig. 5-75)

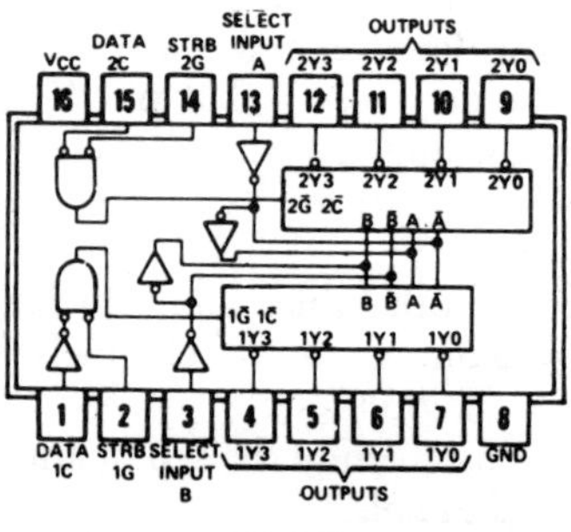

74155

Fig. 5-75.

Schematic Diagram of Circuit (Fig. 5-76)

Step 1

Wire the circuit as shown. You will attempt to detect a series of pulses produced by the pulser at the output channels. The lamp monitor will serve as the "detector."

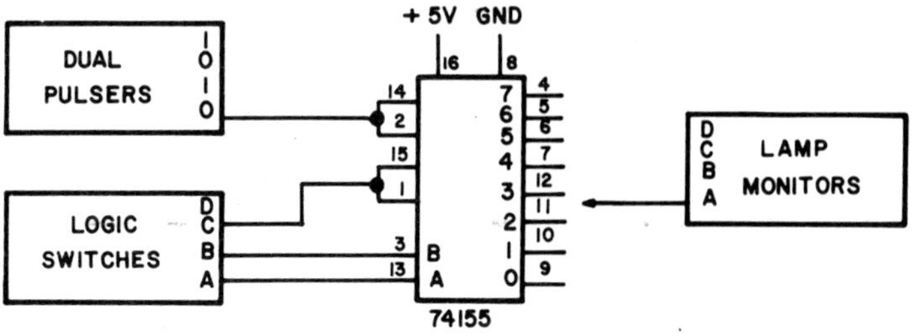

Fig. 5-76.

Step 2

Set CBA = 000, repeatedly press and release the pulser, and determine at which output channel the "data" appears. The data should appear at channel 0, corresponding to pin 9.

Step 3

Set CBA = 001 and repeat the experiment in Step 2. Where does the data appear now?

Channel 1.

Step 4

Set CBA = 010 and repeat Step 2 a third time. Does the data appear at channel 2? It should.

Step 5

Fill in the truth table below. Just enter the channel number in the appropriate column at which the data appears.

C	B	A	Channel Number
0	0	0	0
0	0	1	1
0	1	0	
0	1	1	
1	0	0	
1	0	1	
1	1	0	
1	1	1	

Questions

1. Compare Experiments Nos. 12 and 13. Why can you, on one hand, make a 3-line-to-8-line decoder and, on the other, make a 1-of-8 demultiplexer? What gimmick do you employ to do so? Do you require an extra input or group of inputs *to make a demultiplexer from a decoder? Explain in detail.*

EXPERIMENT NO. 14

Purpose

The purpose of this experiment is to demonstrate how any 2-input positive gate can be constructed using the 74153 4-line-to-1-line data selector/multiplexer.

Pin Configuration of Integrated-Circuit Chip (Fig. 5-77)

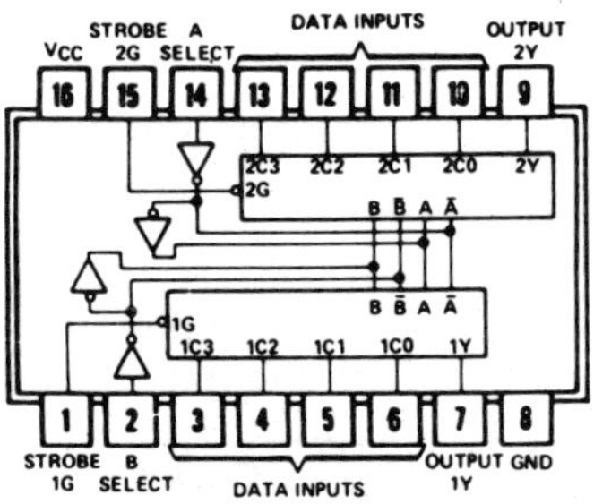

74153

Fig. 5-77.

Schematic Diagram of the Circuit (Fig. 5-78)

Step 1

Using a set of logic switches, a display, and a breadboard, wire the circuit given in the schematic diagram. Note that you will use only two of the four available logic switches. Use short lengths of wire at the four data input pins; these wires will be connected either to +5 Volts or to GROUND as desired during the experiment. You are only using one of the two available data selectors/multiplexers on the 74153 integrated circuit.

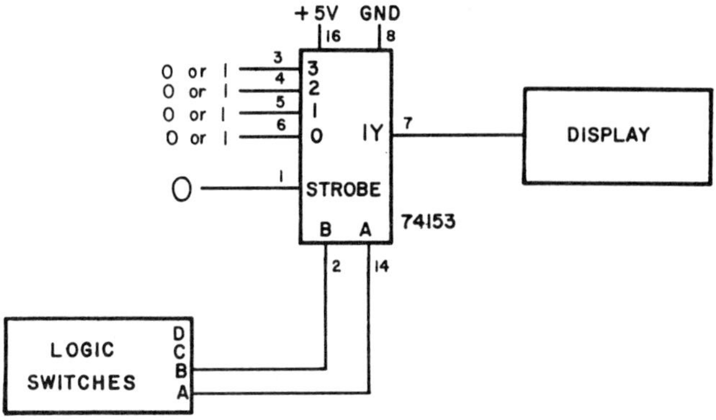

Fig. 5-78.

Step 2

Carefully study the truth table given below for information on how the "select inputs" actually select one of the four data input pins.

B	A	Input Selected	Input Pin Number
0	0	0	6
0	1	1	5
1	0	2	4
1	1	3	3

Thus, if A = B = 0, you select the input at pin number 6 on the 74153 chip. If A = 0 and B = 1, you select the input at pin number 4. You have four different inputs that you can select using a combination of A and B.

Step 3

First wire a 2-input positive AND gate. Set pins 4, 5, and 6 to a "logic 0" state and pin 3 to a "logic 1" state, as shown in the schematic diagram.

Connect power to the breadboard and experimentally determine the following truth table by monitoring the output, at pin 7, on a display.

B	A	Q
0	0	0
0	1	
1	0	
1	1	

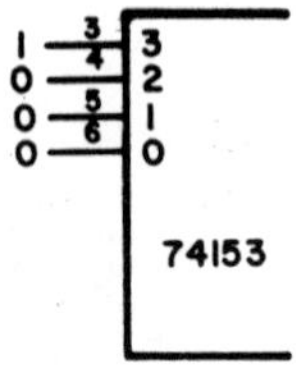

If you wired up this experiment correctly, you should observe a truth table for a 2-input positive AND gate.

Step 4

Next, wire a 2-input positive OR gate. Set pin 6 at logic 0 and pins 3, 4, and 5 to a "logic 1" state, as shown next.

Apply power to the breadboard and fill in the following truth table for a 2-input positive OR gate.

B	A	Q
0	0	0
0	1	
1	0	
1	1	

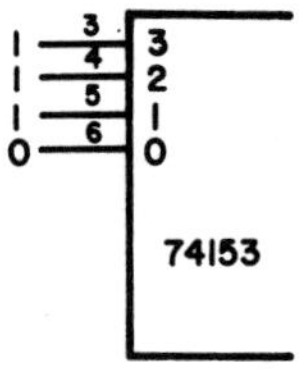

If you have wired this experiment correctly, then the A = B = 0 state should produce a unique output, Q = 0.

Step 5

As a third example, wire a 2-input Exclusive-OR gate. Follow the schematic diagram given next for the input pins 3 through 6.

Fill in the following truth table based on your experimental results.

B	A	Q
0	0	
0	1	
1	0	
1	1	

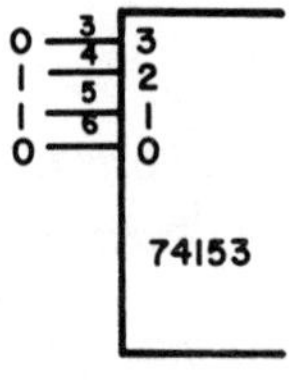

Step 6

In the space following, complete the wiring diagram for the 74153 input pins that correspond to this next unusual truth table.

B	A	Q
0	0	0
0	1	0
1	0	
1	1	0

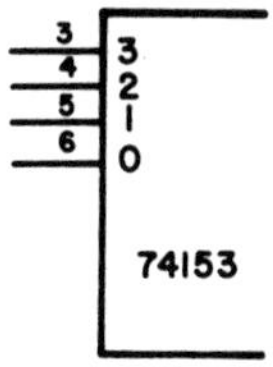

Just fill in the logic states required at pins 3 through 6.

Step 7

In a similar manner, complete the wiring diagram for a 2-input positive NAND gate. You need not perform the experiment.

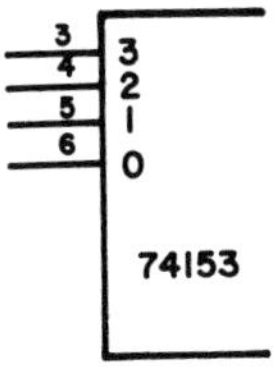

By this point, you should have concluded that it is possible to "create" any of sixteen different type "gates" (see Table 5-27) using the 74153 4-line-to-1-line data selector/multiplexer. Five of these "gates" are well known to you: AND, NAND, OR, NOR, and the 2-input Exclusive-OR. However, the remaining eleven 2-input gates shown in Table 5-27 are not very common and you will rarely be required to "create" any of them.

Table 5-27. Composite Truth Table

B	A	Output, Q, for Sixteen Possible 2-Input "Gates"															
0	0	0	1	0	1	0	1	0	1	0	1	0	1	0	1	0	1
0	1	0	0	1	1	0	0	1	1	0	0	1	1	0	0	1	1
1	0	0	0	0	0	1	1	1	1	0	0	0	0	1	1	1	1
1	1	0	0	0	0	0	0	0	0	1	1	1	1	1	1	1	1

Questions

1. In Table 5-27, we listed the sixteen different types of "gates" that are possible using only two inputs, each input being either at a "logic 0" or a "logic 1" state. In the table, please identify the following gates:

   - 2-input positive AND gate
   - 2-input positive NAND gate
   - 2-input positive OR gate
   - 2-input positive NOR gate
   - 2-input Exclusive-OR gate
   - a gate with an output that is the complement of a 2-input Exclusive-OR gate

   All of these different gates can be "created" using a 74153 data selector/multiplexer integrated circuit.

2. What is a "data selector"? Try to answer this question using the insights that you have gained in this experiment.

3. Define, in your own words, the term "multiplexer."

4. In this experiment, what would happen if the STROBE input at pin 1 were set to a "logic 1" state?

5. Could you suggest how you could create any 2-input gate using up to two inverters and a 2-line-to-1-line data selector/multiplexer? This is a rather difficult and subtle question, so do not try to answer it unless you have the time to do so.

EXPERIMENT NO. 15

Purpose

The purpose of this experiment is to create a 2-input 2-wide AND-OR-INVERT gate using a 74150 16-line-to-1-line data selector/multiplexer integrated circuit.

Pin Configuration of Integrated-Circuit Chip (Fig. 5-79)

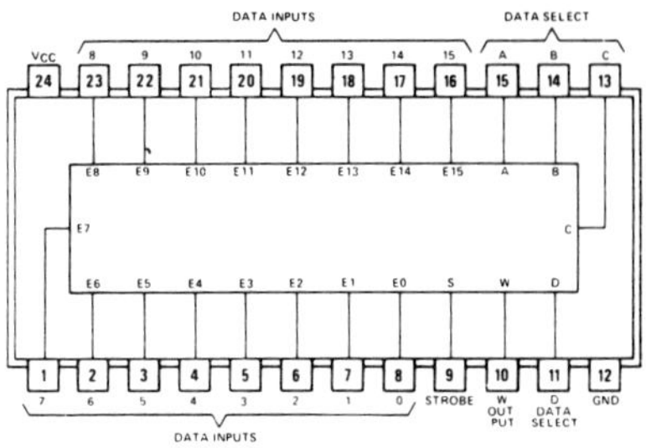

74150

Fig. 5-79.

Schematic Diagram of Circuit (Fig. 5-80)

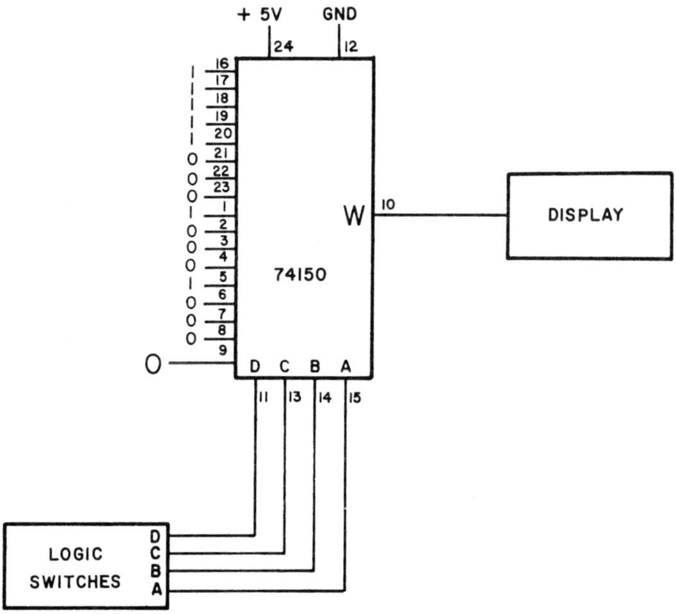

Fig. 5-80.

Step 1

With the power disconnected from the breadboard, wire the circuit shown in the schematic diagram. Use any display that you have available.

Step 2

Apply power to the breadboard and vary the settings of the logic switches. Note the display logic state in each case and fill in the following truth table.

D	C	B	A	Q
0	0	0	0	
0	0	0	1	
0	0	1	0	
0	0	1	1	
0	1	0	0	
0	1	0	1	
0	1	1	0	
0	1	1	1	
1	0	0	0	
1	0	0	1	
1	0	1	0	
1	0	1	1	
1	1	0	0	
1	1	0	1	
1	1	1	0	
1	1	1	1	

If you have wired this experiment correctly, you should observe that the above truth table is the correct one for a 2-input 2-wide AND-OR-INVERT gate. If you do not observe such a result, please check your wiring and repeat the experiment. If your wiring is all right, but your results are not, try another 74150 chip.

Questions

The questions that we shall ask below are somewhat difficult. Answer them only if you have some spare time and would like to challenge your own understanding of multiplexers and the way that they can be used to create truth tables.

1. In the two schematic diagrams shown in Figs. 5-83 and 5-84, we have shown two different ways whereby you can construct an AND-OR-INVERT gate that is 2-wide and has two inputs to each AND gate. We have used the circuit at the bottom of this experiment. Note in the circuit at the top that when DCBA = 0000, the output, Q, is logic 1, and when DCBA = 1111, the output is logic 0. The inputs at pins 1 through 8 and 16 through 23 are precisely the outputs that you would expect from a 2-wide 2-input AND-OR-INVERT gate (if you keep in mind that input channel 0 corresponds to binary DCBA = 0000 and input channel 15 corresponds to binary DCBA = 1111, and that every channel in between these two limits has a corresponding DCBA binary equivalent).

The question is: Use a 74150 chip and set the sixteen input channels to either logic 0 or logic 1 so that the output simulates a 4-input NAND gate. Use the following space for your answer, which should be a circuit diagram similar to the ones shown in Figs. 5-81 and 5-82.

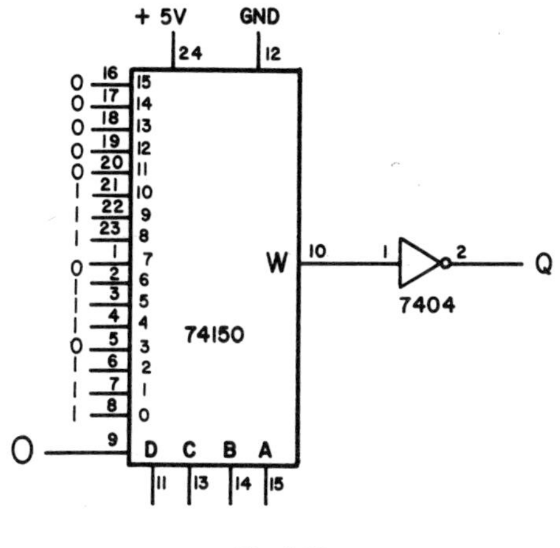

Fig. 5-81.

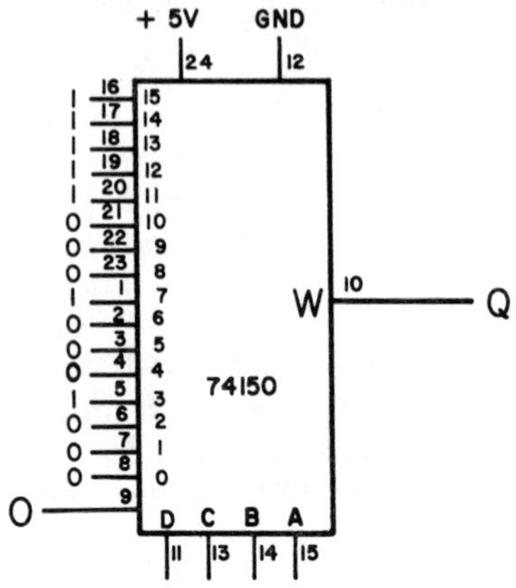

Fig. 5-82.

2. Is it possible to "create" any 2-input, 3-input, or 4-input gate using the 74150 integrated-circuit chip and two, three, or four of the data select inputs at pins 11, 13, 14, and 15? If your answer is yes, please explain how you would do so.

3. Write a truth table for the logic switch inputs A, B, C, and D and determine what type of 4-input gate is represented by the schematic diagram in Fig. 5-83. The pin configuration for the 74151 chip is shown in Fig. 5-84. This is an example of the "folding" technique for creating gates from multiplexer chips.

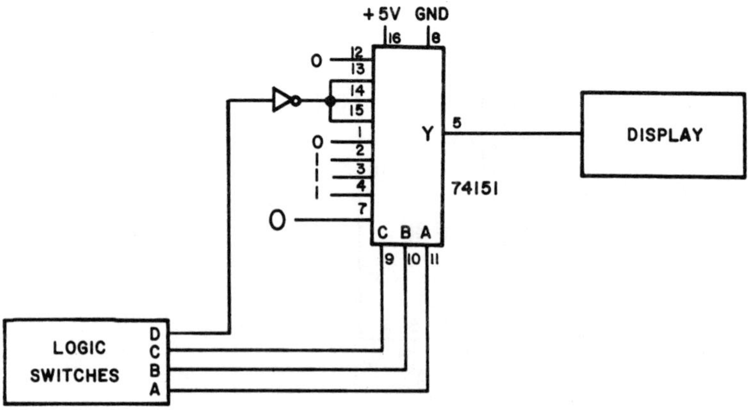

Fig. 5-83.

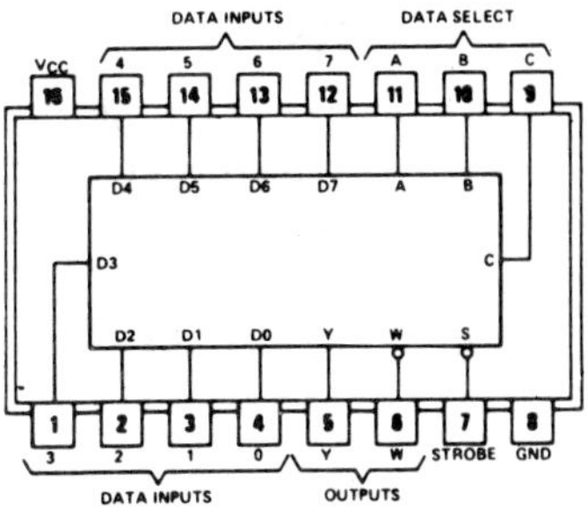

74151

Fig. 5-84.

D	C	B	A	Q
0	0	0	0	
0	0	0	1	
0	0	1	0	
0	0	1	1	
0	1	0	0	
0	1	0	1	
0	1	1	0	
0	1	1	1	
1	0	0	0	
1	0	0	1	
1	0	1	0	
1	0	1	1	
1	1	0	0	
1	1	0	1	
1	1	1	0	
1	1	1	1	

TEST

This test probes your understanding of the digital electronic concepts that we have described in this chapter. Please write your answers on a separate piece of paper.

1. In your own words, describe the five commonly used digital codes:

   binary
   octal
   hexadecimal
   binary coded decimal
   half-ASCII

   Make certain that you explain in what way all of the above codes are similar to each other.
2. In your own words, explain the difference between the following types of devices:

   multiplexer
   demultiplexer
   sequencer
   programmable sequencer
   decoder

3. In your own words, explain the difference between a decoder and a decoder/driver.

Your performance on this test will be acceptable if you can answer all of the above questions correctly in a one-hour closed book examination. You have encountered all of the above concepts in this chapter. The programmable sequencer has been described in the experiments.

WHAT HAVE YOU ACCOMPLISHED IN THIS CHAPTER?

Review of Objectives

We have stated in the Introduction to this chapter that at the end you should be able to do the following:

- Pass a test in which you describe the five commonly used digital codes, including *binary, octal, hexadecimal, binary coded decimal,* and *half-ASCII.*
  You have been asked to do this in question No. 1 on the test.
- Pass a test in which you distinguish between the following types of devices:

  multiplexer
  demultiplexer
  sequencer
  programmable sequencer
  decoder

  You have been asked to do this in question No. 2 on the test.
- Pass a test in which you explain the difference between a *decoder* and a *decoder/driver.*
  You have been asked to do this in question No. 3 on the test.
- Construct decoders from 7442, 74154, and 74155 integrated-circuit chips.
  You have done this in Experiments Nos. 1, 7, 10, and 12.
- Construct sequencers and programmable sequencers from 7442 and 74154 integrated-circuit chips using 7490 and 7493 counters, respectively.
  You have done this in Experiments Nos. 3, 4, 5, and 7.
- Construct demultiplexers from 7442, 74154, and 74155 integrated-circuit chips.
  You have done this in Experiments Nos. 2, 8, 9, 11, and 13.
- Construct multiplexers from 74150 and 74153 integrated-circuit chips.
  You have done this in Experiments Nos. 14 and 15.

appendix

References

BOOKS

1. Rudolf G. Graf, *Modern Dictionary of Electronics,* Howard W. Sams & Co., Inc., Indianapolis, Indiana 46268, 1972.
2. J. Blukis and M. Baker, *Practical Digital Electronics,* Hewlett-Packard Company, 5301 Stevens Creek Blvd., Santa Clara, California 95050, 1974.
3. M. Bird and R. Schmidt, *Practical Digital Electronics: Laboratory Workbook,* Hewlett-Packard Company, Santa Clara, California 95050, 1974.
4. H. V. Malmstadt and C. G. Enke, *Digital Electronics for Scientists,* W. A. Benjamin, Inc., New York, 1969.
5. H. V. Malmstadt, C. G. Enke, and S. R. Crouch, *Instrumentation for Scientists Series,* W. A. Benjamin, Inc., Menlo Park, California, 1973 and 1974.
6. D. E. Lancaster, *TTL Cookbook,* Howard W. Sams & Co., Indianapolis, Indiana 46268, 1974.
7. A. James Diefenderfer, *Principles of Electronic Instrumentation,* W. B. Saunders Company, Philadelphia, Pennsylvania, 1972.
8. J. D. Lenk, *Handbook of Logic Circuits,* Reston Publishing Company, Inc., Reston, Virginia, 1972.
9. F. M. Mims, III, *LED Circuits & Projects,* Howard W. Sams & Co., Inc., Indianapolis, Indiana 46268, 1973.

PERIODICALS

If you would like to maintain a continuing awareness of new advances in digital electronics, or in the field of electronics in general, you will most likely be interested in one or more of the following periodicals.

Electronics, McGraw-Hill Publishing Co., 1221 Avenue of the Americas, New York, New York 10020.

Electronic Design, 50 Essex Street, Rochelle Park, New York 07662.

Computer Design, Circulation Department, Box A, Winchester, Massachusetts 01890.

Kilobaud, 1001001 Inc., Peterborough, New Hampshire 03458.

Radio-Electronics, Gernsback Publications, Inc., 200 Park Avenue South, New York, New York 10003.

Ham Radio Magazine, Communications Technology, Inc., Greenville, New Hampshire 03048.

Index

I

L

M

N

O

T

U

V

W

Z

TO THE READER

This book is one of an expanding series of books that will cover the field of basic electronics and digital electronics from basic gates and flip-flops through microcomputers and digital telecommunications. We are attempting to develop a mailing list of individuals who would like to receive information on the series. We would be delighted to add your name to it if you would fill in the information below and mail this sheet to us. Thanks.

1. I have the following books:

____

____

____

____

____

____

____

____

____

____

____

____

2. My occupation is: ☐ student ☐ teacher, instructor ☐ hobbyist

☐ housewife ☐ scientist, engineer, doctor, etc. ☐ businessman

☐ Other: ____________________

Name (print): ______________________________

Address ______________________________

City ____________________ State __________

Zip Code ____________________

Mail to:

Books
P.O. Box 715
Blacksburg, Virginia 24060